DEVELOPMENTAL
TOXICOLOGY

TARGET ORGAN TOXICOLOGY SERIES

Series Editors

A. Wallace Hayes, John A. Thomas, and Donald E. Gardner

Developmental Toxicology, Third Edition. *Deborah K. Hansen and Barbara D. Abbott, editors, 408 pp., 2008*

Adrenal Toxicology. *Philip W. Harvey, David J. Everett, and Christopher J. Springall, editors, 336 pp., 2008*

Cardiovascular Toxicology, Fourth Edition. *Daniel Acosta, Jr., editor, 712 pp., 2008*

Toxicology of the Gastrointestinal Tract. *Shayne C. Gad, editor, 384 pp., 2007*

Immunotoxicology and Immunopharmacology, Third Edition. *Robert Luebke, Robert House, and Ian Kimber, editors, 676 pp., 2007*

Toxicology of the Lung, Fourth Edition. *Donald E. Gardner, editor, 696 pp., 2006*

Toxicology of the Pancreas. *Parviz M. Pour, editor, 720 pp., 2005*

Toxicology of the Kidney, Third Edition. *Joan B. Tarloff and Lawrence H. Lash, editors, 1200 pp., 2004*

Ovarian Toxicology. *Patricia B. Hoyer, editor, 248 pp., 2004*

Cardiovascular Toxicology, Third Edition. *Daniel Acosta, Jr., editor, 616 pp., 2001*

Nutritional Toxicology, Second Edition. *Frank N. Kotsonis and Maureen A. Mackey, editors, 480 pp., 2001*

Toxicology of Skin. *Howard I. Maibach, editor, 558 pp., 2000*

Neurotoxicology, Second Edition. *Hugh A. Tilson and G. Jean Harry, editors, 386 pp., 1999*

Toxicant–Receptor Interactions: Modulation of Signal Transductions and Gene Expression. *Michael S. Denison and William G. Helferich, editors, 256 pp., 1998*

Toxicology of the Liver, Second Edition. *Gabriel L. Plaa and William R. Hewitt, editors, 444 pp., 1997*

Free Radical Toxicology. *Kendall B. Wallace, editor, 454 pp., 1997*

Endocrine Toxicology, Second Edition. *Raphael J. Witorsch, editor, 336 pp., 1995*

Carcinogenesis. *Michael P. Waalkes and Jerrold M. Ward, editors, 496 pp., 1994*

Developmental Toxicology, Second Edition. *Carole A. Kimmel and Judy Buelke-Sam, editors, 496 pp., 1994*

Nutritional Toxicology. *Frank N. Kotsonis, Maureen A. Mackey, and Jerry J. Hjelle, editors, 336 pp., 1994*

Ophthalmic Toxicology. *George C. Y. Chiou, editor, 352 pp., 1992*

Toxicology of the Blood and Bone Marrow. *Richard D. Irons, editor, 192 pp., 1985*

Toxicology of the Eye, Ear, and Other Special Senses. *A. Wallace Hayes, editor, 264 pp., 1985*

Cutaneous Toxicity. *Victor A. Drill and Paul Lazar, editors, 288 pp., 1984*

DEVELOPMENTAL TOXICOLOGY

THIRD EDITION

Edited by

Deborah K. Hansen

National Center for Toxicological Research
Jefferson, Arkansas, USA

Barbara D. Abbott

U.S. Environmental Protection Agency
Research Triangle Park, North Carolina, USA

CRC Press
Taylor & Francis Group
Boca Raton London New York

CRC Press is an imprint of the
Taylor & Francis Group, an **informa** business

CRC Press
Taylor & Francis Group
6000 Broken Sound Parkway NW, Suite 300
Boca Raton, FL 33487-2742

First issued in paperback 2018

ISBN-13: 978-1-4200-5437-8 (hbk)
ISBN-13: 978-1-138-37244-3 (pbk)

Library of Congress Cataloging-in-Publication Data

Developmental toxicology. – 3rd ed. / edited by Deborah K. Hansen, Barbara D.
  Abbott.
  p. ; cm. – (Target organ toxicology series ; 27)
  Includes bibliographical references and index.
  ISBN-13: 978-1-4200-5437-8 (hardcover : alk. paper) ISBN-10: 1-4200-5437-6
  (hardcover : alk. paper)
  1. Developmental toxicology. 2. Fetus--Effect of drugs on. agents. I. Hansen,
Deborah Kay, 1952– II. Abbott, Barbara D. 3. Teratogenic III. Series.

  [DNLM: 1. Abnormalities, Drug-Induced. effects. 2. Embryonic Development–drug
  3. Fetal Development--drug effects. 4. Risk Assessment. 5. Teratogens. QS 679 D4893
  2008]
  RA1224.45.D47 2008 618.3 2--dc22   2008026172

Visit the Taylor & Francis Web site at
http://www.taylorandfrancis.com

and the CRC Press Web site at
http://www.crcpress.com

Preface

This Third Edition of *Developmental Toxicology* comes 14 years after the publication of the Second Edition. Much has happened in the field of developmental toxicology over that period of time. Many of the advances have been in the areas of mechanistic examinations of developmental toxicants as well as in risk assessment, so these are the two general areas that we have chosen to focus on in this volume.

The first two editions serve as a strong foundation on which to build. Many basic issues in the field of developmental toxicology were covered in those editions, and the information provided in them still remains relevant today. However, significant strides have been made in the area of mechanistic information. Potential mechanisms of developmental toxicants that are covered in this edition include apoptosis, alterations in signal transduction pathways, as well as effects on nutrition and epistasis. The role of the placenta in the transfer of developmental toxicants, nutrients, and wastes, as well as its role in metabolism of compounds is also covered. New techniques being used to address mechanistic questions include targeted gene disruption, cell-culture methods such as whole-embryo culture and the use of embryonic stem cells and genomic approaches. *Xenopus* is being investigated as an alternative to mammalian animal models in an effort to decrease the use of higher animal models as well as to shorten the time involved in preclinical testing. Progress has also been made on the development of physiologically based pharmacokinetic models as a way to extrapolate across species. Interpretation of large amounts of information, especially from newer high-throughput techniques, has required new approaches to analyzing and compiling the information leading to the development of bioinformatic systems and the new field of computational toxicology. How this information can be used in assessing potential risk to humans as well as to specific subpopulations of humans is covered in additional chapters. Finally, methods used to identify developmental toxicants in humans are discussed.

Prevention of birth defects is the goal of all developmental toxicologists. The best way to tackle this important task is an approach in which animal research and information from clinical settings are taken together in an effort to identify which compounds may pose a risk as well as whether certain subpopulations may have an increased risk. The first two editions of *Developmental Toxicology* focused on this integrated approach for characterizing outcome, and we have tried to maintain that approach in this edition too. We hope that we have succeeded in our goal.

Deborah K. Hansen
Barbara D. Abbott

Acknowledgments

We offer many thanks to all of the authors who contributed to this volume and are continuing to contribute to the forefront of developmental toxicology. We would also like to thank the representatives from Taylor and Francis and Informa Healthcare who helped a couple of rookie editors through the process.

Contents

Contributors

Barbara D. Abbott   Reproductive Toxicology Division (MD67), National Health and Environmental Effects Research Laboratory (NHEERL), Office of Research and Development, U.S. Environmental Protection Agency, Research Triangle Park, North Carolina, U.S.A.

Karen A. Augustine-Rauch   Discovery Toxicology, Pharmaceutical Candidate Optimization, Bristol-Myers Squibb Company, Pennington, New Jersey, U.S.A.

Kimberly C. Brannen   Discovery Toxicology, Pharmaceutical Candidate Optimization, Bristol-Myers Squibb Company, Pennington, New Jersey, U.S.A.

Gregg D. Cappon   Pfizer Global Research and Development, Groton, Connecticut, U.S.A.

Christina D. Chambers   Departments of Pediatrics and Family and Preventive Medicine, University of California, San Diego, La Jolla, California, U.S.A.

Jeffrey H. Charlap   Preclinical Services, Charles River Laboratories, Horsham, Pennsylvania, U.S.A.

James J. Chen   Division of Personalized Nutrition and Medicine, FDA/National Center for Toxicological Research, Jefferson, Arkansas, U.S.A.

George P. Daston   Miami Valley Innovation Center, The Procter & Gamble Company, Cincinnati, Ohio, U.S.A.

Robert G. Ellis-Hutchings   Reproductive Toxicology Division, National Health and Environmental Effects Research Laboratory, U.S. Environmental Protection Agency, Research Triangle Park, North Carolina, U.S.A.

Deborah K. Hansen   Division of Genetic and Reproductive Toxicology, FDA/National Center for Toxicological Research, Jefferson, Arkansas, U.S.A.

Phillip Hartig   Reproductive Toxicology Division, National Health and Environmental Effects Research Laboratory, U.S. Environmental Protection Agency, Research Triangle Park, North Carolina, U.S.A.

Sid Hunter   Reproductive Toxicology Division, National Health and Environmental Effects Research Laboratory, U.S. Environmental Protection Agency, Research Triangle Park, North Carolina, U.S.A.

Mark E. Hurtt   Pfizer Global Research and Development, Groton, Connecticut, U.S.A.

Jim Kaput   Division of Personalized Nutrition and Medicine, FDA/National Center for Toxicological Research, Jefferson, Arkansas, U.S.A.

Robert J. Kavlock   National Center for Computational Toxicology (B205–01), Office of Research and Development, U.S. Environmental Protection Agency, Research Triangle Park, North Carolina, U.S.A.

Thomas B. Knudsen   National Center for Computational Toxicology (B205–01), Office of Research and Development, U.S. Environmental Protection Agency, Research Triangle Park, North Carolina, U.S.A.

Kannan Krishnan   Département de Santé Environnementale et Santé au Travail, Faculté de Médecine, Université de Montréal, Montréal, Québec, Canada

Philip Mirkes   Center for Environmental and Rural Health, Texas A&M University, College Station, Texas, U.S.A.

Jorge M. Naciff   Miami Valley Innovation Center, The Procter & Gamble Company, Cincinnati, Ohio, U.S.A.

Terence R. S. Ozolinš   Developmental and Reproductive Toxicology Center of Emphasis, Pfizer Drug Safety Research and Development, Groton, Connecticut, U.S.A.

Julieta M. Panzica-Kelly   Discovery Toxicology, Pharmaceutical Candidate Optimization, Bristol-Myers Squibb Company, Pennington, New Jersey, U.S.A.

John M. Rogers   Reproductive Toxicology Division, National Health and Environmental Effects Research Laboratory, U.S. Environmental Protection Agency, Research Triangle Park, North Carolina, U.S.A.

William Slikker, Jr.   Office of the Director, FDA/National Center for Toxicological Research, Jefferson, Arkansas, U.S.A.

Mathieu Valcke   Département de Santé Environnementale et Santé au Travail, Faculté de Médecine, Université de Montréal, Montréal, Québec, Canada

Role of Apoptosis in Normal and Abnormal Development

Philip Mirkes

Center for Environmental and Rural Health, Texas A&M University,
College Station, Texas, U.S.A.

INTRODUCTION

Although many reviews on the topic of apoptosis have appeared in the last 5 years, most of these have focused on apoptosis in the context of cancer (1), normal development (2), the immune system (3), or neurodegenerative diseases (4). In contrast, the role of apoptosis in developmental toxicology has largely been ignored. Thus, this review will focus on apoptosis and its role in abnormal development, particularly abnormal development induced by teratogens. To understand and appreciate what is known about the role of apoptosis in abnormal development, it is necessary to understand what is now known about apoptosis in general. Thus, this chapter is organized into five basic sections: the morphology of apoptosis, the genetics of apoptosis, the biochemistry of apoptosis, the role of apoptosis in normal development, and the role of apoptosis in teratogen-induced abnormal development.

MORPHOLOGY OF APOPTOSIS

Although cell death was widely known to occur during normal development and the etiology of many diseases, little was known about how this cell death occurred prior to 1972. The 1972 publication by Kerr, Wyllie, and Currie (5) marks the beginning of a leap forward in our understanding of the mechanisms of cell death.

Kerr et al. noted that one of the earliest changes in cellular morphology, associated with what they initially called "shrinkage necrosis," involved condensation of the cytoplasm and nuclear chromatin and the subsequent aggregation of this condensed chromatin beneath the nuclear envelope. At later stages, the dying cell fragmented into small round bodies consisting of membrane-bound masses of condensed cytoplasm, some of which contained fragments of the nucleus. Finally, these round bodies were phagocytosed by resident macrophages. The morphology of this cell death process was designated "apoptosis," which rapidly supplanted a variety of other descriptors, such as shrinkage necrosis and heterophagy. Of interest, Kerr was also one of the first to describe teratogen-induced apoptosis, that is, mesenchymal cell death in the developing vertebral arches induced by 7-hydroxymethyl-12-methylbenz(a)anthracene (6). Subsequent to 1972, studies confirmed that the morphology of cell death defined as apoptosis by Kerr and his colleagues occurred in a variety of cell types and tissues and under a variety of conditions; however, a mechanistic understanding of apoptosis had to await the genetic studies in the nematode, *C. elegans*, which were published beginning in 1983.

GENETICS OF DEVELOPMENTAL CELL DEATH

The nematode, *C. elegans*, proved to be an excellent model for the elucidation of the underlying mechanisms of cell death because (1) the nematode adult has fewer than 1000 somatic cells; (2) the generation of 959 somatic cells during development is accompanied by deaths of 131 cells [programmed cell deaths (PCD)]; (3) development of this transparent worm involves invariant patterns of cell division, migrations, and deaths that can easily be monitored; (4) *C. elegans* has a short generation time (3 days at 20°C); and (5) the worm lends itself to genetic analysis. Using ethyl methanesulfonate mutagenesis of *C. elegans*, Horvitz and his colleagues isolated a number of single-gene mutations that control specific events in developmental cell deaths, and for the first time demonstrated that apoptosis is an active process requiring the function of specific genes (7).

Two of the key genes, initially described in 1986 (8), were *ced-3* and *ced-4*, both of which are required for developmental cell deaths in *C. elegans*. Subsequently, another gene called *ced-9* was discovered that regulates the activity of *ced-3* and *ced-4*. *Ced-9* gain-of-function mutations prevent cells from dying, whereas mutations that inactivate *ced-9* are lethal. This and subsequent studies showed that *ced-9* inhibits the activity of the *ced-4*, which in turn functions to activate *ced-3*. Studies in the late 1990 s identified *egl-1* mutations that also affected cell death in *C. elegans* (9). Gain-of-function mutations in *egl-1* caused hermaphrodite-specific neurons to undergo cell death, whereas a loss-of-function *egl-1* mutation prevented most, if not all, somatic cell deaths in this roundworm. Genetic analyses showed that *egl-1* acts upstream of or in parallel to *ced-4* and *ced-3* and requires *ced-9* to exert its effect on developmental cell death. Subsequent studies demonstrated that EGL-1 binds to and inhibits the activity of

Table 1 Vertebrate Homologs of CED-3, CED-4, CED-9, and EGL-1

C. elegans	EGL-1	CED-9	CED-4	CED-3
	Proapoptotic	*Antiapoptotic*	Apaf-1	*Group 1-cytokine*
	BH1-3	BH1-3		*processing*
	Bax	Mcl-1		Caspase-1
	Bak	Bfl-1/A-1		Caspase-4
	Bok/MTD	BH1-4		Caspase-5
	BH2–3	Bcl-2		Caspase-11
	Bcl-G	Bcl-XL		Caspase-12
	Bfk	Bcl-w		Caspase-13
	BH-3 only	BOO/DIVA		Caspase-14
	Bad	NRH/NR-3		*Group 2-initiator*
	Bid			*caspases*
	Bim/Bod			Caspase-2
	Bik/Blk/Nbk			Caspase-8
	Bmf			Caspase-9
	Bnip3			Caspase-10
	Hrk/DP5			*Group 3-executioner*
	Nix			*caspases*
	Noxa			Caspase-3
	Puma/BBC3			Caspase-6
	SPIKE			Caspase-7

VERTEBRATES

CED-9 by disrupting the association between CED-9 and CED-4 (9,10). Thus, studies using *C. elegans* suggest a linear pathway of apoptosis in which CED-4 activates CED-3, leading to cell death. This cell death machinery is held in check by CED-9, which binds and inactivates CED-4. Finally, EGL-1 inhibits the inhibitory activity of CED-9 and in the process allows CED-4 to activate CED-3 (Table 1). The cell death pathway elucidated in *C. elegans* served as the catalyst for a concerted effort over the past 25 years to understand the detailed mechanisms by which vertebrate, particularly human,cells die.

BIOCHEMISTRY OF CELL DEATH

The discovery of these key cell death genes in the roundworm led to a frenzy of research activity focused on isolating vertebrate (and nonvertebrate) homologs. Whereas there are only two genes, *egl-1* and *ced-9*, that regulate *ced-4* activity in the roundworm, there are more than 20 known vertebrate homologs (Table 1) belonging to the Bcl-2 family, so named because the first member discovered was isolated from a B cell lymphoma/leukemia 2 (11). Homologs of the antiapoptotic CED-9 included Bcl-2, Bcl-X$_L$, Bcl-W, Bcl-B, Boo/DIVA, NR-13, Mc-1, and A1, whereas homologs of the proapoptotic EGL-1 include Bax, Bmf, Bid, BNIP3, Bad, Bak, Hrk/Dp5, Bik, Noxa, Bim, PUMA/Bbc3, Bok Bcl-X$_s$, and Blk.

Similarly, whereas there is only one *ced-3* gene in *C. elegans*, there are more than 10 known vertebrate homologs (Table 1), called caspases (cysteine aspartate proteases). The name caspase is derived from the fact that caspases are proteases that specifically cleave proteins between cysteine and aspartate residues. In contrast, there is only one vertebrate homolog of *ced-4*, called Apaf-1 (Apoptosis protease activating factor-1).

In addition to defining the vertebrate homologs of *C. elegans* ced genes, other studies focused on the mechanisms by which these gene products regulate the apoptotic process. From these studies, it is now clear that there are two major apoptotic pathways, the intrinsic mitochondrial and the extrinsic, receptor-mediated apoptotic pathways.

Intrinsic Mitochondrial Cell Death Pathway

The intrinsic mitochondrial apoptotic pathway (Fig. 1) is activated by a wide variety of drugs, chemicals, and physical agents [e.g., radiation, hyperthermia (HS)]. As will be discussed subsequently, a variety of teratogens has also been shown to activate this pathway in mammalian embryos. In ways not yet completely elucidated, apoptotic stimuli converge on the mitochondria, resulting in changes in the inner mitochondrial membrane, opening of the mitochondrial permeability transition pore, loss of mitochondrial transmembrane potential,

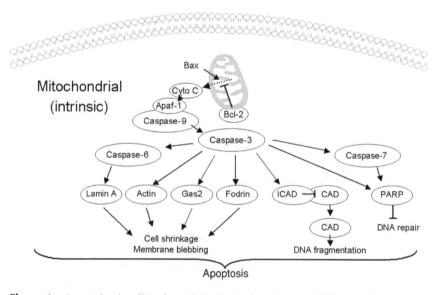

Figure 1 Apoptotic signaling through the intrinsic, mitochondrial apoptotic pathway. Apoptotic stimuli converge on mitochondria to induce cytochrome c, which then activates downstream events in the pathway.

and the release of proapoptotic proteins from their normal location within mitochondria into the cytoplasm (12).

Cytochrome c

One of these proapoptotic proteins, cytochrome c binds and activates Apaf-1 (homolog of CED-4) in the presence of dATP (13). The Apaf-1/cytochrome c then forms a docking complex by oligomerizing into a heptameric structure resembling a wheel the core of which contains seven N-terminal **CARD**s (**ca**spase **r**ecruitment **d**omains), forming the so-called apoptosome (14,15). The apoptosome then recruits and activates multiple initiator procaspase-9 molecules through binding of procaspase-9 CARD domains to Apaf-1 CARD domains. Bringing individual procaspase-9 molecules into close proximity facilitates their activation, although the exact mechanisms regulating the activation of procaspase-9 remain unclear (15). Although apparently not required for activation, one of the consequences of procaspase-9 activation is the cleavage of the proenzyme into lower molecular weight subunits [Fig. 2 (A)] that can readily be detected in Western blots using available antibodies.

Once activated, caspase-9 can then activate so-called downstream effector caspases: caspase-3, -6, and -7. Like caspase-9, these effector caspases normally exist in cells as proenzymes. Activation of these effector procaspases by caspase-9 involves cleavage between specific cysteine/aspartate residues to generate active subunits [Fig. 2 (B)]. Using Western blots, activation of effector caspases can be detected as the disappearance of the proenzyme and/or the appearance of specific subunits.

Effector caspases, in turn, target a variety of intracellular proteins for proteolytic cleavage. According to a recent census, approximately 400 caspase substrates have now been identified (16). Cleavage of caspase substrates can activate or inactivate particular proteins. For example, caspase-mediated cleavage of rho-associated kinase 1 activates this protein by removing the C-terminal autoinhibitory region of this molecule (17). In contrast, caspase-3-mediated cleavage of the inhibitor (ICAD) of a caspase-activated DNase (CAD) results in the inactivation of ICAD and the resultant activation of CAD (18). In this case, cleavage of ICAD inactivates the inhibitor, thereby activating CAD. Although cleavage of most caspase substrates cannot be linked to specific events that coordinate the death of cell, cleavage of rho-associated kinase has been implicated in the membrane blebbing that occurs during the later stages of apoptosis, and the activation of CAD by the inactivation of ICAD has been linked to the well-described internucleosomal degradation of DNA (DNA laddering) that occurs late in apoptosis (17,18). One of the most well-studied caspase substrates is poly (ADP-ribose) polymerase-1 (PARP-1), an enzyme known to play a role in DNA repair. Caspase-3 cleavage of PARP converts the 116-kDa active enzyme into inactive fragments, an 85-kDa fragment containing the catalytic domain, and a 25-kDa N-terminal fragment containing the DNA-binding domain (19). The disappearance of the active enzyme and

(A)   Caspase-9 activation

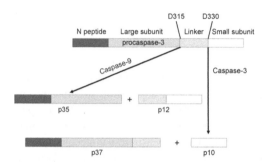

(B)   Capase-3 activation

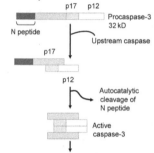

(C)   Poly (ADP-ribose) polymerase (PARP) cleavage

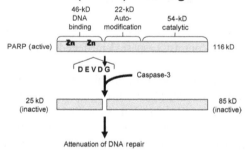

Figure 2   (**A**) Activation of caspase-9 is associated with cleavage of the proenzyme at specific aspartate residues to generate subunits. Autocatalysis results in the formation of p35 and p12 subunits, whereas subsequent cleavage by activated caspase-3 generates p37 and p10 subunits. (**B**) Procaspase-3 cleavage and activation by upstream initiator caspases (caspase-2, -8, and -9). (**C**) Caspase-3 targeted cleavage and inactivation of poly (ADP-ribose) PARP.

the appearance of the inactive 89-kDa fragment can be detected using a Western blot approach [Fig. 2 (C)].

Other Mitochondrial Proapoptotic Proteins

Although cytochrome c is the most well-studied mitochondrial proapoptotic protein released from mitochondria in response to an apoptotic stimulus, several other mitochondrial proteins have been shown to be released from mitochondria and implicated as proapoptotic proteins. These include Smac/Diablo (second mitochondria-derived activator of caspase/direct IAP-binding protein with low pI), HtrA2/Omi (high temperature requirement A2), AIF (apoptosis-inducing factor), CAD (caspase-activated DNAse), and Endo-G (endonuclease-G). Smac/Diablo and HtrA2 activate the caspase-dependent mitochondrial apoptotic pathway by inhibiting IAPs (inhibitors of apoptosis proteins) activity (20,21). AIF translocates to the nucleus, where it induces DNA fragmentation into 50- to 300-kb pieces and condensation of peripheral nuclear condensation (22), referred to as stage 1 condensation (23). Endo-G also translocates to the nucleus, where it attacks chromatin to produce oligonucleosomal DNA fragments (24). CAD also translocates to the nucleus, where it is activated by caspase-3, and then generates oligonucleosomal DNA fragments and advanced chromatin condensation (25), referred to as stage 2 condensation (23). Despite the data suggesting that these mitochondrial proteins, as well as others, play a proapoptotic role in apoptosis, more recent studies using transgenic and knockout mice suggest that these proteins may not play an essential role (26).

Extrinsic, Receptor-mediated Cell Death Pathway

In contrast to the intrinsic mitochondrial apoptotic pathway, the extrinsic, receptor-mediated pathway (Fig. 3) is initially activated by the interaction of specific ligands with death domain (DD-) containing receptors of the tumor necrosis factor (TNF) receptor gene superfamily, the two most well-studied examples being the fatty acid synthetase ligand binding to the Fas receptor (FasR) and tumor necrosis factor alpha (TNFα) binding to the TNF receptor (TNFR). Binding of these ligands to their respective receptors initiates the recruitment of DD-containing adaptor molecules, that is, the Fas-associated death domain protein (FADD) binds to the cytoplasmic DD of FasR whereas the TNF receptor-associated death domain protein binds to the cytoplasmic DD domain of TNFR (27). TNF receptor-associated death domain protein subsequently also recruits FADD. FADD, bound to either the FasR or the TNFR, then recruits procaspase-8, the initiator caspase in the receptor-mediated pathway through interactions of the death effector domains in procaspase-8, and FasR/TNFR (28). Binding of procaspase-8 completes the assembly of the death-inducing signaling complex, resulting in the autocatalytic activation of procaspase-8 (29). Thus, the initiator caspases for both the intrinsic

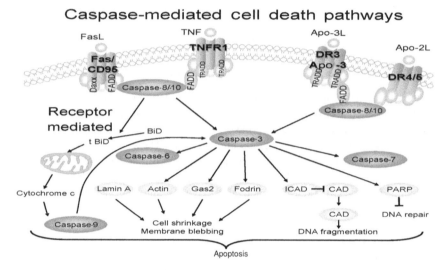

Figure 3 Integration of intrinsic, mitochondrial, and extrinsic, receptor-mediated apoptotic pathways through caspase-8-mediated cleavage and activation of Bid.

and the extrinsic apoptotic pathways are activated by the assembly of a multi-protein-tethering complex within the cell.

Once activated, caspase-8 can then activate the downstream effector caspases in two ways (Fig. 3). First, caspase-8 can activate downstream effector caspases directly, thus activating all of the downstream events described earlier for the mitochondrial apoptotic pathway (30,31). Alternatively, caspase-8 can cleave a BH3-interacting death agonist (Bid), a member of the Bcl-2 family of proteins, at aspartic acid 59 to form a truncated Bid (tBid) (32). Subsequently, tBid is able to migrate to the mitochondria and effect cyctochrome c release (33). It should be noted that in some apoptotic scenarios, cleavage of Bid does not occur; nonetheless, Bid still migrates to the mitochondria. In this case, Bid has been shown to bind to Bax, an event that triggers a change in Bax conformation, which is associated with the release of cytochrome c (34). The release of cytochrome c then activates Apaf-1 and all of the events in the mitochondrial apoptotic pathway described earlier. A more detailed review of the role of Bcl-2 family proteins in regulating apoptotic pathways is provided in the next section.

Regulators of the Cell Death Pathways

The fact that cell death plays such an integral role in normal cell homeostasis, central nervous system (CNS) function, immune system function, and normal development suggest that cells must have mechanisms for regulating cell death that preclude the development of cancer, neurodegenerative diseases, abnormal immune system function, and birth defects unless these mechanisms are somehow subverted. As documented earlier, the essential components of the apoptotic pathway (e.g., Apaf-1, procaspases, proapoptotic Bcl-2 family proteins) are conserved among metazoans and, more important, are constitutively expressed in cells under "normal" growth conditions. The proapoptotic activities of these components are held in check by a variety of prosurvival proteins. As might be expected, the mechanisms available for the regulation of cell death are complex, and a review of all mechanisms is beyond the scope of this chapter. Therefore, this review will focus on several well-studied proteins known to play a role in regulating the intrinsic mitochondrial and extrinsic, receptor-mediated apoptotic pathways, that is, Bcl-2 family proteins, IAP family proteins, p53 family proteins, and heat shock family proteins.

Bcl-2 Family Proteins

Table 1 lists members of the Bcl-2 family, which contains both anti- and proapoptotic members that choreograph a complex interplay of checks and balances that determine, in part, whether a cell lives or dies (35–37). The Bcl-2 family contains three subroups, defined by the presence of one or more Bcl-2 homology (BH) domains, which correspond to a helices that dictate structure and function. The BH multidomain antiapoptotic subgroup contains seven members having either the BH1–3 (MCL-1 and BFL-1/A1) or BH1–4 (Bcl-2, Bcl-XL, Bcl-w, BOO/DIVA, and NRH/NR-3) domains. Expression of Bcl-2, the founding member of this subgroup, blocks cell death induced by a variety of apoptotic stimuli (38,39), including all of the hallmark events described by Kerr et al. (5). What proved to be especially interesting is that Bcl-2 was shown to be localized to the mitochondrion (40), the first indication that mitochondria play an important role in regulating apoptosis.

The BH multidomain proapoptotic subgroup contains five members: the BH1-3 (Bax, Bak, Bok/MTD) and BH2-3 (Bcl-G, Bfk) domain members. The founding member of this subgroup is Bax, which was identified by its interaction with Bcl-2 (41). Cells deficient in either Bax or Bak retain normal response to apoptotic stimuli; however, cells deficient in both Bax and Bak are resistant to apoptotic stimuli, suggesting that these two proapoptotic members are essential to activation of the intrinsic apoptotic pathway linked to either the mitochondrion (42) or the endoplasmic reticulum (43).

The remaining 11 proapoptotic members contain only the BH3 domain and thus belong to so-called BH3-only subgroup (Bad, Bid, Bim/Bod, Bik/Blk/NBK, Bmf, BNIP3, Hrk/DP5, Nix, NOXA, PUMA/Bbc3, and Spike). Accumulated evidence suggests that different apoptotic stimuli activate specific, and sometimes overlapping, sets of BH3-only proteins (44). As already discussed, ligand–receptor interaction, for example, Fas/FasR induces the activation of caspase-8, caspase-8-mediated activation (cleavage) of Bid, and then activation of the intrinsic mitochondrial apoptotic pathway. Thus, Bid serves to link the extrinsic and intrinsic pathways and thereby to amplify apoptotic signaling in response to ligand–receptor binding.

In addition to Bid, Bad is another example of a BH3-only protein that is activated by posttranslational modifications, in this case in response to cytokine and/or growth factor deprivation. When Bad is phosphorylated on serine 136 and/or 112 by the serine–threonine kinase, Akt, it then associates with 14–3-3, a multifunctional phosphoserine-binding protein, leading to the sequestration (inactivation) of Bad. In response to cytokines/growth factors, Bad is dephosphorylated, which leads to its dissociation from 14–3-3, translocation to mitochondria, and activation of the mitochondrial apoptotic pathway (45,46).

Bim and Bmf play a role in transmitting signals from the cytoskeleton to the apoptotic machinery. Studies suggest that regulation of Bim activity is complex and may involve not only posttranslational modifications such as phosphorylation on serine 69 (47) and ubiquitylation (48) but also binding to dynein-light chains associated with microtubules (49) and transcriptional upregulation (50). Like Bim, Bmf can be regulated by dynein-light chain binding (51) and by transcriptional upregulation (52). Additional research is required to determine the importance of these modes of regulation, whether different modes of regulation are cell specific, and whether different combinations of BH3-only proteins are required to effect apoptosis.

PUMA (**p**53 **u**pregulated **m**odulator of **a**poptosis) and *NOXA* play a key role in regulating apoptosis, are p53 target genes containing functional p53-binding sites, and are transcriptionally upregulated in response to DNA damage (p53-dependent) by hypoxia, serum deprivation, and glucocorticoids (p53-independent). It is assumed that PUMA and NOXA, once transcribed and translated, do not require posttranslational modifications to be activated, rather they directly translocate to the mitochondrion to engage the apoptotic machinery. Whether PUMA and NOXA are essential for the regulation of apoptosis appears to be cell context (e.g., p53 status), exposure, and/or cell-type dependent (53).

Other less well-studied BH3-only members include Bik/Blk/Nbk, Hrk/DP5, Nix, Spike, and Mule/Arf-Bp1. Bik is apparently another p53-dependent BH3-only protein that is transcriptionally upregulated by genotoxic stresses (54) and posttranslational phosphorylation on threonine 33 and serine 35 (55). At least in the mouse, Bik deficiency does not confer resistance to apoptosis (56). Hrk is transcriptionally upregulated in response to nerve growth factor withdrawal and β-amyloid treatment (57,58). Bnip3 is also transcriptionally upregulated in response to hypoxia, reoxygenation, exposure to the calcium ionophore A23187, and the protein kinase C activator phorbol myristic acid (59) and mediates cell death by activating Bax and/or Bak (60). Nix/Bnip3 L is transcriptionally upregulated by erythropoietin in erythrocytes (61) and by phorbol myristic acid and PKCa in cardiac myocytes (62). Finally, Spike is a novel BH3-only protein that regulates apoptosis not only by binding to antiapoptotic Bcl-2 proteins in mitochondria but also by binding to an endoplasmic protein, Bap31, an adapter protein for procaspase-8 and Bcl-XL that appears to play a role in Fas-induced cell death (63).

Given that there are at least 7 antiapoptotic and 16 proapoptotic members of the Bcl-2 family, the obvious question is how do members of these two groups ultimately regulate cytochrome c release and the downstream activation of the mitochondrial apoptotic pathway. The answer to this question is that, although we now know a lot about how this family of proteins interacts to regulate cytochrome c release and activate the mitochondrial apoptotic pathway, there is much yet to learn. What we do know is that proapoptotic multidomain proteins (MDPs), particularly Bax and Bak, possess intrinsic cell death–inducing activity; however, this activity requires oligomerization to trigger cytochrome c release. On the other hand, antiapoptotic MDPs, particularly Bcl-2 and Bcl-XL, oppose the intrinsic death-inducing actions of proapoptotic MDPs. Although much studied, the mechanism by which antiapoptotic proteins suppress the activity of proapoptotic proteins remains unclear. Early studies demonstrated that antiapoptotic proteins physically interact with proapoptotic proteins, mutually opposing each other (41); however, subsequent mutagenesis studies suggest that antiapoptotic proteins can suppress the activity of proapoptotic proteins without binding (64,65). In addition, the activation of proapoptotic MDPs is made even more complicated by the activities of BH3-only proteins. Available data show that BH3-only proteins differentially interact with both pro- and antiapoptotic proteins to regulate cytochrome c release (Fig. 4). These differential interactions, and other data, have led to the development of two models for Bax/Bak activation [see recent review (36) for detailed overview]. The Direct Activation model (66,67) proposes that there are two "flavors" of BH3-only proapoptotic proteins, "sensitizer/derepressor" and "activator." "Sensitizer/derepressor" BH3-only proteins (Bad, Bik, Bmf, Hrk, and Noxa) bind only to antiapoptotic MDPs and induce apoptosis by displacing BH-3 proapoptotic proteins bound to prosurvival proteins. The displaced BH-3 proapoptotic proteins then directly engage and activate Bax/Bak. In contrast, "activator" BH3-only proteins (Bim, tBid, and Puma) have the capacity to directly activate Bax/Bak. In the

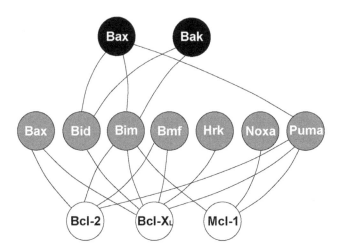

Figure 4 Differential binding partners among anti- and proapoptotic members of the Bcl-2 family.

Prosurvival Neutralization model (68), antiapoptotic MDPs inhibit proapoptotic MDPs, perhaps by direct interactions, and BH3-only proteins induce apoptosis by neutralizing the inhibiting activity of the antiapoptotic MDPs. Once neutralized, Bax/Bak activation occurs spontaneously. Whether one or the other of these models is correct or both are correct, but in different cell death scenarios (cell-type, apoptotic stimulus), remains to be determined.

p53 Family Proteins

Another important family of proteins that regulate cell proliferation and apoptosis, particularly apoptosis induced in response to DNA damage, hypoxia, oncogene expression, and nucleotide depletion, is the p53 family, consisting of p53, p63, and p73 (69,70). The way in which p53 family proteins regulate cell proliferation and apoptosis is also complex because (1) all three members of this family encode multiple isoforms generated by multiple splicing, alternative promoters, and alternative sites of initiation of translation (9 protein isoforms for p53, 6 for p63, and 29 for p73); (2) isoforms can have distinct biological activities; (3) isoforms can interact in different ways to modulate biological activity; and (4) one member of the family, p53, requires activation by posttranslational modifications on multiple sites (phosphorylation, acetylation, methylation, ubiquitination, sumoylation, and neddylation). Of interest in the context of this review, loss of p53 in the mouse leads to a variable rate of exencephaly (71–74), loss of p63 results in animals born with craniofacial malformations and limb truncations (75,76), whereas mice deficient in all p73 isoforms develop hippocampal dysgenesis and hydrocephalus (76). Whereas p53, p63, and p73 play a role in regulating apoptosis, the majority of studies published focus on the mechanisms by which p53 activates the intrinsic mitochondrial apoptotic pathway.

Given the important role that p53 plays as guardian of the genome, it is not surprising that p53 is subject to tight regulatory control. Under normal conditions, that is, lack of stress, p53 levels are low because of rapid turnover, which is regulated by mdm2, an E3 ubiquitin ligase that binds to and ubiquitinates p53, thus targeting it for proteosomal degradation. In response to DNA damage, hypoxia, oncogene expression, and nucleotide depletion, mdm repression of p53 is relieved, p53 levels accumulate, p53 migrates into the nucleus, and p53-dependent activation/repression of p53 target genes ensues (Fig. 5). This activation of p53 is associated with a series of posttranslational modifications, primarily in the N-terminus, a region involved in the transactivation capacity of p53 and interactions with mdm2, and in the C-terminal regulatory region. Posttranslational modifications, phosphorylation, acetylation, methylation, ubiquitination, sumoylation, and neddylation, have been reported at 24 different sites within the p53 molecule (69). Phosphorylations, affected by a number of protein kinases such as ATM (mutated in ataxia-telangiectasia), ATR (A-T and Rad-3 related), check point kinases (Chk1/2), JNK (Jun NH2-terminal kinase), p38, and others, are the most prevalent modifications observed [Fig. 6; for details concerning the complex

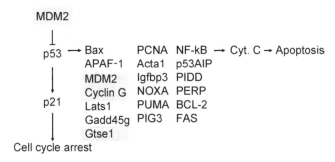

Cell cycle arrest

Figure 5 Key target genes transactivated by p53 that play a role in p53-mediated apoptosis and/or cell cycle arrest.

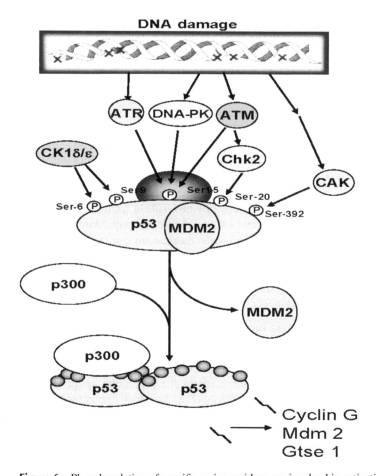

Figure 6 Phosphorylation of specific serine residues are involved in activating p53.

nature of p53 activation, see the excellent review by Lavin and Gueven (77)]. The most frequently observed phosphorylation occurs at serine 15 (serine 18 in the mouse), which, at least in some scenarios, is required for the activation of the mitochondrial apoptotic pathway (78,79).

Activated p53 is thought to induce the mitochondrial apoptotic pathway by transcriptionally upregulating the expression of proapoptotic genes such as *Apaf-1*, *Bax*, *NOXA*, and *PUMA* and by downregulating the expression of antiapoptotic genes such as *Bcl-2* and IAPs (80) (Fig. 5). Although less is known about how p63 and p73 activate the mitochondrial apoptotic pathway, Melino et al. (81) have shown that p73 induces apoptosis by transcriptionally upregulating the expression of PUMA.

Accumulated evidence [see review by Moll et al. (82)] provides convincing evidence that p53 can also activate the mitochondrial apoptotic pathway through transcription-independent mechanisms. In one study (83), Haupt et al. showed that apoptosis could be induced in HeLa cells expressing a truncated p53 (p53dl214) that was completely inactive as a transactivator of p53-responsive genes. Schuler et al. (84) showed that p53 activates the mitochondrial apoptotic pathway through the induction of the release of cytochrome c. In another study, direct targeting of p53 to the mitochondria has been reported to induce apoptosis in ML-1 and RKO cells exposed to campothecin and hypoxia (85,86). Localization of p53 to mitochondria occurs within 1 hour of p53 activation and precedes changes in mitochondrial membrane potential, cytochrome c release, and procaspase-3 activation. More recently, Erster et al. have shown that p53 translocation to the mitochondria occurs in vivo (various mouse tissues) before p53-mediated transcription of proapoptotic genes (87). In addition, they showed that mitochondrial p53 accumulation occurs in radiosensitive tissues (thymus, spleen testis, and brain), but not in radioresistant tissues (liver and kidney). More recently, Marchenko et al. suggest that monoubiquitylation of p53 is the mechanism that targets p53 to the mitochondrion (88). Although the mechanism (s) by which cytosolic p53, once targeted to the mitochondria, directly activates the mitochondrial apoptotic pathway is not completely understood, one mechanism seems to be a direct p53-mediated activation of cytosolic Bax (89–91).

IAP Family-/IAP-Binding Proteins

Although Bcl-2 and p53 family members play primary roles in regulating the release of cytochrome c and activation of the mitochondrial apoptotic pathway, cells have also evolved secondary mechanisms for preventing apoptosis. One such mechanism involves a family of IAPs (Fig. 7), which are known to bind both initiator and executioner caspases. IAPs, first identified in baculoviruses, are characterized by one or more baculovirus IAP repeat (BIR) domains. Studies have shown that the BIR2 domain of XIAP is involved in the inhibition of caspase-3 and -9, whereas the BIR3 domain functions to inhibit caspase-9 (92,93). In addition, some IAPs contain RING and CARD domains. The RING domain of XIAP has E3 ubiquitin ligase activity and therefore the potential to promote

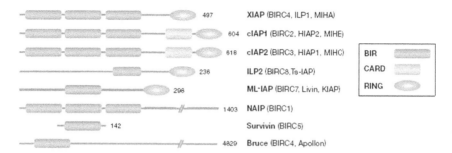

Figure 7 (*See color insert*) Comparison of conserved motifs (BIR, CARD, and RING) among members of the IAP family.

proteosomal degradation of target proteins (94). Overexpression of IAPs protects cells from, whereas silencing IAPs sensitizes cells to, apoptotic stimuli (95). Given that IAPs are known to bind both initiator and executioner caspases, it has been assumed that the antiapoptotic effects of IAPs are related to IAP-induced inhibition of caspases. Recent studies, however, suggest that whereas XIAP is a potent inhibitor of caspase activity, the other IAPs, although they may bind to caspases, do not inhibit their activity (96,97). Given that the non-XIAP IAPs do not inhibit caspase activity, how do they exert their antiapoptotic potential? One possibility is that IAPs serve as protein sinks, thereby binding proteins, such as Smac, and preventing them from activating caspases (98–100). Given that some IAPs can ubiquitinate target proteins (94), another possibility is that IAPs may ubiquitinate caspases and thereby target them for proteosomal degradation (101). Additional studies, conducted in different cellular contexts, will be required to understand the full range of antiapoptotic mechanisms of IAP family members.

Heat Shock Family Proteins

Heat shock proteins (Hsps) constitute a highly conserved family of chaperone proteins that assist in the correct folding of nascent and stress-damaged proteins, thereby preventing their aggregation. Some Hsps, for example, Hsp90, are constitutively expressed whereas others, for example, Hsp70-1/-3, exhibit little if any constitutive expression but can be rapidly induced by a variety of chemical and physical stresses. By a variety of mechanisms, Hsps protect cells/tissues/organs/organisms from the negative effects of these stresses (102). Among this family of Hsps, overexpression of Hsp27, 60, 70, and 90 have been shown to protect against apoptosis induced by a variety of apoptotic stimuli (103). The mechanisms underlying this protection are varied and this section will focus on interactions between Hsps and proteins involved in the intrinsic and extrinsic apoptotic pathways (Table 2). Hsp27 can inhibit both the intrinsic and the extrinsic apoptotic pathways. With respect to the intrinsic pathway, Hsp27 has been shown to block cytochrome c (104) and Smac (105) release, thus blocking downstream events in this pathway. In situations where an apoptotic stimulus results

Table 2   HSPs: Inhibitors of Apoptotic Pathways

HSP	Site of inhibition	References
Hsp27	Cytochrome c release	(104)
	Cytochrome c binding	(106,107)
	Smac release from mitochondria	(105)
	Procaspase-3 activation	(108,109)
	Bid translocation to mitochondria	104
	Binding to Daxx	110
Hsp60	Binding to Bax, Bak, and Bcl-XL	(111,112)
Hsp70	Translocation of Bax to mitochondria	(113–115)
	JNK-mediated Bid cleavage	121
	Procaspase-9 recruitment to Apoptosome	(117,118)
	Procaspase-3/-7 activation	(119)
	P53 inhibition	(116)
	Antagonizes AIF	(120)
	Inhibits stress-activated kinases	(125)
	Binding to Death Receptors 4 and 5	(122)
Hsp90	Apaf-1 binding	(108)

in the release of cytochrome c, Hsp27 has been shown to inhibit the activation of procaspase-9 (106) by binding to cytochrome c and thereby disrupting apoptosome formation (107). Other data indicate that Hsp27 binds to the prodomain of caspase-3, thereby inhibiting procaspase-3 activation (108,109). Hsp27 can also inhibit apoptosis initiated through the extrinsic apoptotic pathway by modulating the translocation of Bid from the cytoplasm to mitochondria (104) or by binding to Daxx, a mediator of Fas-induced apoptosis (110).

Hsp60, a predominantly mitochondrial protein important for folding proteins after import into the mitochondria, appears to exert its antiapoptotic effects in myocytes by binding to Bax and Bak (111), as well as Bcl-XL (112). Shan et al., also show that Hsp60 is associated with increased ubiquitination of Bax, presumably leading to its degradation and a decrease in the ubiquitination of Bcl-XL.

Members of the Hsp70 family can inhibit apoptosis by interfering with events upstream and downstream of cytochrome c release from mitochondria. Identification of important upstream events includes studies using melanocytes (113), macrophages (114), Hela cells (114), COS-7 cells (114), and T cells (115). These studies show that Hsp70, alone or in partnership with Hsp40 members, blocks apoptotic stimuli-induced translocation of Bax to the mitochondria. The mechanism for this Hsp70-induced inhibition of Bax translocation remains unclear and may (114) or may not (115) involve direct interactions between Hsp70 and Bax.

Other studies have demonstrated that mortalin, a novel member of the Hsp70 family, binds to a C-terminal region (amino acid residues 312–352) of p53 that includes its cytoplasmic sequestration domain (116). Although details are lacking, presumably the Hsp70-sequestered p53 is prevented from migrating into the nucleus and assuming its proapoptotic transactivation activity. Downstream events regulated by Hsp70 include inhibiting procaspase-9 recruitment to the apoptosome (117,118), inhibiting procaspase-3/-7 activation (119), and antagonizing AIF (120).

Hsp70 can also interact with factors in the extrinsic receptor-mediated apoptotic pathway. Gabai et al. have reported that Hsp70 blocks TNF-induced apoptosis by inhibiting JNK-mediated Bid cleavage (121). Although the mechanism for JNK-mediated Bid cleavage and activation is unclear, data support the suggestion that JNK regulates Bid cleavage by caspase-8 or other proteases or by regulating the activation of these proteases. In addition, the studies of Guo et al. show that Hsp70 inhibits the assembly and activity of the Apo-2 L/TRAIL-induced death-inducing signaling complex formation and apoptosis by binding to death receptors 4 and 5 (122).

Finally, Hsp90 has also been reported to inhibit apoptosis by blocking the mitochondrial apoptotic pathway. Pandey et al. showed that Hsp90-mediated inhibition of the mitochondrial apoptotic pathway involves binding to Apaf-1 (123). Rodina et al. confirmed that Hsp90 forms a complex with Apaf-1; however, they showed that disruption of this complex only partly explains the activation of the mitochondrial apoptotic pathway (124). At least in SCLC cells, full activation of the mitochondrial apoptotic pathway requires a disruption of the Hsp90/Akt complex leading to the degradation of Akt, reduction in Bad phosphorylation, and cytochrome c release.

Genetic Analysis of the Apoptotic Pathway: Insights into the Role of Apoptosis in Normal and Abnormal Development

In an effort to learn more about the role of "apoptotic proteins," many of the genes for the proteins discussed in the previous sections have been "knocked out," primarily by homologous recombination (Table 3). Although the major focus of studies using these knockouts was to probe the role of specific proteins in apoptosis, these studies also provided information about whether specific proteins were essential for any of the PCD that occurs during development. Whereas the deletion of many of the "apoptotic genes" listed in Table 3 show no abnormal phenotype and presumably no effect on PCD, although this was seldom assessed, the loss of some "apoptotic genes" did affect PCD resulting in abnormal development. For example, loss of *caspase-9* and *Apaf-1* results in reduced apoptosis in the developing brain and embryos exhibiting brain defects (e.g., exencephaly) that die around E11–12.5 (*caspase-9*) or E16.5 (*Apaf-1*). These studies indicate that these two genes are essential for at least some of the PCD occurring in the developing brain.

Table 3 Developmental Phenotype in Mice with Targeted Deletions of "Apoptotic" Proteins

Genes	Developmental phenotype	References
Caspase-2	Excess germ cells; accelerated motor neuron cell death	(126)
Caspase-3	Neuronal hyperplasia; perinatal lethality	(127,128)
Caspase-6	Normal, external phenotype at birth	(129)
Caspase-7	Normal, external phenotype at birth	(128)
Caspase-3/-7	10% exencephaly; 100% postnatal mortality	(128)
Caspase-8	Embryonic lethality (E11-12.5); heart defects	(130)
Caspase-9	Abnormal brain development due to decreased apoptosis; embryonic lethality	(131,132)
Apaf-1	Exencephaly; rostral expansion of forebrain; craniofacial malformations; embryonic lethality (E16.5)	(133,134)
Fadd	Embryonic lethality (E11.5); abnormal heart development	(135,136)
Bcl-2	Normal, external phenotype at birth, but growth retarded; early postnatal mortality; polycystic kidneys	(137)
Bcl-XL	Embryonic lethality (E13); extensive apoptosis in developing CNS	(138)
Bcl-w	Normal, external phenotype at birth; Postnatal testicular degeneration	(139,140)
Mcl-1	Peri-implantation embryonic lethality	(141)
A-1	Normal, external phenotype at birth	(142)
Bax	Normal, external phenotype at birth; abnormal retinal development	(143,144)
Bak	Normal, external, and internal development at birth	(145)
Bax/Bak	Interdigital webs on fore and rear paws; imperforate anus; increased neurons in multiple regions of the brain	(145)
Bad	Normal, external, and internal development at birth	(146)
Bid	Normal, external phenotype at birth	(147)
Bik/NBK	No reports of Bik deficient mice located	
Bim	No abnormal phenotype reported; some embryonic lethality before E10	(148)
Bmf	No reports of Bmf deficient mice located	
Bnip3	No reports of Bnip3 deficient mice located	
Hrk/DP5	Null fetuses showed no gross abnormalities	(149)
Noxa	No developmental abnormalities reported	(150,151)
Puma	No developmental abnormalities reported	(151,152)
Cytochrome c	Exencephaly and rostral expansion of forebrain in embryos destined to die prenatally; hydrocephalus in postnatal pups	(153)

(Continued)

Table 3 Developmental Phenotype in Mice with Targeted Deletions of "Apoptotic" Proteins (*Continued*)

Genes	Developmental phenotype	References
AIF	Attempts to derive AIF deficient mice using AIF deficient ES cells failed	(154)
Smac/Diablo	No developmental abnormalities reported at birth	(155)
HtrA2/Omi	No developmental abnormalities reported at birth; postnatal neurodegenerative disorder	(156)
Endonuclease G	No developmental abnormalities reported at birth	(157)
CAD	No developmental abnormalities reported at birth	(158)
XIAP	No developmental abnormalities reported at birth	(159)
cIAP	No reports of cIAP deficient mice located	
survivin	No reports of survivin deficient mice located	
P53	Approximately 10% of null fetuses are exencephalic and all are female	(71,160)
P63	No skin development; perinatal death	(161,162)
P73	Hippocampal dysgenesis; hydrocephalus	(76)
Hsp25/HSPB1	No developmental abnormalities reported at birth	(163)
Hsp70	No developmental abnormalities reported at birth	(125,164,165)
Hsp90	No reports of Hsp90 deficient mice located	

Although loss of Bax or Bak individually had no effect on normal development, the loss of both Bax and Bak resulted in interdigital webbing in the paws, clearly indicating that Bax and Bak are essential for interdigital limb PCD. What is intriguing is that while *caspase-9* and *Apaf-1* are essential for at least some "brain PCD," loss of either of these genes had no effect on other episodes of PCD, for example, limb PCD. Similarly, Bax and Bak are essential for limb PCD but appear to play no essential role in "brain PCD." One explanation for this apparent paradox is that apoptotic proteins act in a tissue-specific fashion. For example, it might be hypothesized that *caspase-9* is essential for "brain PCD" whereas *caspase-8* is essential for limb PCD. A key test of this hypothesis would be to determine the effects of *caspase-8* deletion on limb PCD. Unfortunately, *caspase-8* null embryos die around E11–12.5, before the main episode of interdigital limb PCD. Another explanation is that there are unidentified "modifiers" of PCD. Precedent for this comes from numerous studies describing strain-specific effects. For example, the original *caspase-3* knockout embryos, on a 129 × B6F1 background, exhibited significant ectopic brain masses resulting from decreased brain PCD (127). Backcrossing these mice onto the C57 BL/6 J background drastically attenuated this neuronal phenotype (129). These results indicate that there is a strain-specific gene or genes that can "substitute" for *caspase-3* and thereby suppress the *caspase-3* phenotype. Clearly, a great deal more needs to be learned about the mechanisms of PCD.

NORMAL PCD

Occurrence

It has long been known that PCD can be observed during the development of most, if not all, tissues/organs (166). Despite the fact that PCD is ubiquitous during normal development and the fact that, as reviewed earlier, much is known about the genetics and biochemistry of apoptotic pathways, much less is known about the relationship between PCD and the extrinsic/intrinsic apoptotic pathways and, more important, how PCD is regulated. Nonetheless, although far from complete, an understanding of the molecular basis of vertebrate limb PCD has begun to emerge (167); therefore, the remainder of this section will highlight what is known about the relationship between PCD and the extrinsic/intrinsic apoptotic pathways and how limb PCD is regulated.

Much of what is known about the occurrence of PCD during limb development comes from studies using chick and mouse embryos. In the chick, PCD first appears at stage 21 in the superficial mesoderm at the anterior edge of the limb bud, the so-called anterior necrotic zone. Later, at stage 24, massive mesenchymal cell death occurs at the posterior junction of the limb bud and the body wall, the posterior necrotic zone. At stage 31, cell death is observed in the interdigital mesenchyme, that is, the interdigital necrotic zones. Although cell death can be observed in anterior necrotic zone and posterior necrotic zone of the mouse limb bud, this cell death is not as pronounced as it is in the chick limb bud. Interdigital cell death in the mouse limb bud [Fig. 8 (A)] begins at E12.5 and reaches a peak at E14.5.

Molecular Basis of Limb PCD

Milligan et al. were the first to provide data showing that caspases may play a role in interdigital limb PCD by showing that caspase inhibitors could rescue interdigitial cells destined to die (168). Although these studies did not identify specific caspases, later studies, using immunohistochemistry and antibodies that recognized only active caspase-3 [Fig. 8 (B)], unequivocally demonstrated that procaspase-3 is activated in normal limb PCD (169). More recently, Hurle and colleagues have used Western blotting and immunohistochemistry to show that caspase-3, -6, and -7 are activated in the interdigital mesenchyme of chick limbs (170). These results confirm that the execution phase of the mitochondrial apoptotic pathway is activated in limb PCD. They have also shown, using Western blot analysis and lysates of interdigital mesoderm cells, that cytochrome c release from mitochondria occurs in these cells during interdigital PCD. Despite the fact that cytochrome c release occurs during PCD, the initiator procaspase-9 does not appear to be activated in interdigital mesenchymal cells that undergo PCD (170,171). These same authors have also reported that the initiator procaspase-8 is not activated in cells undergoing PCD. Thus, at this time, it is unclear how the execution phase of the mitochondrial apoptotic pathway is activated.

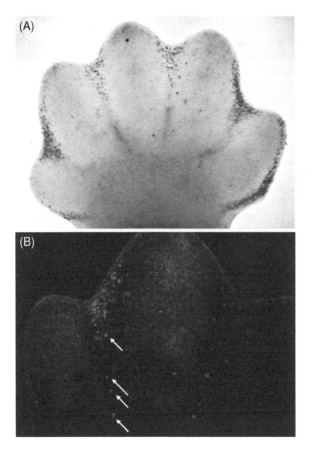

Figure 8   (*See color insert*) (**A**) E13 mouse embryo limb bud stained with Neutral Red showing interdigital PCD. (**B**) E13 mouse limb bud immunohistochemically stained showing caspase-3 is activated in apoptotic cells within the interdigital mesenchyme.

One possibility is procaspase-2, a little studied initiator caspase that has been reported to activate caspase-3, -6, and -7 (172). Using an anti-active caspase-2 antibody, Zuzarte-Luis et al. show that preapoptotic interdigital cells are intensely stained (170). In addition, these authors have shown that FGF2, a potent survival factor for limb mesoderm, downregulates the expression of the *caspase-2* gene. Likewise, downregulation of *caspase-2* by siRNA causes a delay in interdigital PCD. These studies, therefore, support the hypothesis that activation of procaspase-2 during PCD triggers the activation of executioner caspases and subsequent downstream events leading to cell death.

Finally, Zuzarte-Luis et al. show that the proapoptotic factor AIF translocates from mitochondria to the nucleus during interdigital PCD (170), suggesting that PCD may also occur via a caspase-independent pathway. Additional support for a caspase-independent mode of PCD comes from studies showing that the aspartate protease cathepsin D may cooperate with caspases to affect PCD (173). Thus, vertebrate embryos appear to have multiple pathways that lead to PCD.

Regulation of Limb PCD

As shown in Figure 8 (A,B), cells in the interdigital mesenchyme of the developing limb undergo apoptosis whereas their immediate neighbors containing digital mesenchyme do not. This observation leads to two obvious questions: (1) what are the signals that instruct interdigital cells to die and (2) once the death signal is received, how is it linked to the intrinsic mitochondrial apoptotic pathway and/or the lysosomal pathway involving Cathepsin D? Answers to these two questions are far from complete; however, some of the players have been identified [see Zuzarte-Luis and Hurle for an excellent review (167)].

One of the major players in regulating PCD is the family of bone morpho-genetic proteins (BMPs) that belong to the transforming growth factor β superfamily. BMPs exert their effects (cell proliferation, differentiation, apoptosis, left-right asymmetry, neurogenesis, mesodermal patterning, and the development of the kidney, gut, lung, teeth, limb, amnion, and testes) by binding to two types of serine threonine kinase receptors, BMPRI and BMPRII (174). BMP ligand/receptor interactions then trigger complex intracellular signaling via Smad and non-Smad pathways (175). Although the evidence that BMPs are causally linked to limb PCD is clear (176–178), the mechanistic links between BMP intracellular signaling and activation of caspase-dependent and/or independent apoptotic pathways await further research.

TERATOGEN-INDUCED CELL DEATH

More than 1200 chemical and physical agents are known to cause structural and/or functional malformations in experimental animals (179). Although the mechanisms by which these agents disrupt normal development are often not well understood, it is known that many teratogens induce cell death in tissues that subsequently develop abnormally and give rise to structural malformations (180,181). That this cell death may be causally linked to malformations is supported by studies showing that mutations in specific genes have been shown to result in elevated levels of cell death and structural malformations. For example, deletion of the ski proto-oncogene leads to excessive apoptosis in the neuroepithelium, which is associated with abnormal development resulting in exencephaly (182). Conversely, loss of gene function, for example, *Apaf-1*, results in decreased cell death that is also associated with abnormal development, for example, forebrain malformations, spina bifida, and eye and ear malformations (133,134). These studies highlight an important point, either too much or too little cell death can disrupt normal development and give rise to malformations.

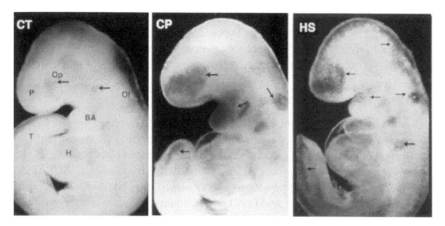

Figure 9 (*See color insert*) Neutral red stained E 9 mouse embryos showing cell death (*arrows*) in untreated (CT), CP-treated or HS-treated embryos.

Examples of Teratogen-Induced Cell Death

Although many teratogens have been shown to induce excessive cell death, the remainder of this section will focus on two teratogens, HS and cyclophosphamide (CP) because (1) CP is a well-studied example of a chemical teratogen, whereas HS is a well-studied physical teratogen, (2) both are animal and human teratogens, and (3) more is known about the biochemistry of HS- and CP-induced cell death than is known about cell death induced by other teratogens. Figure 9 shows day 9 mouse embryos stained with Neutral Red to visualize cell death.

Relationship of Teratogen-Induced Cell Death to PCD

Although little appreciated and not well studied, several investigators have pointed out that teratogens often induce cell death (183–187) or attenuate cell death in areas of normal PCD (188,189). These observations have led to the suggestion that there is a mechanistic link between PCD and teratogen-induced cell death. What the link or links may be remains largely a mystery; however, potential links have been reported. For example, Cheema et al. have shown that ethanol, known to increase cell death in the developing mouse CNS (187), can upregulate the FasR in cerebral cortex cultures (190). If this also occurs in the developing mouse brain, ethanol may induce cell death in the embryonic CNS by upregulating death receptors in "potential" PCD populations already primed to die by a receptor-mediated apoptotic pathway. Alternatively, McAlhany et al. have shown that ethanol selectively induces the activation of c-jun N-terminal-kinase (JNK) (191), a mitogen-activated protein kinase known to play a critical role in naturally occurring cell death during development (192). If ethanol also selectively activates JNK in the developing mouse brain, ethanol may induce cell death in the embryonic CNS by activating downstream signaling in the JNK-mediated apoptotic pathway in "potential" PCD populations already primed to die through activation of the mitochondrial apoptotic pathway. Further elucidation of the links between PCD and teratogen-induced cell death awaits the results of future studies.

Biochemistry of Teratogen-Induced Cell Death

HS and 4-hydroperoxycyclophosphamide (4CP) induce apoptosis in early postimplantation rodent embryos by activating the mitochondrial apoptotic pathway. Activation of this pathway is characterized by the release of cytochrome c and the subsequent activation of caspase-3, -6, -7, and -9; cleavage of poly ADP-ribose polymerase; and DNA fragmentation (171,193–196) [Fig. 10 (A–E)]. Other studies have shown that retinoic acid (197) and cadmium (198) also induce activation of procaspe-3. Thus, at least for this small sampling of teratogens, teratogen-induced apoptosis in early postimplantation mouse embryos involves activation of the mitochondrial apoptotic pathway.

The rapid induction of the mitochondrial apoptotic pathway in teratogen-sensitive neuroepithelial cells and the failure to activate this pathway in teratogen-resistant heart cells suggest that the embryo must possess factors that regulate the efflux of cytochrome c and thereby the activation of the mitochondrial apoptotic pathway. To begin to identify proteins and signaling pathways that regulate cytochrome c release, Mikheeva et al. used DNA microarray gene expression profiling to compare gene expression patterns in HS or 4CP-treated and untreated mouse embryos before and during the activation of the mitochondrial apoptotic

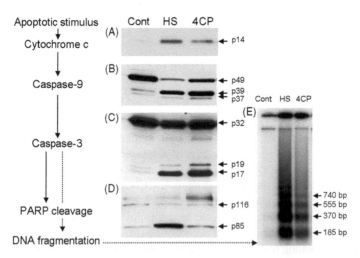

Figure 10   Western blot analysis showing HS- or 4CP-induced cytochrome c release (**A**), activation of caspase-9 (**B**), activation of caspase-3 (**C**), and PARP cleavage (**D**) in day 9 mouse embryos. **E** is a gel assay showing teratogen-induced DNA fragmentation (*DNA laddering*).

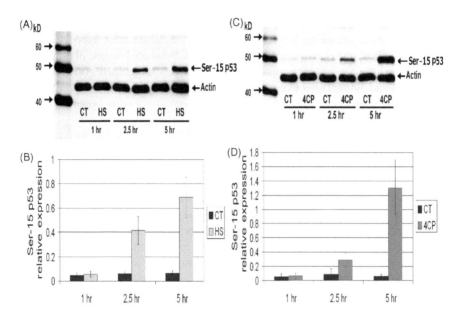

Figure 11 Western blot analysis showing the time-course of HS-induced (**A**) or 4CP-induced (**C**) phosphorylation of serine 15 in p53 in day 9 mouse embryos. Quantitation of increased serine-15 p53 induced by HS (**B**) and 4CP (**D**). *Source*: From Ref. 201.

pathway (199). Their studies identified five candidate "apoptosis-related" genes. Three of these genes, *Mdm2*, *Gtse1*, and *Cyclin G*, are coordinately upregulated by both HS and 4CP during the first 5 hours after embryos are exposed to these teratogens. Because these three genes are all p53-regulated genes (Fig. 5), this suggests that HS and 4CP both activate p53.

Called the "guardian of the genome," p53 is known to play a key role in regulating whether a cell will arrest, undergo apoptosis, senesce, or differentiate in response to various stresses. To achieve this regulatory role, p53 must be activated by a variety of posttranslational modifications, for example, phosphorylation, acetylation, and ubiquitination (200). Using a panel of phospho-specific p53 antibodies directed against human ser-6, -9, -15, -20, -37, -46, and -392 p53, Hosako et al. have shown that p53 is phosphorylated at ser-15 after exposure to HS or 4CP (200) [Fig. 11 (A–D)]. Phosphorylation at this serine is known to regulate apoptosis because p53-mediated apoptosis is significantly impaired when serine 15 is mutated to alanine (78,79). Using an antibody that recognizes phosphorylated and nonphosphorylated p53 (pan-p53), they also showed that the increase in ser-15 p53 is correlated with an increase in total p53. Finally, using ser-15 p53 antibodies and immunohistochemistry, they have shown that activated p53 localizes to the nucleus of stained cells. Together, these results show that HS and 4CP, two teratogens that induce apoptosis in day 9 mouse embryos, also activate p53, a known regulator of apoptosis.

Data also show that p53 is rapidly activated by both HS and 4CP, with activation occurring between 1 hour and 2.5 hours after exposure (201) (Fig. 11). Thus, p53 is activated before HS- and 4CP-induced release of mitochondrial cytochrome c and activation of the caspase cascade, which occur between 2.5 hours and 5 hours after exposure to these two teratogens (193–195). Although not definitive, the kinetics of p53 activation are consistent with a regulatory role for p53 in teratogen-induced apoptosis.

Although data are consistent with a regulatory role for p53 in teratogen-induced apoptosis, the mechanism(s) by which p53 activates the mitochondrial apoptotic pathway in the day 9 mouse embryo are unclear. One known mechanism, elucidated by cell culture studies, is the transcription-dependent expression of proapoptotic genes. Two of the major downstream targets of p53-mediated apoptosis are *Noxa* and *Puma*, transcripts of proapoptotic genes belonging to the Bcl-2 family. At least in some settings, NOXA protein is an essential mediator of p53-dependent apoptosis (202) and activates the mitochondrial apoptotic pathway by interacting with Bcl-2 family members resulting in the release of cytochrome c and the activation of caspase-9 (203,204). Other studies suggest that PUMA protein interacts with BCL-2 and BCL-XL and thereby induces mitochondrial membrane potential change, cytochrome c release, and caspase activation (205–207).

Although results indicate that HS and 4CP both induce increased expression of *Noxa* and *Puma* mRNAs, the increased expression of these mRNAs is not coupled with an increased expression of NOXA and PUMA proteins (201). Thus, the simplest hypothesis is that transcriptional upregulation of Noxa and/or Puma is not required to activate the mitochondrial apoptotic pathway in embryos exposed to HS or CP. Although upregulation of NOXA and PUMA proteins may not be required, these proteins may still play a role in the activation of the mitochondrial apoptotic pathway. Of interest, data show that NOXA and PUMA proteins are constitutively expressed in the day 9 embryos in the absence of any teratogenic exposure (201). The function of these proteins in mouse development is unknown, but do not appear to be required for normal development because *Noxa* and *Puma* null mice are born at the expected frequency and exhibit a normal phenotype (151). How the proapoptotic activity of NOXA and PUMA is blocked is also unknown; however, it may be that these proteins are sequestered in an inactive form that is then activated in response to appropriate apoptotic stimuli. The constitutive expression of proapoptotic Bcl-2 family members that are sequestered and then activated in response to various apoptotic stimuli is well documented. Examples described earlier include binding to other proteins (BIM and BMF binding to dynein motor complex), cleavage (inactive BID cleaved to active tBID by caspase-8), and phosphorylation-induced binding of BAD to 14–3-3 (45,49,51,208,209). However, data showing that NOXA or PUMA proteins are sequestered in the absence of an apoptotic stimulus and then activated after an appropriate cell death signal have not been reported.

Even if p53-mediated upregulation of NOXA and PUMA proteins does not play a role in activating the mitochondrial apoptotic pathway in teratogen-exposed mouse embryos, p53 is known to upregulate other proapoptotic proteins, for example, BAX, p53AIP, and PIGs. Whether any of these or other p53 target genes play a role in HS- and 4CP-induced activation of the mitochondrial pathway is unknown; however, studies have shown that there is no significant increase in BAX protein levels in mouse embryos exposed to HS or 4CP (unpublished data). Although p53 may regulate teratogen-induced apoptosis in the mouse embryo by transcriptionally upregulating "proapoptotic" target genes, available data do not support this possibility. Alternatively, as documented earlier, recent evidence has uncovered a transcription-independent role for p53 in the regulation of apoptosis. Whether p53 mediates teratogen-induced apoptosis by a transcription-independent pathway remains to be determined.

Recent data also show that HS and 4CP induce the upregulation of cyclin-dependent kinase p21 mRNA and protein (201). Moreover, results from immuno-histochemical analysis indicate that p21 protein is upregulated in most, if not all, cells of the day 9 mouse embryo after exposure to HS. Because p21 is a known p53 target that plays a central role in arresting the cell cycle after various geno-toxic stresses (69,210), these results suggest that cells of the day 9 mouse embryo have activated the cell cycle arrest arm of the p53 pathway in response to terato-genic exposures. Although it is not known whether HS induces cell cycle arrest in early postimplantation rodent embryos, published studies have shown that phos-phoramide mustard, the major teratogenic metabolite of 4CP, induces alterations in the cell cycle in postimplantation rat embryos (211,212). In addition, Chernoff et al. (1989) and Francis et al. (1990) have shown that CP induced a dose-dependent increase in the percentage of limb bud cells in the S phase of the cell cycle (213,214). Together, these results demonstrate that CP/4CP induce alterations in the cell cycle in early postimplantation mouse embryos exposed in vitro or in vivo.

Published data consistently show that teratogens induce apoptosis in some cells of the embryos and not others (215–219). Using vital dyes and TUNEL staining, studies have shown that teratogen-induced cell death is cell specific, that is, some cells in the mouse embryo die, particularly in areas of normal PCD, while other cells, often neighboring cells, survive (169,193). For example, cells of the embryonic nervous system (neuroepithelial cells) are particularly sensitive to teratogen-induced cell death, mesenchymal cells surrounding the neuroepithe-lium are less sensitive, and cells of the embryonic heart are completely resistant (193) [Fig. 12 (A,B)]. In addition, hallmarks of apoptosis (cytochrome c release, activation of caspases, PARP cleavage, and DNA fragmentation) are not activated in cells of the heart (169,193,194) [Fig. 13 (A–E)]. These results indicate that the mitochondrial apoptotic pathway is blocked in heart cells at the level of the cytochrome c release from mitochondria or at some point upstream of cytochrome c release. More recent data show that heart cell resistance is also associated with significant attenuation of the activation of p53 in heart cells (201) [Fig. 14 (A–D)]. Despite the attenuated activation of p53 in heart cells in response to teratogenic exposures, data indicate that both HS and 4CP induce increased levels of p21

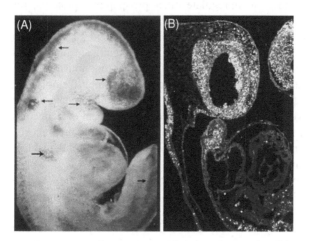

Figure 12 (*See color insert*) Whole mount (**A**) and parasagittal section (**B**) of a day 9 mouse embryo 5 hours after exposure to HS and then stained for activated caspase-3 (*red fluoresence*) and DNA fragmentation (*green fluorescence*). Although occasional cells are stained only red, indicating activation of caspase-3 but not DNA fragmentation, or green, indicating DNA damage but no caspase-3 activation, the majority of cells are yellow indicating that these cells have activated caspase-3 and undergone DNA fragmentation.

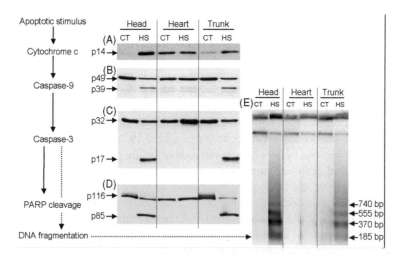

Figure 13 Western blot analysis showing cytochrome c release (**A**), caspase-9 activation (**B**), caspase-3 activation (**C**), and PARP cleavage (**D**) in isolated heads and trunks (sensitive to HS-induced apoptosis) compared to hearts (resistant to HS-induced cell death). **E** is a gel assay comparing DNA fragmentation (*DNA laddering*) in heads, hearts, and trunks isolated from day 9 mouse embryos.

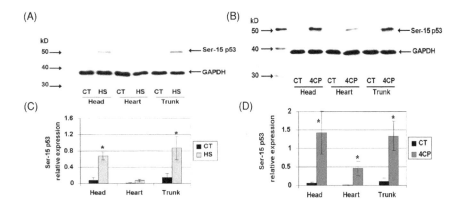

Figure 14 Western blot and associated quantitation of serine 15-p53 induced by HS (*left*) or 4CP in heads, hearts, and trunks isolated from day 9 mouse embryos. *Source*: From Ref. 201.

in heart cells [Fig. 15 (A–D)]. These results suggest that p53 is activated in the heart and when activated subsequently up regulates the expression of p21, thereby arresting heart cells. One caveat, however, is that p21 expression can also be induced via p53-independent mechanisms (220).

Results showing robust activation of p53 in cells sensitive to teratogen-induced apoptosis and attenuated activation of p53 in cells resistant to teratogen-induced apoptosis lead to the hypothesis that high levels of activated p53 induce apoptosis, whereas low levels of activation lead to cell cycle arrest. This hypothesis is supported by studies showing that high levels of ectopic p53 induce apoptosis, whereas lower levels result in cell cycle arrest (221–223). More recently, Speidel et al. showed that low levels of UV irradiation, which led to a relatively low-level activation of p53, induced temporary cell cycle arrest, whereas high levels of UV irradiation, which induced a more robust activation of p53, led to apoptosis (91). Although results from studies using embryos exposed to HS or CP are consistent with the cell culture data, that is, that low levels of p53 activation culminate in cell cycle arrest whereas more robust activation of p53 results in apoptosis, additional research will be required to determine whether the sensitivity/resistance of specific cells to teratogen-induced apoptosis in the day 9 mouse embryo is determined by the extent to which p53 is activated.

SUMMARY

Evidence presented in this review clearly shows that apoptosis is an integral part of normal development and, when apoptosis is induced or inhibited by altered gene function or teratogens, can lead to abnormal development and birth defects. Nonetheless, many gaps remain in our understanding of the mechanisms of PCD and teratogen-induced cell death. Although we now know that at least some

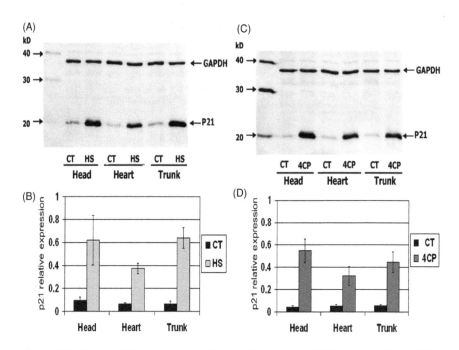

Figure 15 Western blot analysis showing the time-course of HS-induced (**A**) or 4CP-induced (**C**) expression of p21 protein in heads, hearts, and trunks isolated from day 9 mouse embryos. Quantitation of p21 protein levels induced by HS (**B**) and 4CP (**D**). *Source*: From Ref. 201.

teratogens induce apoptosis by activating the mitochondrial apoptotic pathway, a role for the receptor-mediated apoptotic pathway in teratogen-induced cell death has not been documented. Thus, there is a need to study a wide variety of teratogens to determine the range of signaling pathways activated that lead to cell death.

In addition, relatively little is known about how different cells in the embryo "decide" to live or die in response to teratogenic exposures. Thus, there is a critical need to study the role of regulatory apoptotic proteins (e.g., members of the Bcl-2, p53, IAP, and Hsp families) to determine which, if any, play a role in this decision. One potentially informative approach to such studies is the use of knockout mice to determine whether loss of a particular gene suppresses or enhances teratogen-induced cell death. Recent results show the loss of p53 and Hsp70, both sensitize embryos to the teratogenic effects of HS (unpublished data). In the case of Hsp70 knockouts, the increased sensitivity to HS-induced exencephaly is correlated with increased apoptosis in the neuroepithelium (unpublished data).

Another gap is in our understanding of the relationship between PCD and teratogen-induced cell death. Why is it that teratogens seem to preferentially induce cell death in areas of normal PCD? Additional insights into the mechanism of PCD, for example, are receptor-mediated pathways used to activate the apoptotic pathway in PCD, may shed light on why cells surrounding cells programmed to die are sensitive to teratogen-induced cell death. A related gap is in our understanding of the difference between cells that die outside of areas of PCD and cells that do not. Is this cell-specific sensitivity to teratogen-induced cell death mechanistically related to cell-specific sensitivity involving cells in areas of PCD?

Answers to all of these questions will provide exciting insights into the mechanisms cells use to "decide" to live or die in the context of normal and teratogen-induced abnormal development. Ultimately, the hope is that this understanding will provide the knowledge needed to prevent alterations in PCD and thereby prevent associated birth defects.

REFERENCES

1. Green DR, Evan GI. A matter of life and death. Cancer Cells 2002; 1:19–30.
2. Twomey C, McCarthy JV. Pathways of apoptosis and importance in development. J Cell Mol Med 2005; 9:345–359.
3. Opferman JT. Apoptosis in the development of the immune system. Cell Death Differ 2008; 15:234–242.
4. Mattson MP. Neuronal life-and-death signaling, apoptosis, and neurodegenerative disorders. Antiox Redox Signaling 2006; 8:1997–2006.
5. Kerr JF, Wyllie AH, Currie AR. Apoptosis: a basic biological phenomenon with wide-ranging implication in tissue kinetics. Br J Cancer 1972; 26:239–257.
6. Crawford AM, Kerr JFR, Currie AR. The relationship of acute mesodermal cell death to the teratogenic effects of 7-OHM-12-MBA in the foetal rat. Br J Cancer 1972; 26:498–503.
7. Horvitz HR, Sternberg PW, Greenwald IS, et al. Mutations that affect neural cell lineages and cell fates during the development of the nematode Caenorhabditis elegans. Cold Spring Harb Symp Quant Biol 1983; 48:453–463.
8. Ellis HM, Horvitz HR. Genetic control of programmed cell death in the nematode C. elegans. Cell 1986; 44:817–829.
9. Conradt B, Horvitz HR. The C. elegans protein EGL-1 is required for programmed cell death and interacts with the Bcl-2-like protein CED-9. Cell 1998; 93:519–529.
10. Del Peso L, Gonzalez VM, Inohara N, et al. Disruption of the CED-9-CED-4 complex by EGL-1 is a critical step for programmed cell death in Caenorhabditis elegans. J Biol Chem 2000; 275:27205–27211.
11. Tsujimoto Y, Finger LR, Yunis J, et al. Cloning of the chromosome breakpoint of neoplastic B cells with the t(14;18) chromosome translocation. Science 1984; 226:1097–1099.
12. Robertson JD, Orrenius S. Role of mitochondria in toxic cell death. Toxicology 2002; 181–182:491–496.
13. Li P, Nijhawan D, Budihardjo I, et al. Cytochrome c and dATP-dependent formation of Apaf-1/caspase-9 complex initiates an apoptotic protease cascade. Cell 1997; 91:479–489.
14. Schafer ZT, Kornbluth S. The apoptosome: physiological, developmental, and pathological modes of regulation. Dev Cell 2006; 10:549–561.
15. Bao Q, Shi Y. Apoptosome: a platform for the activation of initiator caspases. Cell Death Differ 2007; 14:56–65.

16. Luthl AU and Martin SJ. The CASBAH: A searchable database of caspase substrates. Cell Death Differ 2007; 14:641–650.

17. Coleman ML, Sahai EA, Yeo M, et al. Membrane blebbing during apoptosis results from caspase-mediated activation of ROCK 1. Nat Cell Biol 2001; 3:3339–3345.

18. Liu X, Zou H, Slaughter C, et al. DFF, a heterdimeric protein that functions downstream of caspase-3 to trigger DNA fragmentation during apoptosis. Cell 1997; 89:175–184.

19. Soldani C, Scovassi AI. Poly (ADP-ribose) polymerase-1 cleavage during apoptosis: An update. Apoptosis 2002; 7:321–328.

20. Du C, Fang M, Li Y, et al. Smac, a mitochondrial protein that promotes cytochrome c-dependent caspase activation by eliminating IAP inhibition. Cell 2000; 102:33–42.

21. Verhagen AM, Ekert PG, Pakusch M, et al. Identification of DIABLO, a mammalian protein that promotes apoptosis by binding to and antagonizing IAP proteins. Cell 2000; 102:43–53.

22. Joza N, Susin SA, Daugas E, et al. Essential role of the mitochondrial apoptosis-inducing factor in programmed cell death. Nature 2001; 410:549–554.

23. Susin SA, Daugas E, Ravagnan L, et al. Two distinct pathways leading to nuclear apoptosis. J Exp Med 2000; 192:571–580.

24. Li LY, Luo X, Wang X. Endonuclease G is an apoptotic DNase when released from mitochondria. Nature 2001; 412:95–99.

25. Enari M, Sakahira H, Yokoyama H, et al. A caspase-activated DNase that degrades DNA during apoptosis, and its inhibotor ICAD. Nature 1998; 391:43–50.

26. Ekert PG, Vaux DL. The mitochondrial death squad: hardened killers or innocent bystanders? Curr Opin Cell Biol 2005; 17:626–630.

27. Bhardwaj A, Aggarwal BB. Receptor-mediated choreography of life and death. J Clin Immunol 2003; 23:317–332.

28. Tibbetts MD, Zheng L, Lenardo MJ. The death effector domain protein family: regulators od cellular homeostasis. Nat Immunol 2003; 4:404–409.

29. Kischkel FC, Hellbardt S, Behrmann I, et al. Cytotixicity-dependent APO-1 (Fas/CD95)-associated proteins form a death-inducing signaling complex (DISC) with the receptor. EMBO J 1995; 14:5579–5588.

30. Srinivasula SM, Ahmad M, Fernandes-Alnemri T, et al. Molecular cloning of the Fas-apoptotic pathway: the Fas/APO-1 protease Mch5 is a CrmA-inhibitable protease that activates multiple ced-3/ICE-like cystein proteases. Proc Natl Acad Sci U S A 1996; 93:14486–14491.

31. Muzio M, C, Salvesen GS, Dixit VM. FLICE-induced apoptosis in a cell- free system. Cleavage of caspase zymogens. J Biol Chem 1997; 272:2952–2956.

32. Esposti MD. The roles of Bid. Apoptosis 2002; 7:433–440.

33. Luo X, Budihardjo I, Zou H, et al. Bid, a Bcl2 interacting protein, mediates cytochrome c release from mitochondria in response to activation of cell surface death receptors. Cell 1998; 94:481–490.

34. Desagher S, Osen-Sand A, Nichols A, et al. Bid-induced conformational change of Bax is responsible for mitochondrial cytochrome c release during apoptosis. J Cell Biol 1999; 144(5):891–901.

35. Danial NN, Korsmeyer SJ. Cell death: Critical control points. Cell 2004; 116:205–219.

36. van Delft MF, Huang DC. How the Bcl-2 family of proteins interact to regulate apoptosis. Cell Res 2006; 16:203–213.

37. Reed JC. Proapoptotic multidomain Bcl-2/Bax-family proteins: mechanisms, physiological roles, and therapeutic opportunities. Cell Death Differ 2006; 13:1378–1386.
38. Vaux DL, Cory S, Adams JM. Bcl-2 gene promotes haemopoietic cell survival and cooperates with c-myc to immortalize pre-B cells. Nature 1988; 335:440–442.
39. McDonnell TJ, Deane N, Platt FM, et al. bcl-2-immunoglobulin transgenic mice demonstrate extended B cell survival and follicular lymphoproliferation. Cell 1989; 57:79–88.
40. Hockenbery D, Nunez D, Milliman C, et al. Bcl-2 is an inner mitochondrial membrane protein that blocks programmed cell death. Nature 1990; 348:334–336.
41. Oltvai ZN, Milliman CL, Korsmeyer SJ. Bcl-2 heterodimerizes in vivo with a conserved homolog, Bax, that accelerates programmed cell death. Cell 1993; 74:609–619.
42. Wei MC, Zong WX, Cheng EH, et al. Pro-apoptotic Bax and Bak: a requisite gateway to mitochondrial dysfunction and death. Science 2001; 292:727–730.
43. Scorrano L, Oakes SA, Opferman JT, et al. Bax and Bak regulation of endoplasmic reticulum Ca2+: a control point for apoptosis. Science 2003; 300:135–139.
44. Strasser A. The role of BH3-only proteins in the immune system. Nat Rev Immunol 2005; 5:189–200.
45. Zha J, Harada H, Yang E, et al. Serine phosphorylation of death agonist BAD in response to survival factor results in binding to 14–3-3 not BCL-X(L). Cell 1996; 87:619–628.
46. Datta SR, Dudek H, Tao X, et al. Akt phosphorylation of Bad couples survival signals to the cell-intrinsic death machinery. Cell 1997; 91:231–241.
47. Ley R, Balmanno K, Hadfield K, et al. Activation of the ERK1/2 signaling pathway promotes phosphorylation and proteasome-dependent degradation of the BH3-only protein, bim. J Biol Chem 2003; 278:18811–18816.
48. Akiyamma T, Bouillet P, Miyazaki T, et al. Regulation of osteoclast apoptosis by ubiquitylation of proapoptotic BH3-only Bcl-2 family member Bim. EMBO J 2003; 22:6653–6664.
49. Puthalakath H, Huang DCS, O'Reilly LA, et al. The pro-apoptotic activity of the Bcl-2 family member Bim is regulated by interaction with the dynein motor complex. Mol Cell 1999; 3:287–296.
50. Dijkers PF, Medeenadagger RH, Lammers JJ, et al. Expression of the pro-apoptotic Bcl-2 family member Bim is regulated by the forkhead transcription factor FKHR-L1. Curr Biol 2000; 10:1201–1204.
51. Puthalakath H, Villunger A, O'Reilly LA, et al. Bmf: a pro-apoptotic BH3-only protein regulated by interaction with the myosin V actin motor complex, activated by Anoikis. Science 2001; 293:1829–1832.
52. Zhang Y, Adachi M, Kawamura R, et al. Bmf contributes to histone deacetylase inhibitor-mediated enhancing effects on apoptosis after ionizing radiation. Apoptosis 2006; 8:1349–1357.
53. Shibue T, Taniguchi T. BH3-only proteins: Integrated control point of apoptosis. Int J Cancer 2006; 119:2036–2043.
54. Mathai JP, Germain M, Marcellus RC, et al. Induction and endoplasmic reticulum location of BIK/NBK in response to apoptotic signaling by E1 A and p53. Oncogene 2002; 21:2534–2544.

55. Verma S, Zhao LJ, Chinnadurai G. Phosphorylation of the pro-apoptotic protein BIK: mapping of phosphorylation sites and effect on apoptosis. J Biol Chem 2001; 276:4671–4676.

56. Coultas L, Bouillet P, Stanley EG, et al. Proapoptotic BH3-only Bcl-2 family member Bik/Blk/Nbk is expressed in hemopoietic and endothelial cells but is redundant for the programmed death. Mol Cell Biol 2004; 24:1570–1581.

57. Inohara N, Ding L, Chen S, et al. harakari, a novel regulator of cell death, encodes a protein that activates apoptosis and interacts selectively with survival-promoting proteins Bcl-2 and Bcl-XL. EMBO J 1997; 16:1686–1694.

58. Imaizumi K, Tsuda M, Imai Y, et al. Molecular cloning of a novel polypeptide, DP5, induced during programmed neuronal death. J Biol Chem 1997; 272:18842–18844.

59. Graham RM, Thompson JW, Wei J, et al. Regulation of Bnip3 death pathways by calcium, phosphorylation, and hypoxia-reoxygenation. Antioxid Redox Signal 2007; 9:1309–1316.

60. Kubli DA, Ycaza JE, Gustafsson AB. Bnip3 mediates mitochondrial dysfunction and cell death through Bax and Bak. Biochem J 2007; 405:407–415.

61. Aerbajinai W, Giattina M, Lee YT, et al. Cloning and characterization of a gene expressed during terminal differentiation that encodes a novel inhibitor of growth. J Biol Chem 2003; 279:1916–1921.

62. Galvez AS, Brunskill EW, Marreez Y, et al. Distinct pathways regulate proapoptotic Nix and Bnip3 in cardiac stress. J Biol Chem 2006; 281:1442–1448.

63. Mund T, Gewies A, Schoenfeld N, et al. Spike, a novel BH3-only protein, regulates apoptosis at the endoplasmic reticulum. FASEB J 2003; 17:696–698.

64. Zha H, Reed JC. Heterodimerization-independent functions of cell death regulatory proteins Bax and Bcl-2 in yeast and mammalian cells. J Biol Chem 1997; 272:31482–31488.

65. Simonian PL, Grillot DAM, Merino R, et al. Bax can antagonize Bcl-XL during etoposide and cisplatin-induced cell death independently of its heterodimerization with Bcl-XL. J Biol Chem 1996; 271:22764–22772.

66. Kuwana T, Bouchier-Hayes L, Chipuk JE, et al. BH3 domains of BH3-only proteins differentially regulate Bax-mediated mitochondrial membrane permeabilization both directly and indirectly. Mol Cell 2005; 17:525–535.

67. Letai A, Bassik MC, Walensky LD, et al. Distinct BH3 domains either sensitize or activate mitochondrial apoptosis, serving as prototype cancer therapeutics. Cancer Cells 2005; 2:183–192.

68. Chen L, Willis SN, Wei A, et al. Differential targeting of prosurvival Bcl-2 proteins by their BH3-only ligands allows complementary apoptotic function. Mol Cell 2005; 17:393–403.

69. Harris SL, Levine AJ. The p53 pathway: positive and negative feedback loops. Oncogene 2005; 24:2899–2908.

70. Murray-Zmijewski F, Bourdon J-C. p53/p63/p73 isoforms: an orchestra of isoforms to harmonise cell differentiation and response to stress. Cell Death Differ 2006; 13:962–972.

71. Donehower LA, Harvey M, Slagle BL, et al. Mice deficient for p53 are developmentally normal but susceptible to spontaneous tumors. Nature 1992; 356:215–21.

72. Sah VP, Attardi LD, Mulligan GJ, et al. A subset of p53-deficient embryos exhibit exencephaly. Nat Genet 1995; 10:175–80.

73. Ko LJ, Prives C. p53: puzzle and paradigm. Genes Dev 1996; 10:1054–1072.

74. Amundson SA, Myers TG, Fornace AJ Jr. Roles for p53 in growth arrest and apoptosis: putting on the brakes after genotoxic stress. Oncogene 1998; 17:3287–3299.

75. Mills AA, Zheng B, Wang XJ, et al. p63 is a p53 homologue required for limb and epidermal morphogenesis. Nature 1999; 398:708–713.

76. Yang A, Walker N, Bronson R, et al. p73 deficient mice have neurological, pheromonal and inflammatory defects, but lack spontaneous tumors. Nature 2000; 404:99–103.

77. Lavin MF, Gueven N. The complexity of p53 stabilization and activation. Cell Death Differ 2006; 13:941–950.

78. Chao C, Hergenhahn M, Kaeser MD, et al. Cell type- and promoter-specific roles of Ser18 phosphorylation in regulating p53 responses. J Biol Chem 2003; 42:41028–41033.

79. Sluss HK, Armata H, Gallant J, et al. Phosphorylation of serine 18 regulates distinct p53 functions in mice. Mol Cell Biol 2004; 24:976–984.

80. Yu J, Zhang L. The transcriptional targets of p53 in apoptosis control. Biochem Biophys Res Commun 2005; 331:851–858.

81. Melino G, Bernassola F, Ranalli M, et al. p73 induces apoptosis via PUMA transactivation and Bax mitochondrial translocation. J Biol Chem 2004; 9:8076–8083.

82. Moll UM, Wolff S, Speidel D, et al. Transcription-independent pro-apoptotic functions of p53. Curr Opin Cell Biol 2005; 17:631–636.

83. Haupt Y, Rowan S, Shaulian E, et al. Induction of apoptosis in HeLa cells by transactivation-deficient p53. Genes Dev 1995; 9:2170–2183.

84. Schuler M, Bossy-Wetzel E, Goldstein JC, et al. p53 induces apoptosis by caspase activation through mitochondrial cytochrome c release. J Biol Chem 2000; 275:7337–7342.

85. Marchenko ND, Zaika A, Moll UM. Death signal-induced localization of p53 protein to mitochondria. A potential role in apoptotic signaling. J Biol Chem 2000; 275:16202–16212.

86. Sansome C, Zaika A, Marchenko ND, et al. Hypoxia death stimulus induces translocation of p53 protein to mitochondria: detection by immunofluorescence on whole cells. FEBS Lett 2001; 488:110–115.

87. Erster S, Mihara M, Kim RH, et al. In vivo mitochondrial p53 translocation triggers a rapid first wave of cell death in response to DNA damage that can precede p53 target gene activation. Mol Cell Biol 2004; 24:6728–6741.

88. Marchenko ND, Wolff S, Erster S, et al. Monoubiquitylation promotes mitochondrial p53 translocation. EMBO J 2007; 26:923–934.

89. Chipuk JE, Maurer U, Green DR, et al. Pharmacologic activation of p53 elicits Bax-dependent apoptosis in the absence of transcription. Cancer Cell 2003; 4:371–381.

90. Chipuk JE, Kuwana T, Bouchler-Hayes L, et al. Direct activation of Bax by p53 mediates mitochondrial membrane permeabilization and apoptosis. Science 2004; 303:1010–1014.

91. Speidel D, Helmbold H, Deppert W. Dissection of transcriptional and non-transcriptional p53 activities in the response to genotoxic stress. Oncogene 2006; 25:940–953.

92. Sun C, Cai M, Gunasekera AH, et al. NMR structure and mutagenesis of the inhibitor-of-apoptosis protein XIAP. Nature 1999; 401:818–822.

93. Srinivasula SM, Hegde R, Saleh A, et al. A conserved XIAP-interaction motif in caspase-9 and Smac/DIABLO regulates caspase activity and apoptosis. Nature 2001; 410:112–116.

94. Vaux DL, Silke J. IAPs, RINGs and ubiquitylation. Nat Rev Mol Cell Biol 2005; 6:287–297.

95. Dean EJ, Ranson M, Blackhall F, et al. Novel therapeutic targets in lung cancer: Inhibitor of apoptosis proteins from laboratory to clinic. Cancer Treat Rev 2007; 33:203–212.

96. Eckelman BP, Salvesen GS, Scott FL. Human inhibitor of apoptosis proteins: Why XIAP is the black sheep of the family. EMBO Rep 2006; 7:988–994.

97. Eckelman BP, Salvesen GS. The human anti-apoptotic proteins cIAP1 and cIAP2 bind but do not inhibit caspases. J Biol Chem 2006; 281:3254–3260.

98. Duckett CS. IAP proteins: sticking it to Smac. Biochem J 2005; 385:e1–2.

99. Shin H, Renatus M, Eckelman BP, et al. The BIR domain of IAP-like protein 2 is conformationally unstable: implications for caspase inhibition. Biochem J 2005; 385:1–10.

100. Vucic D, Franklin MC, Wallweber HJ, et al. Engineering ML-IAP to produce an extraordinarily potent caspase 9 inhibitor: implications for Smac-dependent anti-apoptotic activity of ML-IAP. Biochem J 2005; 385:11–20.

101. Huang H, Joazeiro CA, Bonfoco E, et al.. The inhibitor of apoptosis, cIAP2, functions as a ubiquitin-protein ligase and promotes in vitro monoubiquitination of caspases 3 and 7. J Biol Chem 2000; 275:26661–26664.

102. Beere HM. Death versus survival: functional interaction between the apoptotic and stress-inducible heat shock protein pathways. J Clin Invest 2005; 115:2633–2639.

103. Gupta S, Knowlton AA. Hsp60, Bax, apoptosis and the heart. J Cell Mol Med 2005; 9:51–58.

104. Paul C, Manero F, Gonin S, et al. Hsp27 as a negative regulator of cytochrome c release. Mol Cell Biol 2002; 22:816–834.

105. Chauhan D, Li G, Hideshima T, et al. Hsp27 inhibits release of mitochondrial protein Smac in multiple myeloma cells and confers dexamethasone resistance. Blood 2003; 102:3379–3386.

106. Garrido C, Bruey JM, Fromentin A, et al. HSP27 inhibits cytochrome c-dependent activation of procaspase-9. FASEB J 1999; 13:2061–2070.

107. Bruey JM, Ducasse C, Bonniaud P, et al. Hsp27 negatively regulates cell death by interacting with cytochrome c. Nat Cell Biol 2000; 2:645–652.

108. Pandey P, Farber R, Nakazawa A, et al. Hsp27 functions as a negative regulator of cytochrome c-dependent activation of procaspase-3. Oncogene 2000; 19:1975–1981.

109. Voss OH, Batra S, Kolattukudy SJ, et al. Binding of caspase-3 prodomain to heat shock protein 27 regulates monocyte apoptosis by inhibiting caspase-3 proteolytic activation. J Biol Chem 2007; 282:25088–25099.

110. Charette SJ, Lavoie JN, Lambert H, et al. Inhibition of Daxx-mediated apoptosis by heat shock protein 27. Mol Cell Biol 2000; 20:7602–7612.

111. Kirchoff SR, Gupta S, Knowlton AA. Cytosolic Hsp60, apoptosis, and myocardial injury. Circulation 2002; 105:2899–2904.

112. Shan YX, Liu TJ, Su HF, et al. Hsp10 and Hsp60 modulate Bcl-2 family and mitochondrial apoptosis signaling induced by doxorubicin in cardiac muscle cells. J Mol Cell Cardiol 2003; 35:1135–1143.

113. Bivik C, Rosdahl I, Ollinger K. Hsp70 protects against UVB induced apoptosis by preventing release of cathepsins and cytochrome c in human melanocytes. Carcinogenesis 2007; 28:537–544.

114. Gotoh T, Terada K, Oyadomari S, et al. hsp70-DnaJ chaperone pair prevents nitric oxide- and CHOP-induced apoptosis by inhibiting translocation of Bax to mitochondria. Cell Death Differ 2004; 11:390–402.

115. Stankiewicz AR, Lachapelle G, Foo CP, et al. Hsp70 inhibits heat-induced apoptosis upstream of mitochondria by preventing Bax translocation. J Biol Chem 2005; 280:38729–38739.

116. Wadhwa R, Yaguchi T, Hasan MK, et al. Hsp70 family member, mot-2/mthsp70/GRP75, binds to the cytoplasmic sequestration domain of the p53 protein. Exp Cell Res 2002; 274:246–253.

117. Beere HM, Wolf BB, Cain K, et al. Heat-shock protein 70 inhibits apoptosis by preventing recruitment of procaspe-9 to the Apaf-1 apoptosome. Nat Cell Biol 2000; 2:469–475.

118. Saleh A, Stinivasula SM, Balkir L, et al. Negative regulation of the Apaf-1 apoptosome by Hsp70. Nat Cell Biol 2000; 2:476–483.

119. Komarova EY, Afanasyeva EA, Bulatova MM, et al. Downstream caspases are novel targets for the antiapoptotic activity of the molecular chaperone hsp70. Cell Stress Chaperones 2004; 9:265–275.

120. Ravagnan L, Gurbuxani S, Susin SA, et al. Heat-shock protein 70 antagonizes apoptosis-inducing factor. Nat Cell Biol 2001; 3:839–843.

121. Gabai VL, Mabuchi K, Mosser DD, et al. Hsp72 and stress kinase c-jun N-terminal kinase regulate the bid-dependent pathway in tumor necrosis factor-induced apoptosis. Mol Cell Biol 2002; 22:3415–3424.

122. Guo F, Sigua C, Bali P, et al. Mechanistic role of heat shock protein 70 in Bcr-Abl-mediated resistance to apoptosis in human acute leukemia cells. Blood 2005; 105:1246–1255.

123. Pandey P, Saleh A, Nakazawa A, et al. Negative regulation of cytochrome c-mediated oligomerization of Apaf-1 and activation of procaspase-9 by heat shock protein 90. EMBO J 19(16):4310–4322.

124. Rodina A, Vilenchik M, Moulick K, et al. Selective compounds define Hsp90 as a major inhibitor of apoptosis in small-cell lung cancer. Nat Chem Biol 2007; 3:498–507.

125. Lee SH, Kim M, Yoon BW, et al. Targeted hsp70.1 disruption increases infarction volume after focal cerebral ischemia in mice. Stroke 2001; 32:2905–2912.

126. Bergeron L, Perez GI, Macdonald G, et al. Defects in regulation of apoptosis in caspase-2-deficient mice. Genes Dev 1998; 12:1304–1314.

127. Kuida K, Zheng TS, Na S, et al. Decreased apoptosis in the brain and premature lethality in CPP32-deficient mice. Nature 1996; 384:368–372.

128. Lakhani SA, Masud A, Kuida K, et al. Caspases 3 and 7: key mediators of mitochondrial events of apoptosis. Science 2006; 311:847–851.

129. Zheng TS, Hunot S, Kuida K, et al. Caspase knockouts: matters of life and death. Cell Death Differ 1999; 6:1043–1053.

130. Varfolomeev EE, Schuchmann M, Luria V, et al. Targeted disruption of the mouse Caspase 8 gene ablates cell death induction by the TNF receptors, Fas/Apo1, and DR3 and is lethal prenatally. Immunity 1998; 9:267–276.

131. Hakem R, Hakem A, Duncan GS, et al. Differential requirement for caspase 9 in apoptotic pathways in vivo. Cell 1998; 94:339–352.
132. Kuida K, Haydar TF, Kuan CY, et al. Reduced apoptosis and cytochrome c-mediated caspase activation in mice lacking caspase 9. Cell 1998; 94:325–337.
133. Cecconi F, Alvarez-Bolado G, Meyer BI, et al. Apaf1 (CED-4 homolog) regulates programmed cell death in mammalian development. Cell 1998; 94:727–737.
134. Yoshida H, Kong YY, Yoshida R, et al. Apaf1 is required for mitochondrial pathways of apoptosis and brain development. Cell 1998; 94:739–750.
135. Yeh WC, Pompa JL, McCurrach ME, et al. FADD: essential for embryo development and signaling from some, but not all, inducers of apoptosis. Science 1998; 279:1954–1958.
136. Zhang J, Cado D, Chen A, et al. Fas-mediated apoptosis and activation-induced T-cell proliferation are defective in mice lacking FADD/Mort1. Nature 1998; 392:296–300.
137. Veis DJ, Sorenson CM, Shutter JR, et al. Bcl-2-deficient mice demonstrate fulminant lymphoid apoptosis, polycystic kidneys, and hypopigmented hair. Cell 1993; 75:229–240.
138. Motoyama N, Wang F, Roth KA, et al. Massive cell death of immature hematopoietic cells and neurons in Bcl-x-deficient mice. Science 1995; 267:1506–1510.
139. Ross AJ, Waymire KG, Moss JE, et al. Testicular degeneration in Bclw-deficient mice. Nat Genet 1998; 18:251–256.
140. Print CG, Loveland KL, Gibson L, et al. Apoptosis regulator bcl-w is essential for spermatogenesis but appears otherwise redundant. Proc Natl Acad Sci U S A 95:12424–12431.
141. Rinkenberger JL, Horning S, Klocke B, et al. Mcl-1 deficiency results in peri-implantation embryonic lethality. Genes Dev 2000; 14:23–27.
142. Hamasaki A, Sendo F, Nakayama K, et al. Accelerated neutrophil apoptosis in mice lacking A1-a, a subtype of the bcl-2-related A1 gene. J Exp Med 1998; 188:1985–1992.
143. Knudson CM, Tung KS, Tourtellotte WG, et al. Bax-deficient mice with lymphoid hyperplasia and male germ cell death. Science 1995; 270:96–99.
144. Péquignot MO, Provost AC, Sallé S, et al. Major role of BAX in apoptosis during retinal development and in establishment of a functional postnatal retina. Dev Dyn 2003; 228:231–238.
145. Lindsten T, Ross AJ, King A, et al. The combined functions of proapoptotic Bcl-2 family members bak and bax are essential for normal development of multiple tissues. Mol Cell 2000; 6:1389–1399.
146. Ranger AM, Zha J, Harada H, et al. Bad-deficient mice develop diffuse large B cell lymphoma. Proc Natl Acad Sci U S A 2003; 100:9324–9329.
147. Yin XM, Wang K, Gross A, et al. Bid-deficient mice are resistant to Fas-induced hepatocellular apoptosis. Nature 1999; 400:886–891.
148. Bouillet P, Metcalf D, Huang DC, et al. Proapoptotic Bcl-2 relative Bim required for certain apoptotic responses, leukocyte homeostasis, and to preclude autoimmunity. Science 1999; 286:1735–1738.
149. Imaizumi K, Benito A, Kiryu-Seo S, et al. Critical role for DP5/Harakiri, a Bcl-2 homology domain 3-only Bcl-2 family member, in axotomy-induced neuronal cell death. J Neurosci 2004; 24:3721–3725.
150. Shibue T, Takeda K, Oda E, et al. Integral role of Noxa in p53-mediated apoptotic response. Genes Dev 2003; 17:2233–2238.

151. Villunger A, Michalak EM, Coultas L, et al. p53- and drug-induced apoptotic responses mediated by BH3-only proteins puma and noxa. Science 2003; 302:1036–1038.

152. Jeffers JR, Parganas E, Lee Y, et al. Puma is an essential mediator of p53-dependent and -independent apoptotic pathways. Cancer Cell 2003; 4:321–328.

153. Hao Z, Duncan GS, Chang CC, et al. Specific ablation of the apoptotic functions of cytochrome C reveals a differential requirement for cytochrome C and Apaf-1 in apoptosis. Cell 2005; 121:579–591.

154. Vahsen N, Candé C, Brière JJ, et al. AIF deficiency compromises oxidative phosphorylation. EMBO J 2004; 23:4679–4689.

155. Okada H, Suh WK, Jin J, et al. Generation and characterization of Smac/DIABLO-deficient mice. Mol Cell Biol 2002; 22:3509–3517.

156. Martins LM, Morrison A, Klupsch K, et al. Neuroprotective role of the Reaper-related serine protease HtrA2/Omi revealed by targeted deletion in mice. Mol Cell Biol 2004; 24:9848–9862.

157. Irvine RA, Adachi N, Shibata DK, et al. Generation and characterization of endonuclease G null mice. Mol Cell Biol 2005; 25:294–302.

158. Kawane K, Fukuyama H, Yoshida H, et al. Impaired thymic development in mouse embryos deficient in apoptotic DNA degradation. Nat Immunol 2003; 4:138–144.

159. Harlin H, Reffey SB, Duckett CS, et al. Characterization of XIAP-deficient mice. Mol Cell Biol 2001; 21:3604–3608.

160. Jacks T, Remington L, Williams BO, et al. Tumor spectrum analysis in p53-mutant mice. Curr Biol 1994; 4:1–7.

161. Candi E, Rufini A, Terrinoni A, et al. Differential roles of p63 isoforms in epidermal development: selective genetic complementation in p63 null mice. Cell Death Differ 2006; 13:1037–1047.

162. Yang A, Schweitzer R, Sun D, et al. p63 is essential for regenerative proliferation in limb, craniofacial and epithelial development. Nature 1999; 398:714–718.

163. Huang L, Min JN, Masters S, et al. Insights into function and regulation of small heat shock protein 25 (HSPB1) in a mouse model with targeted gene disruption. Genesis 2007; 45:487–501.

164. Huang L, Mivechi NF, Moskophidis D. Insights into regulation and function of the major stress-induced hsp70 molecular chaperone in vivo: analysis of mice with targeted gene disruption of the hsp70.1 or hsp70.3 gene. Mol Cell Biol 2001; 21:8575–8591.

165. Hunt CR, Dix DJ, Sharma GG, et al. Genomic instability and enhanced radiosensitivity in Hsp70.1- and Hsp70.3-deficient mice. Mol Cell Biol 2004; 24:899–911.

166. Glucksmann A. Cell deaths in normal vertebrate ontogeny. Biol Rev Camb Philos Soc 1951; 26:59–86.

167. Zuzarte-Luis V, Hurle JM. Programmed cell death in the embryonic vertebrate limb. Semin Cell Dev Biol 2005; 16:261–269.

168. Milligan CE, Prevette D, Yaginuma H, et al. Peptide inhibitors of the ICE protease family arrest programmed cell death of motoneurons in vivo and in vitro. Neuron 1995; 15:385–393.

169. Umpierre CC, Little SA, Mirkes PE. Co-localization of active caspase-3 and DNA fragmentation (TUNEL) in normal and hyperthermia-induced abnormal mouse development. Teratology 2001; 63:134–143.

170. Zuzarte-Luis V, Berciano MT, Lafarga M,et al. Caspase redundancy and release of mitochondrial apoptotic factors characterize interdigital apoptosis. Apoptosis 2006; 11:701–715.
171. Huang C, Hales BF. Role of caspases in murine limb bud cell death induced by 4-hydroperoxycyclophosphamide, an activated analog of cyclophosphamide. Teratology 2002; 66:288–299.
172. Grossmann J, Walther K, Artinger M, et al. Apoptotic signaling during initiation of detachment-induced apoptosis ("anoikis") of primary human intestinal epithelial cells. Cell Growth Differ 2001; 12:147–155.
173. Zuzarte-Luis V, Montero JA, Kawakami Y, et al. Lysosomal cathepsins in embryonic programmed cell death. Dev Biol 2007; 301:205–217.
174. Balemans W, Van Hul W. Extracellular regulation of BMP signaling in vertebrates: a cocktail of modulators. Dev Biol 2002; 250:231–250.
175. Herpin A, Cunningham C. Cross-talk between bone morphogenetic protein pathway and other major signaling pathways results in tightly regulated cell-specific outcomes. FEBS J 2007; 274:2977–2985.
176. Ganan Y, Macias D, Duterque-Coquilland M, et al. Role of TGFβs and BMPs as signals controlling the position of the digits and the areas of interdigital cell death in the developing chick limb autopod. Development 1996; 122:2349–2357.
177. Yokouchi Y, Sakiyama J-I, Kameda T, et al. BMP-2/-4 mediate programmed cell death in chicken limb buds. Development 1996; 122:3725–3734.
178. Guha U, Gomes WA, Kobayashi T, et al. In vivo evidence that BMP signaling is necessary for apoptosis in the mouse limb. Dev Biol 2002; 249:108–120.
179. Shepard TH. Catalog of teratogenic agents. 10th ed. The Johns Hopkins University Press, Baltimore, MD, 2001.
180. Knudsen TB. Cell death. In Drug Toxicity in Embryonic Development I, Kavlock RJ, Daston GP, eds. Springer, Berlin, 1997; 124(1):211–244.
181. Scott WJ. Cell death and reduced proliferateive rate. In Handbook of Teratology, Wilson JG, Fraser FC, eds. Plenum, New York, NY, 1997; 2:81–98.
182. Berk M, Desai SY, Heyman HC, et al. Mice lacking the ski proto-oncogene have defects in neurulation, craniofacial, patterning, and skeletal muscle development. Genes Dev 1997; 11:2029–2039.
183. Menkes B, Sandoe S, Ilies A. Cell death in teratogenesis. In Advances in Teratology, Woollam DH, ed. Academic Press, New York, 1970; 169–215.
184. Milaire J, Rooze M. Hereditary and induced modifications of the normal necrotic patterns in the developing limb buds of the rat and mouse: Facts and hypothesis. Arch Biol (Bruxelles) 1983; 94:459–490.
185. Sulik KK, Cook CS, Webster WS. Teratogens and craniofacial malformations: Relationships to cell death. Dev Suppl 1988; 103:213–231.
186. Alles AJ, Sulik KK. Retinoic-acid-induced limb-reduction defects: perturbation of zones of programmed cell death as a pathogenetic mechanism. Teratology 1989; 40:163–171.
187. Dunty WC Jr, Chen SY, Zucker RM, et al. Selective vulnerability of embryonic cell populations to ethanol-induced apoptosis: Implications for alcohol-related birth defects and neurodevelopmental disorder. Alcohol Clin Exp Res 2001; 25:1523–1535.

188. Nakamura N, Fufioka M, Mori C. Alterations in programmed cell death and gene expression by 5-bromodeoxyuridine during limb development in mice. Toxicol Appl Pharmacol 2000; 167:100–106.

189. Kise K, Nakagawa M, Okamoto N, et al. Teratogenic effects of bis-diamine on the developing cardiac conduction system. Birth Defects Res (Part A) 2005; 73:547–554.

190. Cheema ZF, West JR, Miranda RC. Ethanol induces Fas/Apo [apoptosis]-1 mRNA and cell suicide in the developing cerebral cortex. Alcohol Clin Exp Res 2000; 24:535–543.

191. McAlhany RE Jr, West JR, Miranda RC. Glial-derived neurotrophic factor (GDNF) prevents ethanol-induced apoptosis and JUN kinase phosphorylation. Brain Res Dev Brain Res 2000; 119:209–216.

192. Borsello T, Forloni G. JNK signalling: a possible target to prevent neurodegeneration. Curr Pharm Des 2007; 13:1875–1886.

193. Mirkes PE, Little SA. Teratogen-induced cell death in postimplantation mouse embryos: differential tissue sensitivity and hallmarks of apoptosis. Cell Death Differ 1998; 5:592–600.

194. Mirkes PE, Little SA. Cytochrome c release from mitochondria of early postimplantation murine embryos exposed to 4-hydroperoxycyclophosphamide, heat shock, and staurosporine. Toxicol Appl Pharmacol 2000; 162:197–206.

195. Little SA, Mirkes PE. Teratogen-induced activation of caspase-9 and the mitochondrial apoptotic pathway in early postimplantation mouse embryos. Toxicol Appl Pharmacol 2002; 181:142–151.

196. Little SA, Kim WK, Mirkes PE Teratogen-induced activation of caspase-6 and caspase-7 in early postimplantation mouse embryos. Cell Biol Toxicol 2003; 19:215–226.

197. Ali-Khan SE, Hales BF. Caspase-3 mediates retinoid-induced apoptosis in organogenesis-stage mouse limb. Birth Defects Res (Part A) 2003; 67:848–860.

198. Fernandez EL, Gustafson A-L, Andersson M, et al. Cadmium-induced changes in apoptotic gene expression levels and DNA damage in mouse embryos are blocked by zinc. Toxicol Sci 2003; 76:162–170.

199. Mikheeva S, Barrier M, Little SA, et al. Alterations in gene expression induced in day-9 mouse embryos exposed to hyperthermia (HS) or 4-hydroperoxycyclophosphamide (4CP): analysis using cDNA microarrays. Toxicol Sci 2004; 79:345–359.

200. Brooks CL, Gu W. Ubiquitination, phosphorylation and acetylation: the molecular basis for p53 regulation. Curr Opin Cell Biol 2003; 15:164–171.

201. Hosako H, Little SA, Barrier M, et al. Teratogen-induced activation of p53 in early postimplantation mouse embryos. Toxicol Sci 2007; 95:257–269.

202. Yakovlev AG, Di Giovanni S, Wang G, et al. BOK and NOXA are essential mediators of p53-dependent apoptosis. J Biol Chem 2004; 279:28367–28374.

203. Oda E, Ohki R, Murasawa H, et al. Noxa, a BH3-only member of the Bcl-2 family and candidate mediator of p53-induced apoptosis. Science 2000; 288:1053–1058.

204. Seo YW, Shin JN, Ko KH, et al. The molecular mechanism of Noxa-induced mitochondrial dysfunction in p53-mediated cell death. J Biol Chem 2003; 278:48292–48299.

205. Chipuk JE, Bouchier-Hayes L, Kuwana T, et al. PUMA couples the nuclear and cytoplasmic proapoptotic function of p53. Science 2005; 309:1732–1735.
206. Yu J, Zhang L, Hwang PM, et al. PUMA induces the rapid apoptosis of colorectal cancer cells. Mol Cell 2001; 7:673–682.
207. Yu J, Wang Z, Kinzler KW, et al. PUMA mediates the apoptotic response to p53 in colorectal cancer cells. Proc Natl Acad Sci U S A 2003; 100:1931–1936.
208. Gross A, Jockel J, Wei MC, et al. Enforced dimerization of BAX results in its translocation, mitochondrial dysfunction and apoptosis. EMBO J 1998; 17:3878–3885.
209. Li H, Zhu H, Xu CJ, et al. Cleavage of BID by caspase 8 mediates the mitochondrial damage in the Fas pathway of apoptosis. Cell 1998; 94:491–501.
210. Taylor WR, Stark GR. Regulation of the G2/M transition by p53. Oncogene 2001; 20:1803–1815.
211. Little SA, Mirkes PE. Effects of 4-hydroperoxycyclophosphamide (4-OOH-CP) and 4-hydroperoxydechlorocyclophosphamide (4-OOH-deCICP) on the cell cycle of post implantation rat embryos. Teratology 1992; 45:163–173.
212. Mirkes PE, Ricks JL, Pascoe-Mason JM. Cell cycle analysis in the cardiac and neuroepithelial tissues of day 10 rat embryos and the effects of phosphoramide mustard, the major metabolite of cyclophosphamide. Teratology 1989; 39:115–120.
213. Chernoff N, Rogers JM, Alles AJ, et al. Cell cycle alterations and cell death in cyclophosphamide teratogenesis. Teratogen Carcinogen Mutagen 1989; 9:199–209.
214. Francis BM, Rogers JM, Sulik KK, et al. Cyclophosphamide teratogenesis: evidence for compensatory responses to induced cellular toxicity. Teratology 1990; 42:473–482.
215. Gao X, Blackburn MR, Knudsen TB. Activation of apoptosis in early mouse embryos by 2'-deoxyadenosine exposure. Teratology 1994; 49:1–12.
216. Mirkes PE. Cyclophosphamide teratogenesis: a review. Teratogen Carcinogen Mutagen 1985; 5:75–88.
217. Mirkes PE. Effects of acute exposures to elevated temperatures on rat embryo growth and development *in vitro*. Teratology 1985; 32:259–266.
218. Mirkes PE, Ellison A, Little SA. Resistance of rat embryonic heart cells to the cytotoxic effects of cyclophosphamide does not involve aldehyde dehydrogenase-mediated metabolism. Teratology 1991; 43:307–318.
219. Thayer JM, Mirkes PE. Programmed cell death and N-acetoxy-2-acetylaminofluorene-induced apoptosis in the rat embryo. Teratology 1995; 51:418–429.
220. O'Reilly MA. Redox activation of p21Cip1/WAF1/Sdi1: a multifunctional regulator of cell survival and death. Antioxid Redox Signal 2005; 7:108–118.
221. Chen F, Oikawa S, Hiraku Y, et al. Metal-mediated oxidative DNA damage induced by nitro-2-aminophenols. Cancer Lett 1998; 126:67–74.
222. Lokshin M, Tanaka T, Prives C. Transcriptional regulation by p53 and p73. Cold Spring Harb Symp Quant Biol 2005; 70:121–128.
223. Ronen D, Schwartz D, Teitz Y, et al. Induction of HL-60 cells to undergo apoptosis is determined by high levels of wild-type p53 protein whereas differentiation of the cells is mediated by lower p53 levels. Cell Growth Differ 1996; 7:21–30.

2

Signal Transduction Pathways as Targets for Teratogens

Barbara D. Abbott

Reproductive Toxicology Division (MD67), National Health and Environmental Effects Research Laboratory (NHEERL), Office of Research and Development, U.S. Environmental Protection Agency, Research Triangle Park, North Carolina, U.S.A.

INTRODUCTION

Control of morphogenetic processes is critical for embryonic development. These processes include proliferation, cell death, extracellular matrix (ECM) and cytoskeletal remodeling, cell–cell and cell–ECM adhesion, cell motility, cell shape modifications, and differentiation to tissue/organ specific cellular phenotypes. Signaling pathways provide the regulation necessary to control these processes during development allowing the critical events to occur at the right time and at the specific locations necessary for an organ to form, mature, and become functional. These signaling events are tightly regulated during development, and disruption of the signaling pathways by exogenous agents can be catastrophic for the embryo. This chapter provides an introduction to signal transduction pathways essential for development and presents examples of teratogenic modes of action that involve disrupting signal transduction.

Signal transduction pathways can be described and grouped by their cellular location (membrane bound, cytoplasmic, and nuclear), ligands, cofactors or signaling intermediates utilized, kinase activities, or targeted genes. The National Research Council evaluated the mechanisms of action of developmental toxicants and focused on identifying signaling pathways that impact development and may be targets of teratogens. Their report "Scientific Frontiers in Developmental

Toxicology and Risk Assessment" [1] listed 17 signaling pathways as important regulators of development. Six pathways were identified as important in early development as in later during organogenesis: (1) the wingless-int (Wnt) pathway which signals via ß-catenin (canonical pathway) or jun N-terminal kinase (JNK) (noncanonical pathway); (2) the receptor serine/threonine kinase pathway which includes families of cell-surface receptors including transforming growth factor ß (TGFß pathway), and bone morphogenetic proteins (BMPs) which signal via Smad transcription factors; (3) the sonic hedgehog pathway (Shh) which signals through binding to the patched receptor (Ptc) and is regulated by smoothened (Smo); (4) the small G protein [Ras] linked receptor tyrosine kinase pathway which includes many growth factors or mitogens that bind to cell-surface receptors including epidermal growth factor (EGF), vascular endothelial growth factor (VEGF), platelet-derived growth factor, fibroblast growth factor (FGF), insulin-like growth factor, and ephrins, and these growth factors form ligand–receptor complexes that signal via small G proteins, protein kinase C (PKC), Ras, Rho, and trigger kinase cascades such as extracellular signal-regulated kinase (ERK), mitogen activated protein kinase (MAPK), jun N-terminal kinase (JNK/p38); (5) the Notch-Delta pathway; and (6) cytokine receptors (receptor-linked cytoplasmic tyrosine kinases) that bind ligands such as growth hormone, erythropoietin, prolactin, thrombopoietin, interleukins, and interferons, and signal via the Janus kinase (JAK)/signal transducer and activator of transcription protein (STAT) pathway. Other pathways considered to be active during organogenesis and later include the interleukin-1 receptor signaling through nuclear factor-Kappa B (NFκB) and inhibitor of NFκB (IκB) pathway; nuclear hormone receptor pathways which include zinc-finger DNA-binding transcription factor receptors including estrogen receptor (ER), glucocorticoid receptor (GR), mineralocorticoid receptor (MR), androgen receptor (AR), prostaglandin receptor (PR), thyroid hormone receptor (TR), vitamin D3 receptor (VDR), retinoic acid receptor (RAR), retinoid X receptor (RXR), peroxisome proliferator-activated receptor (PPAR); the apoptosis pathway which typically involves activation of caspase proteolytic enzymes subsequent to signaling and modulation via tumor necrosis factor, Fas, BAX, Bcl2, FADD, and TRADD; receptor phosphotyrosine phosphatase pathway which regulates signaling of other pathways by dephosphorylation of receptors and intermediates. Seven pathways are considered to be important in differentiated cells: (1) the receptor guanylate cyclase pathway which affects signaling via c-Fos, JunB, cyclic AMP response element-binding protein (CREB), activator protein 1 (AP-1), and ion channels; (2) the nitric oxide receptor pathway which involves a cytoplasmic enzyme that binds NO at a heme group converting GTP to cyclic GMP and affecting transcription via c-Fos; (3) the G-protein coupled receptor (large G proteins) pathway which includes a very broad range of ligands (proteins, peptides, and small molecules) which bind cell-surface receptors and affect a broad range of events (transcription, metabolism, motility, secretion, and activity of other kinase pathways); the (4) integrin; (5) cadherin and (6) gap junction pathways, which are involved in cell-to-cell signaling and cell-environment signaling and affect adhesion, motility, and passage of ions, metabolites and signaling molecules between cells; and (7) ligand-gated cation channel pathways which include several receptors and ligands (acetylcholine, glutamate, NMDA, and GABA) and affect membrane potentials and calcium-dependent events and are important in neuronal and myocardial signaling. All of these pathways have important roles during development and many are required throughout life. There is an extensive literature for each of these pathways and their roles in normal and abnormal development. Simple diagrams for many of these pathways can be found on the Biocarta website (www.biocarta.com) and detailed network diagrams for some of the pathways are on the Signal Transduction Knowledge Environment website in the "Connections Maps Pathways" section of the "Database of Cell Signaling" (www.sciencemag.org).

Prior to discussing the role that some of these pathways have in mediating the response of the embryo to a developmental toxicant, it may be useful to briefly discuss some general concepts regarding the roles of signaling pathways in morphogenesis, briefly describe some common features of receptor-mediated signaling pathways, and consider some aspects that affect the complexity of regulation of signal transduction.

SIGNALING IN MORPHOGENESIS

Development of the embryo depends on regulation of thousands of different gene combinations with expression of specific sets of genes at specific times and places. Signaling between cells and from the cell surface to the nucleus is a major factor in coordination of the developmental plan. Signaling pathways are used repeatedly during development of a morphological structure and the same pathways will be expressed to form very different tissues and organs. Specific pathways must be turned on and off precisely at the correct developmental stages in each tissue and must act in concert with other signals in order to form structures correctly. In response to signaling, genes in cells are turned on or off, cells proliferate, migrate, change shape or phenotype, or interactions between cells or between the cell and the ECM may be affected. For example, there are common themes in morphogenesis that can be directed by coordinated expression of several signaling pathways, but that ultimately lead to morphogenesis of organs with very different final structures and functions. The morphogenesis of kidney, lung, tooth, mammary gland, and hair or feathers all begin with formation of a placode. The placode enlarges to form a bud and the bud undergoes branching. Budding and branching involve interactions between epithelial and mesenchymal cell types [2]. Cell-type transformations may also be involved with transformation of a mesenchymal cell to an epithelial phenotype, as in kidney glomerulus formation. This series of events (form a placode, develop a bud, branch, and differentiate) occurs in tissues with ectodermal (hair, teeth, feathers, scales, beaks, nails, eccrine glands), endodermal (lung, pancreas), or mesodermal (kidney) origins [3]. Signaling regulates the placodes and buds formation and the direction of growth, as well as the shape and size of the branches. Branching is an iterative process that can progress to form tree-like structures (as in kidney, mammary gland, and lung), but the basic process and signaling can be modified to form different types and patterns of branches, and different relations between branch diameter and branch "generation" for the different organs.

During morphogenesis there may exist "signaling centers" from which signals are initiated and controlled. Placodes are signaling centers for bud formation, and there are both positive and negative signals to determine where placodes form. Ectodysplasin, a member of the tumor necrosis factor pathway, and its receptor EDAR are among the earliest markers of placode formation and are required for formation of all placodes of ectodermal origin [3]. Other genes that promote placode formation include Wnt, Shh, Ptc, FGF, and TGFß2 [4]. Inhibition of Wnt signaling inhibits formation of many ectodermal derivatives, including hair, teeth, and mammary gland. Negative regulators such as BMP and TGFß1 determine the location of a forming placode and restrict its lateral borders. FGFs and BMPs antagonize each other and differentially regulate the same genes. Outgrowth of the bud from the placode and later branching of the bud are also under the control of a signaling center. In the tooth bud that signaling center is the enamel knot and signaling molecules are locally expressed in that zone and regulate morphogenesis and growth [4]. The function of the enamel knot is regulated by lymphoid enhancer factor 1 (LEF1) and EDAR. EDAR expression is confined to the enamel knot. Ectodysplasin binds EDAR and activates NFκB. LEF1 mediates Wnt signaling resulting in upregulation of FGF, which then binds FGF receptors and promotes regional proliferation. Epithelial BMPs upregulate mesenchymal BMP4 via muscle segment homeobox genes, Msx1 and Msx2, and the mesenchymal BMP4 further upregulates LEF1 in the signaling center. In addition, Shh, expressed in the enamel knot acts via mesenchyme to regulate regional epithelial growth, and lunatic fringe contributes to modulate signaling in the lateral regions. The interactions between these multiple signaling pathways regulate tooth bud development and illustrate the sequential and reciprocal signaling between epithelial and mesenchymal cells and the complex signaling interactions that are required to regulate morphogenesis.

No promoter regions have been identified in signaling network genes that direct expression specifically to form teeth or any other organ. All of the signaling networks regulating tooth morphogenesis also govern development in other organs. Similar morphogenetic events and signaling sequences occur during formation of the kidney and lung. Formation of the lung involves budding, tubule formation, branching, and epithelial–mesenchymal inductions and interactions [5]. Lung buds grow out from the ventral foregut into mesenchyme and repeatedly branch to form primary, secondary, and tertiary bronchi. Branching patterns are under developmental control with a specific relationship between branch generation and diameter. The branching structures express specific markers and require differential gene expression patterns for the final phenotype from primary bronchi to alveoli [6]. FGF18 appears to be involved in signaling differentiation of support structures such as cartilage, smooth muscle, and blood vessels [7]. FGF10 has a critical role in early lung branching, and the expression patterns of the growth factor and its receptor are complex and dynamic [8]. Epithelial lung buds grow toward FGF10 expressed in mesenchyme. Shh is a feedback signal that shuts off FGF10 expression in the mesenchyme near the growing tips of the buds, splitting the FGF10 expression domain into two smaller subdomains. The bud bifurcates and each new tip grows toward one of the FGF10 subdomains, thus producing a branched bud structure. Shh and Ptc have dynamic expression patterns in the mesenchyme and the downstream genes Gli1, Gli2, and Gli3 are also expressed in mesenchyme. Another regulator of this symphony of events, FGF9, inhibits responses to Shh and is expressed in mesothelium [5].

In summary, similar morphogenetic events occur to form functionally diverse organs. Many of the same signaling pathways are used over and over during morphogenesis. Wnts, FGFs, BMPs, Shh, Ptc, and TGFßs are among the signaling pathways involved in formation of ectodermal, endodermal, and meso-dermal structures, e.g., tooth, lung, and kidney. Precise regulation of the expression patterns is critical to determine the location and ultimate morphology of the struc-tures formed. In this way, common morphogenetic and signaling themes lead to structural and functional diversity.

FEATURES OF RECEPTOR-MEDIATED SIGNALING PATHWAYS

Membrane-Bound Receptors

Receptor-mediated signal transduction is a multistage process that typically involves transfer of information from outside the cell or from the cell surface to the cytoplasm and the nucleus. In the case of pathways with membrane-bound receptors, ligand binds to the ligand-binding domain on the extracellular region of the receptor. The receptor enters an active conformation or state, may recruit additional partners or coactivators, and activates the next intermediate in the sig-naling pathway, which subsequently results in a cascade of activations (Fig. 1). The receptor may phosphorylate itself, or another intermediate, changing its activation state. Intermediary signaling peptides in turn phosphorylate the next protein in the pathway, continuing the cascade of signal transduction until ultimately the original signal translates to cellular responses. Cellular processes in which intermediary proteins are engaged may also become altered. Intermediates and targets of the sig-naling cascade may be involved in transcription, translation, cell-cycle regulation, cell movement, differentiation, or members of other signaling pathways. Some of the genes regulated by the signaling cascade may result in feedback inhibition of the signaling pathway. In general, each of the pathways with membrane-bound receptors can be described as having unique sets of signal transducing proteins. For example, TGFß pathways utilize Smad's to transduce signals, receptor tyro-sine kinase growth factor pathways trigger MAPK, ERK, or JNK/p38 cascades of phosphorylation, and cytokine pathways signal via JAK and STAT. Disrup-tion of signaling via these pathways can lead to alterations in cellular processes

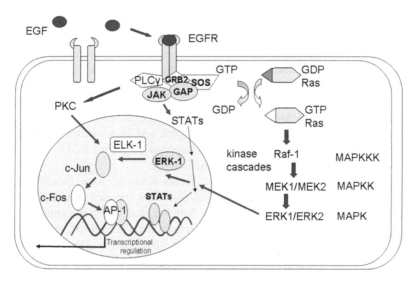

Figure 1 The EGFR/MAPK signaling pathway regulates cell proliferation and is an example of a membrane-bound tyrosine kinase receptor that transduces signals through a cytoplasmic kinase cascade. EGFR binds ligand and dimerization and autophosphorylation follow. Proteins associated with the receptor and involved in signaling include growth factor receptor-bound protein 2 (GRB2), Son of Sevenless (SOS), a GTP/GDP exchange factor protein and guanosine triphosphate-activating protein (GAP). EGFR activates Ras and the MAP kinase pathway, ultimately resulting in phosphorylation of ERK, ELK, and transcription factors, such as c-Jun and c-Fos which form the AP-1 complex. Signaling can also occur through activities of PLCγ and PKC or through the JAK–STAT pathway. The EGFR pathway has extensive cross talk with other pathways and this diagram is a simple representation of a complex signaling pathway which interacts extensively with other signaling systems.

(e.g., proliferation, differentiation, migration, and cell death) that are essential to embryonic development.

Nuclear and Cytoplasmic Receptors

The nuclear hormone receptor pathway and the nitric oxide receptor pathway have intracellular receptors with ligands that pass readily through the cell membrane to interact with the receptor. Nuclear hormone receptors may be found in the cytoplasm or nucleus, and may interact with chaperone proteins, small peptides and proteins that promote inhibition or activation of the receptor. The superfamily of steroid hormone nuclear receptors that act as ligand-induced transcription factors includes ER, TR, RAR, RXR, AR, GR, VDR, MR, PR, and PPAR. Members of the family share common structural features and exhibit high levels of homology. There is typically a variable N-terminal region (A/B domain), a conserved DNA-binding domain (C), a hinge region (D), a conserved ligand-binding domain (E), and a variable C-terminal region (F). In the unliganded state, the receptors may be found in the cytoplasm or nucleus depending on the specific receptor. The receptors may exist as stabilized complexes with heatshock proteins (HSP) as chaperone proteins and numerous other small proteins may participate in these complexes. Ligands for these receptors are passively transported across the cell membrane and bind to the receptor. In the presence of ligand, the receptors dimerize, e.g., ER forms homodimers while RAR, TR, and PPAR heterodimerize with RXR. If not already in the nucleus, the receptor–ligand complex moves into the nucleus where it binds to specific DNA sequences (response elements) to regulate expression of target genes. The DNA-binding domains of the receptors in this superfamily have a zinc-finger domain that interacts with the DNA response element (DRE) upstream of the regulated gene. These receptors can be repressors or activators of gene expression. Each of the nuclear receptor pathways is vulnerable to disruption by exogenous chemicals (natural and manufactured), and agonism or antagonism of their signaling can lead to developmental toxicity.

REGULATION OF NUCLEAR RECEPTOR SIGNAL TRANSDUCTION

Membrane-bound receptors transduce signals through a complex cascade of modifications to cytoplasmic intermediary signaling proteins (Fig. 1). Steroid hormone and PPAR receptors are located in the cytoplasm; signaling via these pathways may thus appear simpler, i.e., receptor binds ligand, moves to the nucleus, interacts with DNA, and changes transcription. However, the control of these receptors can be very complex with multiple options for repression or activation. The PPAR signaling pathway (Fig. 2) provides a good example of the complexity that can exist in regulation of signaling by nuclear receptor family members.

Three isoforms of PPAR have been characterized (PPARα, PPARβ/δ, and PPARγ). Each of these has unique tissue expression patterns, physiological roles, and ligand specificity [9,10]. PPARs control energy homeostasis and are important regulators of adipogenesis, lipid metabolism, inflammatory responses, and hematopoiesis [11]. PPARα and PPARγ are implicated in chronic diseases such as diabetes, obesity, and atherosclerosis [9]. PPARβ appears to have roles in embryo implantation, tumorigenesis in the colon, cholesterol transport, and skin wound healing. The PPAR family of transcriptional factors also has critical roles in reproduction and development. During development, PPARα, PPARβ, and PPARγ exhibit specific patterns of expression in the embryo, extra-embryonic membranes, uterus, and placenta that indicate roles in implantation of the embryo, development of the embryo, maintaining pregnancy, and initiation of labor at term [12–16]. PPARγ and PPARβ are required for placenta formation and function. The knockout of PPARγ is lethal in utero around GD10 (gestation days) due to the failure of the labyrinthine layer of the placenta to develop [17]. PPARβ/δ deficiency results in placental defects and frequent mid-gestation lethality [14]. PPARα protein was detected immuno-histochemically in the mouse blastocyst as early as GD5 and on GD11 it was found in the tongue, liver, digestive tract, heart, and vertebrae [10]. PPARα mRNA was also detected using in situ hybridization in the rat fetus on GD13.5 (roughly equivalent developmentally to GD11.5 in the

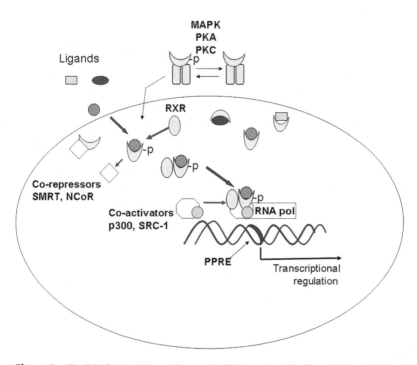

Figure 2 The PPAR signaling pathway provides an example of a cytoplasmic, ligand activated receptor that translocates to the nucleus and binds specific regulatory DNA sequences (PPRE) in promoter regions of genes to regulate transcription. Ligands (endogenous fatty acids, drugs such as fibrates, and commercial compounds such as the perfluorinated alkyl acid PFOA) bind the receptor. A conformational change results and PPAR forms a heterodimer with RXR. Different ligands influence the specific conformational change and produce different effects on gene regulation. The ligand–PPAR–RXR complex recruits coactivators (p300 and SRC-1), associates with the PPRE, and forms an RNA polymerase transcriptional activation complex. Corepressors (SMRT and NCoR) can be associated with PPAR and prevent binding to DNA. In the absence of ligand these complexes can inhibit gene expression. Phosphorylation of PPAR by MAPK, and PKA, PKC activities can either enhance (PPARα) or inhibit (PPARγ) activity.

mouse) in CNS, tongue, digestive tract, vertebrae, liver, and heart [18,19]. PPARα also plays a role in the mode of action of the developmental toxicant, perfluorooctanoic acid (PFOA). PFOA is a member of a family of perfluorinated chemicals that has a variety of commercial applications. PFOA persists in the environment and is found in wildlife and humans. In mice, PFOA is developmentally toxic producing mortality, delayed eye opening, growth deficits, and altered pubertal maturation. PFOA activates PPARα and expression of PPARα is required for the induction of mortality that occurs in neonatal mice after exposure to PFOA during pregnancy [20].

As with other members of the nuclear receptor family, PPAR isoforms consist of several domains, the N-terminal A/B domain, the DNA-binding domain (C), a hinge region (D), and C-terminal ligand-binding domain (E/F), which contains the ligand-dependent activation function 2 (AF-2). Phosphorylation of the A/B domain enhances transcriptional activity, for example, insulin enhances transcriptional activity of PPARα via phosphorylation of MAP-kinase A/B domain serine 12 and 21 sites, but phosphorylation of serine 112 of PPARγ lowers transcriptional activity [21]. The reduced activity reflects interference with ligand binding, revealing potential regulation of activity in adjacent domains. In the canonical pathway for transcriptional regulation, PPAR binds ligand, forms a heterodimer with RXR which binds to the peroxisome proliferator response element (PPRE) in the promoter region of regulated genes. Cofactors and coactivators are recruited to the ligand–receptor complex, and heterodimerization with RXR is required for the transcriptional complex to bind to the PPRE. Ligand binding to the C-terminal domain induces conformational changes that also involve the AF-2 helix. The AF helix becomes folded against the ligand-binding domain forming a hydrophobic cleft allowing interaction with the steroid receptor coactivator-1 (SRC-1). Coactivator recruitment depends on the allosteric alterations in the AF-2 helical domain [22]. SRC-1 also interacts with the A/B domain and contacts another cofactor, CREB-binding protein (CBP/p300). SRC-1 also interacts in a ligand-dependent manner with other steroid hormone receptors to act as a coactivator. The recruitment of coactivators is an essential component of gene regulation by nuclear receptors, and numerous coactivators have been identified. Selective inhibition of SRC-1, PPARγ coactivator-1α (PGC-1α), and PPAR binding protein demonstrated that the activity of PPAR depends on recruitment of specific coactivator complexes [23]. Binding of different PPAR ligands (various natural and synthetic ligands are known for all three isoforms) appears to result in somewhat different conformational changes to the receptor and this in turn appears to affect cofactors, which participate in the activational complex [23]. The combination of ligand-specific conformational changes and cofactor recruitment may offer a mechanism for the differential gene expression observed following activation by different ligands [22].

PPAR can be either an activator or a repressor of gene activity. Association of the unliganded PPARγ with corepressors, such as silencing mediator for retinoid and thyroid hormone receptors (SMRT) or the nuclear corepressor (NCoR), prevents binding to DNA. Upon ligand binding, the corepressor is released. PPARß, however, can associate with NCoR/SMRT, form a heterodimer with RXR, and suppress expression of PPARα and PPARγ target genes by binding to their PPREs, repressing gene transcription [24,25]. Structural and X-ray crystallography examination of the PPAR interactions with cofactors suggest that there are differences in the binding modes of coactivators and corepressors. In addition, the association between PPAR and RXR may stabilize the AF-2 helix in such a way that favors recruitment of coactivators even in the absence of a bound ligand [22]. This supports the observation that the PPAR/RXR complex can be activated by RXR agonists alone.

The complexity of the interactions of the receptors with heterodimers, coactivators, or corepressors and the possibility of activity in the unliganded state may all contribute to the regulation of signaling via these pathways. This complex regulation of signaling through interactions with activators, repressors, and partner proteins may also provide opportunities for toxicant disruption of signaling other than via direct interaction at the ligand-binding domain.

DISRUPTION OF SIGNAL TRANSDUCTION AS A TERATOGENIC MODE OF ACTION

All of the pathways mentioned in the introduction have important roles during development, and many are required throughout life. Clearly, it is beyond the scope of this chapter to expand upon the roles of each of these pathways in the developing embryo. Also, it is not possible in this chapter to explore the instances in which toxicants are known or suspected to disrupt cell signaling as a potential mode of action. However, several examples will be presented for which there is strong experimental evidence to support a mode of action in which the toxicant directly interferes with signal transduction, leading to malformations or death of the embryo. These examples will illustrate several of the potential interactions between toxicants and various target sites in signaling pathways and link the disruption of the signaling to cellular consequences which ultimately lead to malformations.

Sonic Hedgehog Signaling and Cyclopamine-Induced Craniofacial Dysplasia

Shh is a morphogen that plays an important role in the developing embryo. This signaling pathway regulates proliferation and differentiation and controls developmental patterning of the head, brain, and limbs. Mutations of the genes in this pathway are associated with holoprosencephaly in humans. Holoprosencephaly is a congenital anomaly that features hypotelorism and deficiencies in the forebrain. The malformation can be caused by a variety of environmental and genetic factors and can be variable in the degree of severity observed. Maternal diabetes, ethyl alcohol, retinoic acid, sonic hedgehog mutations, and defects in cholesterol biosynthesis all are considered causal factors in humans and a wide range of teratogenic agents has been shown to be capable of producing holoprosencephaly in animal models [26]. The steroidal alkaloids, cyclopamine and jervine, are derived from the desert plant *Veratrum californicum* and have been extensively studied for their ability to produce holoprosencephaly and cyclopia in mammalian and chick embryos. In the 1950s an epidemic of congenital deformities in sheep was linked to consumption of *V. californicum* [27] and subsequent studies confirmed that cyclopamine and jervine were potent teratogens, capable of producing cyclopia in extreme cases [28]. Since those initial studies, these compounds have served as

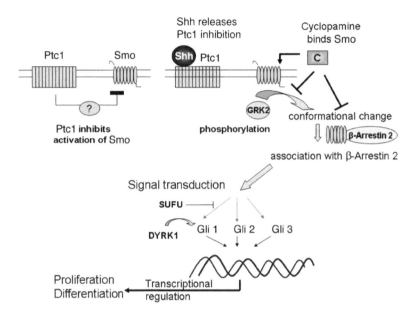

Figure 3 (*See color insert*) In the Shh signaling pathway, Ptc1 inhibits Smo. The manner in which the inhibition of Smo occurs is unclear. Shh binds Ptc1 and releases the inhibition allowing Smo to be activated by GRK2. Activated Smo undergoes conformational change and associates with ß-Arrestin 2. This initiates a series of transcriptional activations via the Gli family. Supressor of fused (SUFU) inhibits and DYRK1 enhances the transcriptional activity of Gli. Cyclopamine binds Smo, likely interfering with conformational change and inhibiting phosphorylation by GRK2 and subsequent association with ß-Arrestin 2, thus shutting down signaling through the Shh pathway.

model compounds for the study of cranial morphogenesis, and it was demonstrated that the effects are a consequence of blocking signaling through the Shh pathway.

The schematic overview of the Shh pathway (Fig. 3) summarizes some of the key features of this pathway. Although, this diagram represents extensive research efforts from multiple investigations over several decades, some regulatory steps remain to be elucidated fully. Briefly, the signaling pathway includes Shh, Ptc1, and Smo. Shh is a secreted protein which is activated by cleavage and binding of a cholesterol moiety. Ptc1 inhibits activation of Smo, and binding of Shh to Ptc1 releases that inhibition. It is yet unclear exactly how Ptc1 inhibition of Smo occurs (thus the oval containing a question mark on the diagram). No direct protein–protein interactions of Ptc1 and Smo have been detected, and it is proposed that the inhibitory effect on Smo conformation is produced indirectly via interactions with small endogenous molecules [29,30]. However, a conformational change is considered to be the primary determinant of Smo activation [31]. After Shh binding to Ptc1 and release of Smo inhibition, Smo is phosphorylated by G protein-coupled receptor kinase 2 (GRK2) and this phosphorylation promotes conformational change and facilitates an association of Smo with ß-Arrestin 2, and subsequent endocytosis in clathrin-coated pits [32]. The pathway downstream of Smo remains somewhat unclear; however, activation of Smo leads to signal transduction via transcriptional activators of the Gli family (Gli1, Gli2, and Gli3). GRK2 was shown to be an integral component of the signaling pathway as reduction in GRK2 expression by short hairpin RNA reduced Gli1 signaling in response to a Smo agonist [33]. Supressor of fused interacts with Gli proteins to repress Shh signaling, while the kinase dual specificity/tyrosine-phosphorylated and regulated kinase 1 (DYRK1) stimulates Gli1 transcriptional activity.

Cyclopamine, jervine, and other members of this alkaloid family have a striking resemblance to the chemical structure of cholesterol. There are unique features of the chemical structure of the alkaloid compounds that influence their potency as teratogens. Greater potency for induction of holoprosencephaly is associated with having the furan E ring at a right angle in relation to the A–D rings as in cyclopamine; changes in polarity and stearic bulk of the ring system also affect potency [34]. It was originally speculated that the mechanism for teratogenicity might be related to disruption of cholesterol transport or homeostasis, or to post-translational modification of Shh, cleavage and binding of a cholesterol moiety to give an active Shh molecule. However, studies revealed that cyclopamine directly interfered with Shh signaling and did not affect cholesterol transport or the processing and activation of Shh [34–36]. Cyclopamine binds to the heptahelical bundle of Smo, likely affecting conformation, and interfering with the phosphorylation of Smo by GRK2 and with the association of Smo and ß-Arrestin 2 [29,32]. Cyclopamine interferes with Shh signaling by binding to Smo but appears to act in a Ptc1 dependent manner. Patched activity modulated the binding of cyclopamine to Smo as increased expression of Ptc1 resulted in increased binding of cyclopamine [29].

The Shh pathway regulates outgrowth and patterning of the mid and upper face. Signaling through this pathway also induces proliferation and differentiation in craniofacial tissues. Disruption of this signaling by cyclopamine results in hypoplasia of midline structures of the face, but is not associated with increased cell death or altered neural crest cell (NCC) migration. Mouse embryos exposed to cyclopamine in whole embryo culture exhibited a mild facial phenotype suggestive of midfacial hypoplasia and this was accompanied by decreased expression of Ptc1 and Gli1 [37]. Although there was no effect on NCC migration from the midbrain toward the nasal region, the number of NCC was reduced by cyclopamine exposure. The reduction in NCC could be due to either an effect on proliferation or programed cell death (PCD). An effect of exposure on proliferation seems more likely; in studies using chick embryos, cyclopamine exposure completely blocked Shh signaling without producing cytotoxicity in chick embryonic neural plate explants and after exposure of the chick embryo in ovo there was no increase in PCD of NCCs [34,38]. As excessive PCD was not present, Cordero et al. [38] proposed that the cyclopamine-produced malformations are a result of a ventral shift in an organizing center (defined by the boundary of Shh/FGF8 expression) that regulates patterning and outgrowth of the frontonasal primordium. Cell types normally induced in the ventral neural tube by Shh are either absent or appear at the ventral midline in chick embryos exposed to cyclopamine [36]. In the absence of Shh signaling, a ventral shift in FGF8 expression was observed in the chick, and it was proposed that such a shift could result in effects on patterning and proliferation. Further studies of the effects of cyclopamine on cultured mouse embryos also revealed defects in vasculogenesis and reduced expression of vascular endothelial growth factor (VEGF) and BMP4, suggesting that the etiology of the effects of cyclopamine may also involve Shh-regulated effects on angiogenesis [39].

In summary, cyclopamine reduces Shh signaling by binding to Smo, inhibiting phosphorylation of Smo and its association with ß-Arrestin with subsequent decreased downstream signaling through Gli transcriptional activators. Aspects of the molecular pathway remain unclear, and an explanation of how this decreased signaling translates to morphological effects is also incomplete. The morphological effects may be a consequence of effects on regulation of cell proliferation, as loss of Shh signaling results in fewer neural cells without increased PCD or effects on NCC migration.

Wnt Signaling and Thalidomide-Induced Limb Dysmorphology

The Wnt signaling pathway involves binding of secreted glycoproteins (i.e., Wnt family members) to cell-surface receptors (such as Frizzled). The Frizzled receptor is related to Smo (the protein necessary for hedgehog signaling) and has seven transmembrane domains and a cysteine-rich N-terminal domain. Wnt regulates expression of many genes by networking with a variety of receptor-mediated signaling pathways. Wnt signaling via the ß-catenin pathway (referred to as the canonical pathway) is an important regulator of transcription, spindle orientation, cell polarity, and possibly cadherin-mediated adhesion and gap junctions. In this pathway, Wnt stabilizes ß-catenin and regulates a diverse array of biological processes. Wnt signaling also occurs via ß-catenin-independent pathways which include activation of Ca^{2+} flux, G proteins, and JNK activation to affect gene expression, cell migration, adhesion, and polarity. Wnt signaling and activation of the JNK pathway are important regulatory events in pattern formation and during organogenesis. Refer to ref. [40] for an overview of Wnt signaling in limb development and skeletal morphogenesis.

During Wnt signaling via the ß-catenin pathway, Wnt stabilizes ß-catenin by controlling its phosphorylation (Fig. 4). In the absence of Wnt, ß-catenin is phosphorylated by GSK-3ß in a complex in the cytosol with axin and adenomatous polyposis coli. The phosphorylated ß-catenin is recognized by beta-transducin repeat containing protein (ß-TrCP) (an F-box protein) that targets it for ubiquitination and degradation [41]. However, when Wnt is present and binds to Frizzled and low density lipoprotein (LDL) receptor related protein 5 or 6 (LRP5/6), dishevelled (Dsh) is activated and GSK-3ß is suppressed, thus rescuing ß-catenin from degradation. The recruitment of Dsh and inhibition of GSK-3ß require the involvement of ß-Arrestin, which is also an important participant in the Shh signaling pathway (previously discussed), and there appear to be many possibilities

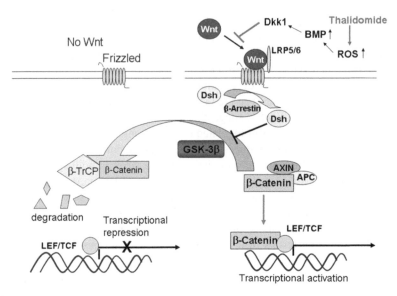

Figure 4 (*See color insert*) In the absence of Wnt, ß-catenin is phosphorylated by GSK-3ß, associates with ß-TrCP, and is targeted for degradation by ubiquitination. If Wnt is available to bind to Frizzled, in association with LRP5/6, Dsh is activated and GSK-3ß is suppressed. Recruitment of Dsh requires participation of ß-Arrestin. Thus in the presence of Wnt, ß-catenin is stabilized, associates with AXIN and adenomatous polyposis coli, moves into the nucleus, and activates transcription via LEF/TCF. In the absence of ß-catenin, LEF/TCF acts to repress transcription. Thalidomide exposure results in oxidative stress, increased ROS, increased BMP signaling, increasing expression of Dkk1, a target gene of BMP. Dkk is a Wnt antagonist and inhibits binding to Frizzled and subsequent signaling via stabilized ß-catenin.

for cross talk between Wnt and other signaling pathways [42,43]. ß-Catenin then converts LEF/TCF (T cell transcription factor) from a transcriptional repressor to an activator [41,44,45]. The noncanonical Wnt signaling that is independent of ß-catenin also uses the Frizzled receptor, but uses a different coreceptor (the proteoglycan protein Knypec) and different domains of Dsh appear to be required. Signaling downstream of Dsh becomes complex and diverse with JNK activation leading to gene transcriptional regulation (refer to review by Yang [40] for more detailed pathway information).

Thalidomide, originally marketed as a sedative, was also given to pregnant women to treat morning sickness. By the early 1960s, thalidomide was linked to birth defects including phocomelia (short limbs), amelia (absence of limbs), ear, eye, heart, and gastrointestinal defects. The therapeutic actions of the chemical are believed to involve an inflammatory response, the immune system, anti-angiogenic properties, and/or increased production of reactive oxygen species (ROS). ROS are known to change the expression of genes and affect signaling pathways [46]. Many mechanisms have been proposed for the teratogenic effects of thalidomide including DNA damage due to increased ROS, effects on DNA transcription, growth factors, angiogenesis, and apoptosis; see Table 1 of Stephens and Fillmore [47]. Current hypotheses propose that thalidomide affects limb outgrowth by inducing oxidative stress and/or by inhibiting angiogenesis.

Recently, Knobloch et al. [48] have linked thalidomide-induction of limb and eye defects to disruption of Wnt signaling through the canonical Wnt ß-catenin pathway. Their study provides evidence to support a Wnt/ß-catenin mediated mechanism for thalidomide teratogenicity. Briefly, this mechanism occurs as (1) thalidomide induces oxidative stress via ROS formation, (2) oxidative stress results in enhanced BMP signaling, (3) the BMP target gene Dickkopf1 (Dkk1) is upregulated, (4) Dkk1, a Wnt antagonist, inhibits Wnt binding to Frizzled and downstream ß-catenin signaling, (5) leading to changes in gene expression that ultimately lead to increased cell death. In support of this mechanism, thalidomide is known to induce ROS in limb bud cells of species susceptible to the teratogenic effects of the compound, but not in thalidomide-resistant mice or rats [49,50]. Also, Wnt signaling is known to be important in limb development [47]. In addition, the expression of Dkk1 is regulated by BMPs and promotes PCD in the limb bud; overexpression of Dkk1 in chicken wing buds results in wing truncations [51–53]. The study of Knobloch et al. [48] showed that thalidomide-induced limb truncation and microphthalmia in chick embryos, and these defects correlated with increased cell death and increased expression of BMP4, BMP5, BMP7, and Dkk1 in the apical ectodermal ridge as well as in the underlying mesenchyme. Blocking the activity of BMP, Dkk1, and GSK-3ß eliminated the cell death and counteracted the teratogenicity reducing the incidence of limb truncations and microophthalmia in the chick embryo. All of these inhibitors act to restore signaling via the canonical Wnt ß-catenin pathway. As presented in the introduction to Wnt signaling, GSK-3ß phosphorylates ß-catenin targeting it for degradation. If Wnt signaling is not occurring (Wnt does not bind the Frizzled receptor), ß-catenin is localized to the cytoplasm; however, once Wnt signaling occurs ß-catenin enters the nucleus. Thalidomide exposure reduced the number of cells with nuclear ß-catenin indicating reduced Wnt signaling. Study of Knobloch et al. also showed that production of ROS was necessary for producing the enhanced BMP expression and suppression of Wnt signaling. In their study, these molecular outcomes were observed in chick embryos, primary chick limb bud cells, and chick and human embryonic fibroblasts exposed to thalidomide, suggesting that the observations in chick would also be relevant to humans. While thalidomide-induced pathologies likely involve a wide spectrum of mechanisms, the evidence is strong that one mechanism involves an increase in ROS leading to reductions in Wnt signaling that results in increased cell death and malformations in the limb bud.

Cation Channel Signaling: Relationship to Hypoxia and Malformations

Chemicals and drugs that interact with membrane receptors that regulate flux of potassium ions in the embryonic heart have the potential to disrupt heart function. The ensuing hypoxic conditions in tissues can result in edema, hemorrhage, cell death, and malformations or death of the embryo. The class III antiarrhythmic drugs, dofetilide, almokalant, and d-sotalol are examples of such agents. These drugs act by lengthening myocardial refractoriness and increasing action potential duration. The repolarization of the myocardium requires activity of the delayed rectifying potassium (K^+) current, and inhibition of the rapidly activating current (I_{kr}) is the target of these drugs. Exposure of pregnant rats to these drugs resulted in embryonic edema, hemorrhage, abnormalities, and fetal death [54]. The abnormalities included cleft palate, short tails, digital hypoplasia, or more severe limb deficiencies, ventricular septal defects, great vessel defects, and urogenital defects [55]. The pattern of defects and observations of edema and hemorrhage in regions that later developed tail and digital shortening or defects suggested that hypoxia was involved and that effects of the drugs on embryonic heart function could be responsible, as reviewed by Webster et al. [56]. The adult rat heart is not dependent on an I_k current, but in the rat embryo the heart is dependent on I_k and the embryo is especially sensitive to the drugs targeting I_{kr} between GD9 and GD14 [57,58]. The I_{kr} inhibitory drugs cause embryo/fetal bradycardia and the decreased heart function leads to hypoxia, edema, hemorrhage, and ultimately malformations. Similar patterns of malformations can be produced by other methods of inducing hypoxia in the embryo, such as clamping uterine vessels [59,60]. Grabowski described the teratogenic effects of hypoxia and its link to hemorrhage and edema as the "fetal oedema syndrome" with examples of teratogens and gene mutations that produce malformations through this pathogenic mechanism in chick and mammalian embryos [61,62]. Similarly, Webster et al. refer to the pattern of defects that occur in response to fetal hypoxia, whether caused by drugs or other agents, as fetal hypoxia syndrome [56]. A similar pattern of defects is observed in the human after prenatal exposure to phenytoin and is referred to as the fetal hydantoin syndrome.

Diphenylhydantoin (DPH, phenytoin) is an anticonvulsant that is used to treat epilepsy and if taken during pregnancy produces facial dysmorphology, distal digital hypoplasia, intrauterine growth retardation (IUGR) and mental retardation, collectively referred to as fetal hydantoin syndrome [63]. The teratogenicity of phenytoin was first detected in mice exposed from GD9 to GD15 with resulting increased cleft lip and palate [64]; for a review of craniofacial dysplasia induced by phenytoin see Webster et al. [56]. Phenytoin inhibits both sodium and calcium channels but is also a weak inhibitor of the I_{kr} current in the embryonic heart. As discussed for the class III antiarrythmic drugs, the inhibition of I_{kr} by phenytoin leads to bradycardia, arrhythmia, and hypoxia with edema and hemorrhage in regions that subsequently exhibit malformations. Concentration-dependent bradycardia is induced in rat embryos exposed to phenytoin in culture, and within hours hypoxia can be observed [56,65,66]. The embryonic hypoxia may result from effects on the mother that lead to reduced placental circulation as well as from the direct effects on fetal heart function.

A number of mechanisms have been proposed as mediating the effects of phenytoin, and quite possibly several could be involved in producing the fetal hydantoin syndrome. Other mechanisms that have been proposed include damage from reactive intermediates produced during DPH metabolism and DPH-induced vitamin K deficiency. Metabolism by cytochrome P450 (CYP450) enzymes or prostaglandin synthetase appears to be required to produce reactive intermediate(s) [67]. The evidence for inhibition of I_{kr} potassium channel signaling by phenytoin supports a role for this pathway in producing bradycardia, arrhythmia, hypoxia, edema, and hemorrhage in tissues that later exhibit malformations. Even if other actions of phenytoin are involved, the disruption of this signaling pathway in the fetal heart is implicated in the teratogenic mechanism leading to the fetal hydantoin syndrome.

Nuclear Receptor Signaling and Dioxin-Induced Cleft Palate

The aryl hydrocarbon receptor (Ahr) is a member of the PER-ARNT-SIM (PAS) family of basic region helix-loop-helix (bHLH) transcription factors and is a ligand-activated receptor that translocates to the nucleus to regulate gene expression (Fig. 5). This family of receptors has many similarities to the steroid hormone nuclear receptors. Ahr forms a complex in the cytosol with HSP90 and several HSP accessory proteins and immunophilin-like proteins (XAP2/ARA9/AIP and p23) [68,69]. After ligand binding, the complex translocates to the nucleus and

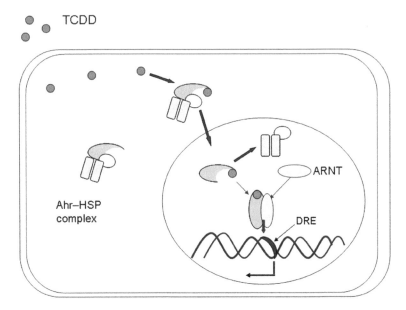

Figure 5 The Ahr–ARNT signaling pathway has similarities to the steroid hormone receptor pathways. Ahr and ARNT are members of the bHLH family of receptors. Ahr associates with HSP90 and other small peptides in the cytoplasm and on binding ligand, translocates to the nucleus. The HSP complex dissociates and Ahr forms a heterodimer with ARNT which binds to the DRE in the promoter region of genes that are transcriptionally regulated by the pathway.

forms a heterodimer with the aryl hydrocarbon nuclear translocator (ARNT/HIF-1ß), another member of the PAS-bHLH family. The ligand–receptor complex binds to a response element (variously referred to as the dioxin-, or xenobiotic- or Ahr response element, DRE, XRE, AhrE) and regulates gene expression. The endogenous ligands for the Ahr and its physiological role remain unclear; however, activation of Ahr results in altered expression of a range of genes affecting diverse biological processes, including xenobiotic metabolizing enzymes (CYP1A1, CYP1A2, and CYP1B1), growth regulators (EGFR, TGFα, epiregulin, TGFß, cyclin-dependent kinases, and CDKs), and steroid hormones (ER and regulation of Ahr itself) [70]. Expression of CYP1A1 is directly regulated by binding of the activated receptor complex to DREs in the promoter region of that gene. Epiregulin, a potent mitogen and ligand for EGFR, also has a DRE in its promoter region and is directly upregulated by activated Ahr [71]. DREs have not been identified in promoter regions for many of the other genes whose expression is changed by Ahr, suggesting that the expression is affected indirectly. Activation of Ahr leads to rapid increases in immediate early genes (c-Jun, c-Fos) leading to increased AP-1. There are also links between Ahr activity and stimulation of kinase activity such as SRC-like tyrosine kinases, MAP kinases, PKA, and PKC [72]. Members of the MAPK pathway, extracellular signal-regulated kinase (ERK), and JNK are modulators of ARNT and Ahr activity and the increased or decreased activity of the receptor complex appears to be tissue specific [73,74].

Ligands for Ahr include a number of exogenous compounds and endogenous compounds, including the environmental pollutants coplanar polychlorinated biphenyls, dioxins, and furans, and an endogenous compound indolo-[3,2,-*b*]-carbazole, as reviewed in Denison et al. [75]. 2,3,7,8-Tetrachlorodibenzo-*p*-dioxin (TCDD) is the most potent of the known ligands and the toxicity of this compound has been extensively studied. TCDD has been used as a tool to study the biology of Ahr and most of the information regarding gene regulation by this pathway was obtained using TCDD in model systems. TCDD is a reproductive and developmental toxicant and in mice causes cleft palate and hydronephrosis. Ahr and its partner protein ARNT are expressed in the developing embryo as early as GD10 and the expression patterns become increasingly tissue specific as gestation progresses [76,77]. Ahr and ARNT are expressed in the developing palate of mouse and human embryos, and exposure to TCDD in vitro results in activation and increased expression of CYP1A1 in both species [78,79]. The induction of isolated clefts of the secondary palate (CP) by TCDD is mediated by signaling through the Ahr pathway, as shown by studies in Ahr knockout mice [80]. After exposure to 25 µg TCDD/kg body weight on GD10, 72% of the wild-type mice have CP, and the incidence in the TCDD-treated Ahr knockout mice (9%) is not significantly different from the controls.

The etiology of the CP in C57 Bl/6 mice was determined to involve a failure of the palatal shelves to fuse. Formation of the secondary palate, or roof of the oral cavity, is the result of outgrowth of palatal shelves from the maxillary arches, elevation of these shelves above the tongue, expansion and contact of the shelves, and ultimately fusion. Just prior to contact of the opposing shelves, the medial epithelium loses its peridermal cell layer as it undergoes PCD, allowing the underlying basal cell layer to make contact and adhere. These basal cells stop proliferating, the basement membrane degrades, and epithelial cells disappear from the midline seam. Processes for removal of the epithelial cells may involve migration of the cells to the surface, transformation from an epithelial to a mesenchymal phenotype, and/or cell death. Evidence is available to support all of these processes, and it is likely that all are involved in fusion of the shelves. In C57Bl/6 mice exposed to TCDD on GD12, palatal shelves form, elevate, and come into contact, but fail to fuse. The medial epithelial cells have altered expression of several growth factors, continue to proliferate at a time that control palatal medial epithelial cells cease proliferation, and the medial edge ultimately forms a stratified, squamous oral-like epithelium, as reviewed by Abbott and Birnbaum [81]. In TCDD-exposed palates, the expression of Ahr, EGFR, EGF, TGFα, TGFβ2, and TGFβ3 is increased in the medial epithelial cells, while TGFβ1 decreases (based on protein detection by immunohistochemistry). In situ hybridization of mRNA for these genes indicates that the mRNA levels increase for EGF, TGFα, and TGFβ1, remain similar to controls for EGFR and TGFβ2, and decrease relative to controls for Ahr and TGFβ3.

In control palates, just prior to contact of the opposing shelves the medial epithelial cells stop expressing EGFR and stop proliferating. In palates exposed to TCDD, the medial cells continue to express EGFR and to proliferate. The ongoing expression of EGFR accompanied by increased expression of EGF and excessive proliferation of the medial epithelial cells appears key to the failure of the TCDD-exposed palates to fuse. This was examined further in EGF knockout and TGFα knockout mice. The EGF knockout mice exposed to 24 μg TCDD/kg body weight do not develop CP although that dose produces a significant induction of CP in WT mice. Even at 50 μg TCDD/kg the knockout mice do not have a significant increase in CP, although doses of 100 μg or higher do produce CP. This indicates that expression of EGF is a major factor in mediating the induction of CP. The response of the TGFα knockout mice is not different from the C57/BL6 wild type. It can be concluded that TGFα expression is not required for the induction of CP by TCDD. It is important to note that the TGFα knockout mice still express EGF, and if the response to TCDD depends on expression of EGF, then these knockout mice would be expected to respond as they do. Although EGF and TGFα are both ligands for the EGFR and expression of both growth factors is increased subsequent to TCDD exposure, it appears that the response to TCDD that leads to CP is dependent on signaling through the EGFR pathway after binding of EGF, but not TGFα (EGFR pathway signaling depicted in Fig. 1).

The experimental evidence using Ahr knockout mice demonstrated that the induction of CP required signaling through that pathway and the evidence in the EGF knockout mice indicated that the EGFR signaling pathway was also important to the induction of CP by TCDD. The requirement for EGF was also examined using palate organ culture which allowed the availability of growth factor to be manipulated. For palatal organ culture the midfacial tissues (explants) of GD12 wild type and EGF knockout embryos were suspended in medium and cultured for 4 days. In this culture model, the palatal shelves grow and elevate and fuse during culture. In defined medium without supplemental growth factor, explants showed poor response to TCDD. However, in medium supplemented with EGF, the EGF knockout mice responded to TCDD with a failure of the palates to fuse [82].

EGF gene transcription does not appear to be directly regulated by Ahr as DREs have not been found in the promoter region. The steps between Ahr activation and increased expression of growth factors remain unclear. Ma and Babish [72] proposed a model of TCDD-mediated dysregulation of signal transduction pathways that control cell proliferation in which the Ahr–ligand complex activates MAPK kinase-mediated signal transduction. Although their model does not refer to EGFR, EGF binding to EGFR initiates ERK signaling through a MAPK cascade (Fig. 1).

Even though the evidence is strong for Ahr and EGF to be required for the induction of CP after an exposure of mice to TCDD, the etiology is likely to be complex, involving multiple genes and regulation by multiple signaling pathways. TGFß is clearly important in palatogenesis and expression of TGFs is affected by TCDD expression. There is potential for an interaction between Ahr and RAR/RXR pathways to exist and contribute to the induction of CP as well. TCDD and all trans-retinoic acid are known to interact to produce a high incidence of CP at doses which alone do not cause CP [83]. Responses to retinoic acid exposure on GD12 are in many ways similar to those observed after TCDD exposure [84]. The Ahr, ARNT, RAR, RXR, EGFR, and TGFß pathways may well be interactive in producing the effects on epithelial cells that result in proliferative responses that ultimately contribute to the biological effects observed after exposure to TCDD.

OVERVIEW

Signal transduction pathways are essential for morphogenesis and differentiation of the embryo. Receptor-mediated signaling pathways can be grouped into general categories according to their cellular location and signaling intermediates. Cellular processes involved in morphogenesis and regulated by cell signaling include proliferation, cell death, ECM and cytoskeletal remodeling, cell–cell and cell–ECM adhesion, cell motility, cell shape modifications, and differentiation to tissue/organ specific cellular phenotypes. Pathways participating in formation of structures of ectodermal, endodermal, and mesodermal origin all use coordinated and integrated signaling through multiple pathways, and the same patterns and pathways of signaling are used repeatedly to form diverse structures. Tight regulation of the timing and location of activation of signaling is critical for determination of the morphogenetic outcome. Whether membrane-bound or cytoplasmic, the receptor signaling pathways are complex, having the potential to interact with coactivators, corepressors, and multiple intermediate signal transducers and transcriptional activators. Developmental toxicants can interfere with signal transduction, blocking or inappropriately activating the pathway, disrupting critical cellular processes, and resulting in malformations. The pathways–teratogens–malformations presented as examples include sonic hedgehog–cyclopamine–holoprosencephaly, Wnt–thalidomide–limb defects, cation channel–phenytoin–fetal hydantoin syndrome, and Ahr–TCDD–cleft palate. This chapter provides an overview of concepts involved in signal transduction, regulation of cellular processes that impact morphogenesis, and disruption of cell signaling as a mechanism of action.

REFERENCES

1. NRC, Scientific Frontiers in Developmental Toxicology and Risk Assessment. National Academy Press, Washington, D.C., 2000.
2. Metzger RJ, Krasnow MA. Genetic control of branching morphogenesis. Science 1999; 284(5420):1635–1639.
3. Pispa J, Thesleff I. Mechanisms of ectodermal organogenesis. Dev Biol 2003; 262(2):195–205.
4. Thesleff I, Mikkola M. The role of growth factors in tooth development. Int Rev Cytol 2002; 217:93–135.
5. Weaver M, Batts L, Hogan BL. Tissue interactions pattern the mesenchyme of the embryonic mouse lung. Dev Biol 2003; 258(1):169–184.
6. Liu Y, Hogan BL. Differential gene expression in the distal tip endoderm of the embryonic mouse lung. Gene Expr Patterns 2002; 2(3–4):229–233.
7. Whitsett JA, Clark JC, Picard L, et al. Fibroblast growth factor 18 influences proximal programming during lung morphogenesis. J Biol Chem 2002; 277(25):22743–22749.
8. Perl AK, Hokuto I, Impagnatiello MA, et al. Temporal effects of Sprouty on lung morphogenesis. Dev Biol 2003; 258(1):154–168.
9. Wahli W. Peroxisome proliferator-activated receptors (PPARs): From metabolic control to epidermal wound healing. Swiss Med Wkly 2002; 132(7–8):83–91.
10. Keller JM, Collet P, Bianchi A, et al. Implications of peroxisome proliferator-activated receptors (PPARS) in development, cell life status and disease. Int J Dev Biol 2000; 44(5):429–442.
11. Hihi AK, Michalik L, Wahli W. PPARs: Transcriptional effectors of fatty acids and their derivatives. Cell Mol Life Sci 2002; 59(5):790–798.
12. Tarrade A, Schoonjans K, Pavan L, et al. PPARgamma/RXRalpha heterodimers control human trophoblast invasion. J Clin Endocrinol Metab 2001; 86(10):5017–5024.
13. Ding NZ, Ma XH, Diao HL, et al. Differential expression of peroxisome proliferator-activated receptor delta at implantation sites and in decidual cells of rat uterus. Reproduction 2003; 125(6):817–825.
14. Ding NZ, Teng CB, Ma H, et al. Peroxisome proliferator-activated receptor delta expression and regulation in mouse uterus during embryo implantation and decidualization. Mol Reprod Dev 2003; 66(3):218–224.
15. Dunn-Albanese LR, Ackerman WE 4th, Xie Y, et al. Reciprocal expression of peroxisome proliferator-activated receptor-gamma and cyclooxygenase-2 in human term parturition. Am J Obstet Gynecol 2004; 190(3):809–816.
16. Berry EB, Eykholt R, Helliwell RJ, et al. Peroxisome proliferator-activated receptor isoform expression changes in human gestational tissues with labor at term. Mol Pharmacol 2003; 64(6):1586–1590.

17. Asami-Miyagishi R, Iseki S, Usui M, et al. Expression and function of PPARgamma in rat placental development. Biochem Biophys Res Commun 2004; 315(2):497–501.
18. Braissant O, Wahli W. Differential expression of peroxisome proliferator-activated receptor-alpha, -beta, and -gamma during rat embryonic development. Endocrinology 1998; 139(6):2748–2754.
19. Michalik L, Desvergne B, Dreyer C, et al. PPAR expression and function during vertebrate development. Int J Dev Biol 2002; 46(1):105–114.
20. Abbott BD, Wolf CJ, Schmid JE, et al. Perfluorooctanoic acid induced developmental toxicity in the mouse is dependent on expression of peroxisome proliferator activated receptor-alpha. Toxicol Sci 2007; 98(2):571–581. doi: 10.1093/toxsci/kfm110.
21. Escher P, Wahli W. Peroxisome proliferator-activated receptors: Insight into multiple cellular functions. Mutat Res 2000; 448(2):121–138.
22. Zoete V, Grosdidier A, Michielin O. Peroxisome proliferator-activated receptor structures: Ligand specificity, molecular switch and interactions with regulators. Biochim Biophys Acta 2007; 1771(8):915–925.
23. Tien ES, Hannon DB, Thompson JT, et al. Examination of ligand-dependent coactivator recruitment by peroxisome proliferator-activated receptor-alpha (PPAR-alpha). PPAR Res 2006; 2006: 1–9. Article ID69612.
24. Dowell P, Ishmael JE, Avram D, et al. Identification of nuclear receptor corepressor as a peroxisome proliferator-activated receptor alpha interacting protein. J Biol Chem 1999; 274(22):15901–15907.
25. Tan NS, Michalik L, Desvergne B, et al. Multiple expression control mechanisms of peroxisome proliferator-activated receptors and their target genes. J Steroid Biochem Mol Biol 2005; 93(2–5):99–105.
26. Cohen MM Jr, Shiota K. Teratogenesis of holoprosencephaly. Am J Med Genet 2002; 109(1):1–15.
27. Binns W, James LF, Shupe JL, et al. Cyclopian-type malformation in lambs. Arch Environ Health 1962; 5:106–108.
28. Keeler RF, Binns W. Teratogenic compounds of *Veratrum californicum* (Durand). I. Preparation and characterization of fractions and alkaloids for biologic testing. Can J Biochem 1965; 44:819–828.
29. Chen JK, Taipale J, Cooper MK, et al. Inhibition of Hedgehog signaling by direct binding of cyclopamine to Smoothened. Genes Dev 2002; 16(21):2743–2748.
30. Taipale J, Cooper MK, Maiti T, et al. Patched acts catalytically to suppress the activity of Smoothened. Nature 2002; 418(6900):892–897.
31. Chen JK, Taipale J, Young KE, et al. Small molecule modulation of Smoothened activity. Proc Natl Acad Sci U S A 2002; 99(22):14071–14076.
32. Chen W, Ren XR, Nelson CD, et al. Activity-dependent internalization of smoothened mediated by beta-arrestin 2 and GRK2. Science 2004; 306(5705):2257–2260.
33. Meloni AR, Fralish GB, Kelly P, et al. Smoothened signal transduction is promoted by G protein-coupled receptor kinase 2. Mol Cell Biol 2006; 26(20):7550–7560.
34. Incardona JP, Gaffield W, Lange Y, et al. Cyclopamine inhibition of sonic hedgehog signal transduction is not mediated through effects on cholesterol transport. Dev Biol 2000; 224(2):440–452.
35. Cooper MK, Porter JA, Young KE, et al. Teratogen-mediated inhibition of target tissue response to Shh signaling. Science 1998; 280(5369):1603–1607.

36. Incardona JP, Gaffield W, Kapur RP, et al. The teratogenic Veratrum alkaloid cyclopamine inhibits sonic hedgehog signal transduction. Development 1998; 125(18)3553–3562.
37. Nagase T, Nagase M, Osumi N, et al. Craniofacial anomalies of the cultured mouse embryo induced by inhibition of sonic hedgehog signaling: An animal model of holoprosencephaly. J Craniofac Surg 2005; 16(1):80–88.
38. Cordero D, Marcucio R, Hu D, et al. Temporal perturbations in sonic hedgehog signaling elicit the spectrum of holoprosencephaly phenotypes. J Clin Invest 2004; 114(4):485–494.
39. Nagase T, Nagase M, Yoshimura K, et al. Defects in aortic fusion and craniofacial vasculature in the holoprosencephalic mouse embryo under inhibition of sonic hedgehog signaling. J Craniofac Surg 2006; 17(4):736–744.
40. Yang Y. Wnts and wing: Wnt signaling in vertebrate limb development and musculoskeletal morphogenesis. Birth Defects Res C Embryo Today 2003; 69(4):305–317.
41. Peifer M, Polakis P. Wnt signaling in oncogenesis and embryogenesis—A look outside the nucleus. Science 2000; 287(5458):1606–1609.
42. Bryja V, Gradl D, Schambony A, et al. Beta-arrestin is a necessary component of Wnt/beta-catenin signaling in vitro and in vivo. Proc Natl Acad Sci USA 2007; 104(16):6690–6695.
43. Chen W, Hu LA, Semenov MV, et al. Beta-Arrestin1 modulates lymphoid enhancer factor transcriptional activity through interaction with phosphorylated dishevelled proteins. Proc Natl Acad Sci U S A 2001; 98(26):14889–14894.
44. Chan SK, Struhl G. Evidence that Armadillo transduces wingless by mediating nuclear export or cytosolic activation of Pangolin. Cell 2002; 111(2):265–280.
45. Bienz M, Clevers H. Armadillo/beta-catenin signals in the nucleus—Proof beyond a reasonable doubt? Nat Cell Biol 2003; 5(3):179–182.
46. Hancock JT, Desikan R, Neill SJ. Role of reactive oxygen species in cell signalling pathways. Biochem Soc Trans 2001; 29(Pt 2):345–350.
47. Stephens TD, Fillmore BJ. Hypothesis: Thalidomide embryopathy-proposed mechanism of action. Teratology 2000; 61(3):189–195.
48. Knobloch J, Shaughnessy JD Jr, Ruther U. Thalidomide induces limb deformities by perturbing the Bmp/Dkk1/Wnt signaling pathway. Faseb J 2007; 21:1410–1421.
49. Parman T, Wiley MJ, Wells PG. Free radical-mediated oxidative DNA damage in the mechanism of thalidomide teratogenicity. Nat Med 1999; 5(5):582–585.
50. Hansen JM, Harris C. A novel hypothesis for thalidomide-induced limb teratogenesis: Redox misregulation of the NF-kappaB pathway. Antioxid Redox Signal 2004; 6(1):1–14.
51. Zuzarte-Luis V, Montero JA, Rodriguez-Leon J, et al. A new role for BMP5 during limb development acting through the synergic activation of Smad and MAPK pathways. Dev Biol 2004; 272(1):39–52.
52. Mukhopadhyay M, Shtrom S, Rodriguez-Esteban C, et al. Dickkopf1 is required for embryonic head induction and limb morphogenesis in the mouse. Dev Cell 2001; 1(3):423–434.
53. Grotewold L, Ruther U. The Wnt antagonist Dickkopf-1 is regulated by Bmp signaling and c-Jun and modulates programmed cell death. Embo J 2002; 21(5):966–975.

54. Danielsson BR, Webster WS. Cardiovascular active drugs. In Handbook of Experimental Pharmacology 124/II: Drug Toxicity in Embryonic Development II: Advances in Understanding Mechanisms of Birth Defects: Mechanistic Understanding of Human Developmental Toxicants, Kavlock RJ, Daston GP, eds. Springer-Verlag, Berlin, 1997; 161–190.

55. Webster WS, Brown-Woodman PD, Snow MD, et al. Teratogenic potential of almokalant, dofetilide, and d-sotalol: Drugs with potassium channel blocking activity. Teratology 1996; 53(3):168–175.

56. Webster WS, Howe AM, Abela D, et al. The relationship between cleft lip, maxillary hypoplasia, hypoxia and phenytoin. Curr Pharm Des 2006; 12(12):1431–1448.

57. Spence SG, Vetter C, Hoe CM. Effects of the class III antiarrhythmic, dofetilide (UK-68,798) on the heart rate of midgestation rat embryos, in vitro. Teratology 1994; 49(4):282–292.

58. Abrahamsson C, Palmer M, Ljung B, et al. Induction of rhythm abnormalities in the fetal rat heart. A tentative mechanism for the embryotoxic effect of the class III antiarrhythmic agent almokalant. Cardiovasc Res 1994; 28(3):337–344.

59. Franklin JB, Brent RL. The effect of uterine vascular clamping on the development of rat embryos three to fourteen days old. J Morphol 1964; 115:273–290.

60. Webster WS, Lipson AH, Brown-Woodman PD. Uterine trauma and limb defects. Teratology 1987; 35(2):253–260.

61. Jaffee OC. The effects of moderate hypoxia and moderate hypoxia plus hypercapnea on cardiac development in chick embryos. Teratology 1974; 10:275–282.

62. Grabowski CT. Embryonic oxygen deficiency—A physiological approach to analysis of teratological mechanisms. In Advances in Teratology, Woollam DHM, ed. Logos Press Limited, London, 1970; 125–167.

63. Hanson JW, Myrianthopoulos NC, Harvey MA, et al. Risks to the offspring of women treated with hydantoin anticonvulsants, with emphasis on the fetal hydantoin syndrome. J Pediatr 1976; 89(4):662–668.

64. Massey KM. Teratogenic effects of diphenylhydantoin sodium. J Oral Ther Pharmacol 1966; 2(5):380–385.

65. Danielsson BR, Azarbayjani F, Skold AC, et al. Initiation of phenytoin teratogenesis: Pharmacologically induced embryonic bradycardia and arrhythmia resulting in hypoxia and possible free radical damage at reoxygenation. Teratology 1997; 56(4):271–281.

66. Azarbayjani F, Danielsson BR. Pharmacologically induced embryonic dysrhythmia and episodes of hypoxia followed by reoxygenation: A common teratogenic mechanism for antiepileptic drugs? Teratology 1998; 57(3):117–126.

67. Wells PG, Kim PM, Nicol CJ, et al. Reactive intermediates. In Drug Toxicity in Embryonic Development, Vol 1., Handbook of Experimental Pharmacology, Vol. 124/I., Kavlock RJ, Daston GP, eds. Springer-Verlag, Heidelberg, 1997.

68. Petrulis JR, Perdew GH. The role of chaperone proteins in the aryl hydrocarbon receptor core complex. Chem Biol Interact 2002; 141(1–2)25–40.

69. Gu YZ, Hogenesch JB, Bradfield CA. The PAS superfamily: Sensors of environmental and developmental signals. Annu Rev Pharmacol Toxicol 2000; 40:519–561.

70. Nebert DW, Roe AL, Dieter MZ, et al. Role of the aromatic hydrocarbon receptor and [Ah] gene battery in the oxidative stress response, cell cycle control, and apoptosis. Biochem Pharmacol 2000; 59(1):65–85.

71. Patel RD, Kim DJ, Peters JM, et al. The aryl hydrocarbon receptor directly regulates expression of the potent mitogen epiregulin. Toxicol Sci 2006; 89(1):75–82.
72. Ma X, Babish JG. Activation of signal transduction pathways by dioxins. In Molecular Biology of the Toxic Response, Puga A, Wallace KB, eds. Taylor & Francis, Philadelphia, 1999; 493–516.
73. Tan Z, Huang M, Puga A, et al. A critical role for MAP kinases in the control of Ah receptor complex activity. Toxicol Sci 2004; 82(1):80–87.
74. Chen S, Operana T, Bonzo J, et al. ERK kinase inhibition stabilizes the aryl hydrocarbon receptor: Implications for transcriptional activation and protein degradation. J Biol Chem 2005; 280(6):4350–4359.
75. Denison MS, Pandini A, Nagy SR, et al. Ligand binding and activation of the Ah receptor. Chem Biol Interact 2002; 141(1–2):3–24.
76. Abbott BD, Birnbaum LS, Perdew GH. Developmental expression of two members of a new class of transcription factors: I. Expression of aryl hydrocarbon receptor in the C57BL/6N mouse embryo. Dev Dyn 1995; 204(2):133–143.
77. Abbott BD, Probst MR. Developmental expression of two members of a new class of transcription factors: II. Expression of aryl hydrocarbon receptor nuclear translocator in the C57BL/6N mouse embryo. Dev Dyn 1995; 204(2):144–155.
78. Abbott BD, Buckalew AR, Diliberto JJ, et al. AhR, ARNT, and CYP1A1 mRNA quantitation in cultured human embryonic palates exposed to TCDD and comparison with mouse palate in vivo and in culture. Toxicol Sci 1999; 47:62–75.
79. Abbott BD, Held GA, Wood CR, et al. AhR, ARNT, and CYP1A1 mRNA quantitation in cultured human embryonic palates exposed to TCDD and comparison with mouse palate in vivo and in culture. Toxicol Sci 1999; 47(1):62–75.
80. Peters JM, Narotsky MG, Elizondo G, et al. Amelioration of TCDD-induced teratogenesis in aryl hydrocarbon receptor (AhR)-null mice. Toxicol Sci 1999; 47(1):86–92.
81. Abbott BD, Birnbaum LS. Dioxins and teratogenesis. In Molecular Biology of the Toxic Response, Wallace K, ed. Taylor and Francis, Washington, D.C., 1998; 439–447.
82. Abbott BD, Buckalew AR, Leffler KE. Effects of epidermal growth factor (EGF), transforming growth factor-alpha (TGFalpha), and 2,3,7,8-tetrachlorodibenzo-p-dioxin on fusion of embryonic palates in serum-free organ culture using wild-type, EGF knockout, and TGFalpha knockout mouse strains. Birth Defects Res A Clin Mol Teratol 2005; 73(6):447–454.
83. Birnbaum LS, Harris MW, Stocking LM, et al. Retinoic acid and 2,3,7,8-tetrachlorodibenzo-p-dioxin selectively enhance teratogenesis in C57BL/6N mice. Toxicol Appl Pharmacol 1989; 98(3):487–500.
84. Abbott BD, Best DS, Narotsky MG. Teratogenic effects of retinoic acid are modulated in mice lacking expression of epidermal growth factor and transforming growth factor-alpha. Birth Defects Res A Clin Mol Teratol 2005; 73(4):204–217.

3

Nutrition in Developmental Toxicology

Deborah K. Hansen

Division of Genetic and Reproductive Toxicology, FDA/National Center for Toxicological Research, Jefferson, Arkansas, U.S.A.

INTRODUCTION

The adverse outcomes of developmental toxicology include embryonic/fetal death, structural abnormalities, growth retardation, and functional defects. Early pregnancy loss is one of the more common adverse outcomes. It has been suggested that as many as 50% of human conceptuses are lost before implantation, and another 15 to 20% may be lost prior to delivery (1). Estimates have also suggested that the cause of approximately 65 to 70% of structural abnormalities cannot be determined (2), and after years of investigation, this estimate remains virtually unchanged. It is believed that many of these abnormalities are the result of interactions between the genetic constitution of an individual and various environmental factors. Nutrition is one of the environmental factors that may play a role in a variety of developmental toxicology outcomes.

In the 1930s and 1940s, reports began to appear in the literature examining the relationship between the maternal diet and the pregnancy outcome. Often the diets were described as good, fair, or poor, and it was not always clear which factor(s) was used to determine the quality of the diet. Nevertheless, premature birth was observed to occur twice as frequently in women with diets rated as poor as in women with better nutrition (3). When examining neonates at birth through 14 days of age, infants in poor to very poor condition (as indicated by being stillborn, dying within a few days of birth, having a congenital abnormality, or having a very low birth weight) were born almost exclusively to women with a diet that was rated poor to very poor (4). Dietary intervention during the last 3 to 4 months of pregnancy resulted in better maternal health and lower incidences of miscarriages, stillbirths, and prematurity (5). Famine in the Netherlands during World War II also resulted in increases in infant mortality, prematurity, and very low birth weights (6).

Caloric restriction in animal models decreased fetal growth and produced structural abnormalities in some species (6). Restriction of particular components of the diet has also been shown to decrease fetal growth. For example, protein restriction results in growth deficiency in mice (7), rats (8), and swine (9). Restriction of micronutrients in the diet can also produce abnormalities. The production of cretinism by iodine deficiency was identified nearly 150 years ago and was perhaps the earliest association of diet and birth outcome. This resulted in one of the first public health interventions with iodine supplementation occurring in the form of iodized salt.

Deficiencies of micronutrients can occur in several different ways. A primary deficiency is due to a low intake of the micronutrient in the diet. In developed countries, a secondary deficiency may be more common; secondary deficiencies can occur with adequate dietary intake of a micronutrient. Secondary deficiencies can be due to interactions between nutrients, genetic factors, or drug/chemical effects (10). Examples of an interaction between nutrients involve possible interference of absorption of zinc by other nutrients (11). Early studies suggested that intestinal zinc absorption could be hindered by the presence of iron (12–14). It was also suggested that folate might interfere with intestinal zinc absorption (14), but a later study disputed this finding (15). A number of drugs and chemicals have been suggested to be embryotoxic via effects on folic acid uptake or metabolism, including trimethoprim, sulfasalazine, or methanol (16,17).

Genetic factors may play a significant role in nutritional requirements (18) and nutrient-induced abnormal development. For example, a single-gene defect in zinc absorption can lead to acrodermatitis enteropathica (19); single-gene defects in 5,10-methylenetetrahydrofolate reductase (*MTHFR*) can produce developmental delay, seizures, as well as motor and gait disturbances (20). Much work has been done recently on the roles of genetic polymorphisms and birth defects such as neural tube defects (NTDs). Some of this work regarding polymorphisms in folate metabolism genes will be reviewed later in this chapter. An exciting new area of nutrition research involves the role of nutrients on epigenetic changes (21, 22) and the long-term consequences of such changes for the individual (23,24).

It is beyond the scope of this chapter to review in detail studies which have indicated that various micronutrient deficiencies or excesses may be associated with abnormal birth outcomes in experimental animals and/or humans. A good general reference for developmental toxicity of nutrients can be found in Schardein (25). This review will focus on two nutrients that may play significant roles in dietary-induced birth defects as well as epigenetic events. The first of these, folic acid, is well known to play a role in normal development of the neural tube and is involved in metabolism of S-adenosylmethionine, which is the primary methyl donor for DNA methylation. There is much less known about the developmental toxicity of the second nutrient to be reviewed, biotin. However, recent work has suggested that marginal deficiency of this vitamin may occur fairly frequently in human pregnancy, and animal models have shown that marginal deficiency is teratogenic. Biotin has also recently been shown to be involved in chromatin structure (26).

FOLIC ACID

The discovery that folic acid could prevent NTDs has been called the greatest discovery in nutrition science during the last 30 years (27). The first evidence that this vitamin might play a role in the normal development of the neural tube came from a number of clinical studies done in Great Britain in the mid-1960s. These trials indicated that altered metabolism of folic acid might be associated with adverse pregnancy outcome, especially NTDs (16). Publication of the MRC study in 1991 (28) demonstrated a significant decrease in the recurrence of NTDs when 4 mg of folic acid was taken per day around the time of conception. The Medical Research Council (MRC) study was the first to specifically implicate folic acid as the preventative vitamin since many of the earlier studies had utilized multivitamin supplements. Further, the MRC study clearly demonstrated that folate could prevent recurrence of NTDs, and later studies indicated that supplementation with the vitamin could also prevent the occurrence of NTDs (29,30). The accumulated evidence led the United States Public Health Service in 1992 to recommend that all women of reproductive age who were capable of becoming pregnant should consume 400 μg of folic acid per day. In March of 1996, the U. S. Food and Drug Administration authorized the addition of folic acid to cereal grains. This fortification was optional until January 1, 1998, when it became mandatory (31). The level of fortification approved was expected to add approximately 100 μg of folate to the average daily diet and to result in daily intakes of 400 μg in about 50% of women of reproductive age (32).

Recent evaluations have indicated that only about 33% of women aged 15 to 49 years are getting the recommended 400 μg of folate per day (33). Using data from National Health And Nutrition Examination Survey (NHANES), 2001–2002, the authors found differences based on race/ethnicity with 40.5% of non-Hispanic whites, 19.1% of non-Hispanic blacks, and 21% of Hispanics taking in at least 400 μg/day. They also found that about 76% of the women within the total group getting at least 400 μg/day were taking supplements. Again, there were differences due to race/ethnicity with more non-Hispanic whites taking supplements. Results from phone surveys indicated that the percentage of women who took a multivitamin with folic acid daily had increased from 25% in 1995 to 29% in 1998 (34), to 40% by 2004 (35), and remained at 40% in 2007 (36). The most recent data also showed little difference in supplement use between whites (40%)/nonwhites (36%) and Hispanics (38%)/non-Hispanics (40%). Some of these discrepancies may be due to the differences in data collection or in the timing of the study (2001–2002 vs. 2007).

Using data from various NHANES collections, folate status as determined by serum and/or red cell folate levels increased dramatically soon after fortification (37–39). Using data from NHANES, 1999, and comparing it to NHANE-SIII (1989–1994), the Centers for Disease Control and Prevention (CDC) found that median serum folate had increased nearly threefold (from 4.8 ng/mL to 14.5 ng/mL) after fortification (37). Red cell folate had nearly doubled from 160 ng/mL to 293 ng/mL. However, more recently, serum folate levels appear to be decreasing (39, 40). Median serum folate decreased from 12.6 ng/mL to 11.4 ng/mL to 10.6 ng/mL in 1999–2000, 2001–2002, and 2003–2004, respectively (40). Red cell folate levels were 255, 260, and 235 ng/mL over the same time frame. These decreases were observed in non-Hispanic whites, non-Hispanic blacks, and Mexican Americans. The reasons for the decreases are not clear. However, after the initial large increases in folate status, various enriched foods were analyzed and found to have been overfortified, with up to 150% of the amount of folate was allowed (41). More recently, Johnston and Tamura (42) have found that the amount of folate in bread has decreased from levels present in 2001, indicating that companies may not be overfortifying to the same extent as in 2000–2001. There may also be differences in the consumption of enriched foods; it was noted that about 27% of women indicated that they had been on low carbohydrate diets (40), which could lead to decreased consumption of enriched products.

Since folate has been added to cereal grains in the United States, several epidemiological analyses have indicated that the incidence of NTDs has decreased (32,43–48). The decreases have been very consistent across studies. Soon after fortification, Honein et al. (32) reviewed birth certificate data from 48 states and found a 23% decrease in the prevalence of spina bifida and an 11% decrease in anencephaly when comparing rates from a prefortification to a postfortification period. Williams et al. (43) used data from 24 population-based surveillance systems and compared NTD rates during prefortification to those postfortification, and found a 31% decrease in spina bifida and a 16% decrease in anencephaly. The CDC (45) compared NTD prevalence for 24-month periods from 1995–1996 to 1999–2000 using data from 23 population-based surveillance systems. They observed a 27% decline in NTDs overall with approximately a 34% decrease in spina bifida and a 15% decrease in anencephaly. Canfield et al. (46) using data from 23 states participating in the National Birth Defects Prevention Network found a 15 to 18% decrease in anencephaly and a 34 to 36% decrease in spina bifida. Botto et al. (47) using data from Atlanta and Texas found approximately 16 to 17% decreases in anencephaly and 27 to 35% decreases in spina bifida. All of these decreases were significant but were less than the 50% anticipated when fortification began. The reasons for this are not clear.

In addition to NTDs, orofacial clefting has been reported to be decreased by folate in some studies (46,49–51) (52). A recent meta-analysis of 5 prospective studies indicated that orofacial clefts were decreased by folate (relative risk = 0.55, CI, 0.32–0.95) (53). The same authors analyzed 12 case-control studies and again found a protective effect of folate (relative risk = 0.78, CI, 0.71–0.85). Another recent study found a statistically significant decrease in the prevalence of orofacial clefting postfortification (54). When comparing birth certificate data on clefting incidence prior to fortification (1990–1996) to postfortification (October, 1998 – December, 2002), the prevalence decreased from 85.2 per 100,000 births to 80.2 per 100,000 births.

Cardiovascular defects have also been reported to be decreased by folic acid in some studies (29,46,52,55–59). Fewer data are available regarding associations between folate supplementation and limb reduction defects (46,60–62), urinary tract defects, or Down's syndrome (63–70). Overall, most data do not support much of a preventative effect of folic acid on any birth defect other than NTDs (47).

Even though it has been clear since 1991 that folic acid can prevent NTDs in humans, the mechanism for this preventative effect remains unknown. Initially, investigators looked for folate deficiency to explain this association; however, only rarely was serum or red cell folate deficiency observed in women with offspring with an NTD (71). While several studies (72–77) have not reported a difference between case and control mothers for serum and/or red cell folate levels, a few studies have observed differences (78–80). Overall, studies do not appear to support a simple folate deficiency as a cause of NTDs.

Since a primary deficiency of folate was not apparent, investigators turned their attention to looking for causes for a secondary deficiency possibly due to a genetic defect in folate uptake and/or metabolism. The majority of these studies have looked for associations between genetic polymorphisms and NTDs, but a few studies have looked at associations with other birth defects. The polymorphism that has received the most attention is the substitution of thymidine for cytosine at position 677 in the *MTHFR* gene. The mutation results in the replacement of alanine by valine at position 222 in the polypeptide, and produces an enzyme that is thermolabile, has decreased enzyme activity and results in higher serum homocysteine levels (81). This enzyme also has a decreased affinity for its flavin cofactor and requires riboflavin (82).

The association of the C677T mutation with NTDs was first described by van der Put et al. (81). Since that time, a number of studies have looked at this association. Some studies have found that the TT homozygous genotype is more often found in offspring with NTDs or their parents, but a number of studies have observed no difference in the incidence of the mutation in NTD offspring or their parents. A part of the discrepancy between studies may lie in the frequency of the polymorphism. Large differences in frequency have been observed across different geographical and ethnic groups. A recent analysis of the frequency of the polymorphism in over 7000 newborn infants from various countries around the world showed the TT genotype to be most frequent in Mexico (32%) and least frequent in blacks in the Atlanta area of the United States (2.7%) (83). Within the Atlanta area, the TT genotype occurred in 17.7% of Hispanics, 10.7% of whites, 3.8% of Asians, and 2.7% of blacks. This paralleled the incidence of NTDs, which occur most frequently in the United States among Hispanics, with an intermediate level in whites and a lower frequency in blacks (83).

Blom et al. (84) recently performed a meta-analysis on the association of C677T and NTDs using 34 published studies. Of the 29 studies with data on the infant's genotype, they found an odds ratio of 1.9 (CI, 1.6–2.2) indicating a significant association of the homozygous mutant genotype (TT) with the presence of an NTD. The heterozygous (CT) genotype was also significantly associated with an NTD (odds ratio = 1.3, CI, 1.1–1.4). When looking at the maternal genotype, the TT genotype was associated with an odds ratio of 1.6 (CI, 1.3–2.0), and the CT genotype was associated with a ratio of 1.1 (CI, 1.0–1.3). Regarding the paternal genotype, the TT genotype was associated with an odds ratio of 1.2 (CI, 1.0–1.6). There were 23 studies with data on maternal genotype, and 13 with data on the paternal genotype.

Polymorphisms in several other genes involved in folate metabolism have been identified and investigated for their association with birth defects, particularly, NTDs. Some of these are listed in Table 1. Because of the low frequencies of many of these polymorphisms and their examination in only a small number of individuals, it is not clear if they play much of a role in NTDs or other folate-preventable birth defects.

Since folate deficiency and/or the *MTHFR* mutation result in increased serum levels of homocysteine, this compound has been thought to be associated with NTDs. Although several studies have demonstrated increased levels of homocysteine in NTD pregnancies (85–87), this finding has not been consistent across studies (80,88,89). Additionally, early work in the chick suggested that homocysteine could produce NTDs (90); however, more recent work in mammalian systems has suggested that homocysteine does not produce NTDs in rodents either in vitro (91,92) or in vivo (93).

Another potential mechanism of folate protection for NTDs was suggested by Rothenberg's group. They found that antibodies to the folate receptor were embryotoxic to rat embryos and produced teratogenesis (94). These embryos were rescued by pharmacological doses of folinic acid. They also observed antibodies to the folate receptor in serum from 9 of 12 women who had a pregnancy complicated by an NTD (95). Recently, such antibodies were also identified in serum from 9 of 11 women who had a child with an orofacial cleft (96). More work will need to be done to determine how much of a role folate receptor antibodies might play in folate-responsive birth defects.

Evidence is clear that folate can prevent NTDs and possibly other birth defects. This does not appear to be due to a primary deficiency of the vitamin, and the search for genes responsible for a secondary deficiency will continue, as will research to determine the mechanism for this protective effect. However, the polymorphisms in folate uptake/metabolism genes identified so far do not appear to be significant factors in the reduction of NTDs by folic supplementation.

Table 1 Polymorphisms in Folate Metabolizing Genes Associated with Neural Tube Defects

Enzymes	Genes	Polymorphisms	Protein change	Associated with NTDs	References
Folyl-γ-glutamate	*GCPII*	C1561T	His475Tyr	No	(157)
carboxypeptidase				No	(158)
				No	(159)
Reduced folate carrier-1	*RFC-1*	G80A	His27Arg	No	(160)
				Yes	(161,162)
				Yes	(158)
				No	(159,163)
				Yes	(164)
Transcobalamin II	*TCII*	C766G	Pro259Arg	No	(165)
transporter				No	(166)
5,10-Methylenetetrahydro	*MTHFR*	C677T	Ala222Val	Yes	(84)
folate reductase					
		A1298C	Glu429Ala	No	(159)
				No	(167)
				Yes	(168)
Methionine synthase	*MTR*	A2756G	Asp919Gly	No	(169)
				Yes	(170)
				Yes	(171) (mom)
				Yes	(172)
Methionine synthase	*MTRR*	A66G	Ile22Met	Yes	(173)
reductase				Yes	(159)
Thymidylate synthase	*TS*	28 bp repeat		Possible	(174)
Cystathionine ß-synthase	*CBS*	844ins68		No	(175)
				No	(176)
				No	(177)
Betaine homocysteine	*BHMT*	G742A	Arg239Gln	No	(178)
methyltransferase				No	(179)
Serine hydroxymethyl	*SHMT*	C1420T	Leu474Phe	No	(180)
transferase					
Trifunctional synthase	*MTHFD*	G1958A	Arg653Gln	Yes	(181)
				Yes	(182)

BIOTIN

Biotin is a water-soluble vitamin in the B complex of vitamins and is a cofactor for several enzymes in pathways involved in gluconeogenesis, fatty acid synthesis, amino acid catabolism, and carbohydrate metabolism. Specifically, biotin serves as a coenzyme for four carboxylases—acetyl-CoA carboxylase, pyruvate carboxylase, propionyl-CoA carboxylase, and 3-methylcrotonyl-CoA carboxylase.

Biotin is added to polypeptides by the action of holocarboxylase synthase (HCS), and a number of mutations in HCS have been described. These mutations lead to multiple carboxylase deficiency in which affected infants demonstrate a skin rash, ketolactic acidosis, difficulty in feeding and breathing, alopecia, developmental delay, and lethargy. The symptoms are responsive to biotin supplementation (97). The recycling of biotin by breakdown of carboxylases is due to the activity of the enzyme, biotinidase. Deficiency of this enzyme also results in multiple carboxylase deficiency, which is responsive to biotin supplementation (98, 99).

Biotin is synthesized by microorganisms and plants and is widely distributed in foods, although at low concentrations compared to other water-soluble vitamins. Because of its wide distribution in foods and its production by intestinal bacteria, spontaneous biotin deficiency is extremely rare. Biotin dietary intake has been estimated to be 35 to 70 µg/day (100). Deficiency can be induced by parenteral feeding without biotin supplementation or in individuals consuming large amounts of raw egg whites (100) as well as in individuals with HCS or biotinidase deficiency. The avidin in egg whites binds to biotin thereby removing it from the diet.

Dietary biotin requirements remain somewhat uncertain due to the lack of validated indicators of biotin status. Recent work has suggested that urinary excretion of biotin or 3-hydroxyisovaleric acid (3-HIA) may be earlier and more sensitive biomarkers of biotin status than the serum biotin concentration (101). Increased amounts of 3-HIA are produced and excreted due to the decrease in activity of 3-methylcrotonyl-CoA carboxylase. Using urinary 3-HIA excretion as a biomarker of biotin status and comparing it to other biomarkers, Mock and coworkers developed a human model of marginal biotin deficiency (101). It is possible that some populations may be more susceptible to becoming biotin deficient, and pregnant women may be one such group of individuals. Although the literature reports are conflicting depending on the biomarker utilized, most studies have demonstrated that biotin levels are decreased during pregnancy (102–107).

Biotin appears to be transported across the placenta by an active transport method (108,109), although fetal accumulation was not noted in these studies. Fetal levels were reported to be higher than maternal levels in another study (110), but the assay used in this study may have inadvertently measured inactive biotin metabolites (106). Watanabe (111) did observe an increase in fetal levels of biotin in mice when compared to the levels in the dam when mice were supplemented with excess biotin.

Epileptics may also be more susceptible to develop biotin deficiency. Several anticonvulsant medications, including phenobarbital, phenytoin, primidone, and carbamazepine, appear to alter biotin status either by inhibiting uptake (112), by altering biotin metabolism (113,114), or by decreasing protein and mRNA expression of pyruvate carboxylase (115). All of these drugs have also demonstrated teratogenicity in experimental animals and humans (25); however, biotin deficiency has not been speculated to be a mechanism of developmental toxicity for any of these drugs.

Some of the earliest reports of the possible teratogenicity of biotin deficiency came from work with chickens (106). Biotin-deficient hens produced eggs with decreased hatchability and chicks with beak and limb deformities. Using mice, Watanabe (116) reported that biotin deficiency produced developmental toxicity. Mice were fed a diet in which casein was replaced as a protein source by spray dried egg white. Another group of mice were fed the biotin-deficient diet with sufficient biotin added to neutralize all of the avidin and to provide sufficient biotin for normal pregnancy. In the absence of evidence of maternal toxicity, there were no differences in the numbers of implants or live fetuses; however, biotin-deficient fetuses weighed less than those in the other two groups. Additionally, approximately 80 to 90% of fetuses in the biotin-deficient litters had micrognathia and/or cleft palate. Additionally, over 40% of biotin-deficient fetuses had micromelia; this defect did not occur in any fetuses in the other groups. Additional skeletal defects in the biotin-deficient group consisted of mandibular hypoplasia, limb hypoplasia, and extra rib.

Watanabe and Endo (117) fed mice chow with three different amounts of avidin added; no maternal toxicity or embryolethality was observed. There were dose-related decreases in fetal body weight and increases in malformed fetuses. The malformations observed were micrognathia, cleft palate, and micromelia. Results from the high-dose avidin group were very similar to the results from their previous study in which egg white was substituted for casein as the protein source in the chow. A more thorough dose–response analysis was performed by Mock et al. (118), who examined multiple biomarkers of biotin deficiency in addition to developmental toxicity. Casein was replaced by egg white at 1.0, 1.3, 2.0, 5, 10, or 25 g/100 g chow (indicated as 1.0%, 1.3%, etc). Two control groups were used: one group was fed the casein diet with 0% egg white added, and the other was fed a standard rodent diet. A third control group was fed the chow with 25% egg white with sufficient biotin added to bind all avidin and supply sufficient biotin for normal growth and development. Biotin deficiency did not produce any evidence of maternal toxicity, adverse clinical signs, or adversely affect implantation or embryolethality at any dose. The high-dose group was the only group to demonstrate a decrease in fetal weight. There was a striking dose-related increase in the number of fetuses with cleft palate, with all diets with at least 2% egg white demonstrating an increased incidence of clefting. Micrognathia was increased at all egg white doses of 3% and above, and microglossia frequency was increased at egg white concentrations of 5% and higher. Forelimb, hind limb, and pelvic girdle hypoplasia were also increased in a dose-responsive manner at all concentrations of egg white of 3% or higher. Virtually all fetuses in the 25% group had cleft palate, micrognathia, and limb hypoplasia. There were no malformations among the fetuses in the group fed 25% egg white with supplemental biotin. With increasing concentrations of egg white, there were increases in the excretion of 3-HIA and decreased excretion of biotin. Maternal hepatic biotin concentration and propionyl-CoA carboxylase activity were decreased only at the highest doses. Fetal hepatic biotin was decreased at 10 and 25% egg white, and fetal hepatic propionyl-CoA carboxylase activity was decreased at 3% egg white and all higher doses. These results demonstrated that an egg white dose (3%) that produced only a minor change in a biotin biomarker (increased 3-HIA excretion) with no changes in maternal biotin excretion or maternal hepatic propionyl-CoA carboxylase activity could decrease fetal hepatic propionyl-CoA carboxylase activity and significantly increase structural abnormalities, suggesting that a marginal biotin deficiency could result in fetal abnormalities.

Biotin-deficient rats did not display any developmental toxicity (119). This was confirmed by Watanabe and Endo (120), who examined the developmental toxicity of biotin deficiency in Jcl:ICR, C57 BL/6 N/Jcl, and A/Jax strain mice as well as Syrian hamsters and Wistar rats. Maternal and fetal liver biotin were also measured in each group of animals. Maternal toxicity was present in ICR and C57 mice, rats, and hamsters, but not in A/Jax mice. There were no differences in the number of implantation sites among any of the strains or species, but biotin deficiency did decrease the number of live fetuses and increased the frequency of resorptions in hamsters. A later study in hamsters observed exencephaly, cleft palate, micromelia, and hemorrhage among hamster fetuses examined on GD 14 suggesting that the increase in resorptions may have been due to malformed embryos that did not survive until the end of gestation (121). Biotin deficiency decreased fetal weight in every group except the A/Jax mice. No malformations were observed among rat fetuses, and only two hamster fetuses were abnormal (both had exencephaly). All three strains of mice demonstrated an increased frequency of abnormal fetuses in the biotin-deficient group. Cleft palate was increased in all strains; micrognathia and micromelia were present only in ICR and C57 mice, but not in A/Jax mice. Maternal liver biotin content was approximately the same among control animals of all strains and species, and the deficient chow decreased hepatic biotin content by about 50% in each strain and species. However, there were large differences in fetal biotin contents. Despite the differences in teratogenic effects, fetal livers from ICR mice, Wistar rats, and Syrian hamsters contained virtually identical levels of biotin in the biotin-deficient animals.

It was questioned whether the results observed in this study were due to biotin deficiency or whether biotin deficiency might have potentiated the teratogenic effects of vitamin A, which was present in excess in this diet (122). Watanabe and Endo (123,124) examined this question in more detail using a diet with spray-dried egg white as the protein source and various amounts of vitamin A added. The results suggested that the teratogenicity observed in the mice was due to deficiency of biotin, which was not potentiated by excess vitamin A.

Excesses of some vitamins can cause developmental toxicity, but this may not be the case for biotin. Excess biotin delivered to mice either in the diet or by subcutaneous injection on GD 0, 6, and 12 had no adverse effects on development, even though serum levels of biotin were increased by over 200-fold at the end of gestation (111). Biotin also appeared to accumulate in the fetus and placenta with increases in placenta of over 150-fold and in fetal liver of about 77-fold, whereas the concentration in maternal liver increased only 2-fold. Excess biotin applied to the head region near the eye of embryonic chick embryos produced structural malformations of the retina that were dose and time dependent (125). These malformations could be prevented by the simultaneous administration of avidin. Excess biotin did not affect proliferation in the eye region but did increase retinal apoptosis.

The mechanism for the embryotoxic effect of biotin deficiency on skeletal growth is unknown. Abnormal lipid metabolism has been suggested since both acetyl-CoA and propionyl-CoA carboxylases are involved in fatty acid elongation. Abnormal lipid profiles have been reported in biotin-deficient rats (126–128) and humans (129). In addition, alterations in skeletal growth of biotin-deficient chicks have been suggested to be due to abnormalities in arachidonic acid metabolism (106). This mechanism has not been investigated in other experimental model systems.

The mechanism for the embryotoxic effect of biotin deficiency on palatal growth is also unknown. Watanabe's group examined development and fusion of the palate using a serum-free organ culture system (130). When palates were removed from GD 12 biotin supplemented mice and placed in a serum-free organ culture system, approximately 80% of the explants fused within 72 hours if biotin was present in the medium. If the media was biotin free, only 27% of the explants fused, indicating that continual biotin exposure is necessary for normal palatal fusion. Only a slight decrease in the frequency of fused palates occurred when avidin was added to the culture system; since tissue biotin is bound to cellular proteins, it would not be available to interact with avidin. When palates were removed from biotin-deficient embryos, approximately one-third of the explants fused in biotin supplemented media, while only 6% fused when cultured in biotin-free media. This also suggested that the timing of biotin exposure was important in that palates which did not have biotin present early in the process of palatal fusion were not capable of fusion if biotin was present later in the fusion process. The addition of the end products of biotin-requiring enzymes did not increase the incidence of fusion. Biotin-deficient palates incorporated less radiolabeled methionine into cellular proteins than did biotin-sufficient palates, suggesting that biotin inhibited protein synthesis, which is required for proliferation and palatal fusion. However, the reason that palatal fusion requires biotin remains unknown.

Although the best-known function of biotin is as a coenzyme for the four carboxylases discussed earlier, biotin can also affect gene expression (131). Some of the first genes observed to be affected by biotin status were those involved in glucose metabolism (132). Later work indicated that biotin deficiency decreased both protein and mRNA levels of HCS in rats (133). They also observed decreased protein levels of pyruvate and propionyl-CoA carboxylases with no changes in their mRNA levels. Solorzano-Vargas et al. (134) observed decreased mRNA for HCS in human cells that was responsive to biotin and cyclic guanine monophosphate (GMP) (cGMP) supplementation. They hypothesized that biotinyl-adenine monophosphate (AMP), which is an intermediate in the synthesis of holocarboxylases, is part of a signaling pathway involving activation of cGMP, which enhances expression of HCS, acetyl-coA carboxylase, and propionyl-CoA carboxylase. Recently, marginal biotin deficiency in mice decreased the abundance of all four of the biotin-requiring holocarboxylases in both maternal and fetal liver (135). Each enzyme was decreased by 40 to 60% in maternal liver but was decreased by about 90 to 95% in fetal liver. There were no changes in mRNA levels of any of the carboxylases in either maternal or fetal liver. Supplementation of humans with biotin demonstrated increased levels of 3-methylcrotonyl-CoA carboxylase in peripheral blood mononuclear cells (136). Biotin can also regulate expression of the sodium-dependent multivitamin transporter, which is responsible for cellular uptake of the vitamin (137). Human cells cultured in biotin-deficient medium responded with increased expression of this transporter (138,139). These data are consistent in indicating that biotin can alter the expression of several of the genes involved in its uptake and utilization.

Microarray analysis has demonstrated that expression of many other genes may also be altered by biotin. Most of this work has been performed either in peripheral blood mononuclear cells or in cultured cells and has indicated that expression of some 2000 human genes may be changed (140). The gene changes in adult cells were not randomly distributed but appeared to be clustered with about 28% involved in cell signaling (141). Transcription factors NF-κB (142) as well as Sp1 and Sp3 (143) are sensitive to biotin regulation as well as several proteins involved in the receptor tyrosine kinase pathway (144). These signaling pathways are also important in embryonic development; however, the roles that they may play in the teratogenic effect of biotin deficiency remain to be determined.

Using purified histones, Hymes et al. (145) first demonstrated that serum biotinidase was able to covalently bind biotin onto lysine residues of histones. However, biotinidase does not appear to be the enzyme that biotinylates histones under physiological conditions; rather HCS has been localized in the nucleus and appears to be able to transfer a biotin moiety to histones (146). Investigators have identified some of the sites of biotinylation in human histones; these include K4, K9, and K18 in histone H3 (147); K8 and K12 in histone H4 (148); and K9 and K13 in histone H2 A (149). Biotinylation levels can be affected by acetylation and phosphorylation of nearby sites (147,148).

Many of the functions of biotinylated histones remain to be determined. Several histones were biotinylated in peripheral blood mononuclear cells in response to a stimulus to proliferate; it appeared that all classes of histones were affected, including H1, H2 A, H2B, H3, and H4 (150). Additionally, fibroblasts from patients with HCS deficiency had much lower levels of histone biotinylation (146), but it is not clear if this effect causes a decrease in proliferation. There are also cell cycle–dependent differences with increased biotin uptake and biotinylation of 3-methylcrotonyl-CoA and propionyl-CoA carboxylases in the G1 phase (151) and possibly also cell cycle differences in biotinylation of histones (152,153). The role of histone biotinylation in repair of DNA damage is uncertain. Biotin supplementation of Jurkat cells in vitro increased the expression of cytochrome P_{450} 1B1 and increased the frequency of single strand breaks in DNA (154); this effect appeared to be the result of the increased expression of cytochrome P_{450} 1B1 and not a direct effect of biotin supplementation. Using a site-specific antibody to K12 biotinylation of histone H4, Kothapalli et al. (155) observed decreased biotinylation of this histone 10 to 20 minutes after treatment of JAr choriocarcinoma cells in vitro with etoposide; double strand breaks in DNA were observed 60 minutes after addition of etoposide to the culture media, although damage had probably occurred earlier. These results suggested that biotinylation of histone H4 in particular might be an early response to DNA damage. Additionally, this particular biotinylated histone appears to be localized to heterochromatin (156), suggesting that it may play a role in chromatin silencing. Further research is needed to determine the biological significance of histone biotinylation and any role it may play in epigenetic changes and developmental toxicity.

In conclusion, it appears that a marginal biotin deficiency may develop during pregnancy, and animal studies have demonstrated that such a deficiency may result in developmental toxicity. Much research remains to be done to determine if biotin deficiency does produce developmental toxicity in humans and the mechanism for this effect.

CONCLUSIONS

The causes of most structural birth defects are unknown but are believed to be multifactorial in nature. Deficiency or excess of several nutrients have been shown to produce birth defects in animal models or in humans. A deficiency may not be due to decreased nutrient intake, but may be due to genetic variability in its uptake, metabolism, or utilization. Although a good deal of work has been done examining associations between various genetic polymorphisms in folic acid uptake or metabolism and birth defects, no polymorphism to date appears to be largely responsible for the 50 to 75% projected decrease in folate-preventable NTDs. Much work remains to be done to determine the mechanism whereby folic acid is able to prevent NTDs as well as possibly other defects. One of those defects, orofacial clefting, is the main teratogenic effect in mice of a marginal deficiency of the water-soluble vitamin, biotin. Such a deficiency may be more prevalent than previously believed and may be responsible, in part, for developmental toxicity among humans. Lastly, a new area of nutrition research in development involves nutrient-induced epigenetic changes and the potential long-term consequences of these alterations on an individual.

REFERENCES

1. Hirschi KK, Keen CL. Nutrition in embryonic and fetal development. Nutrition 2000; 16:495–499.
2. Holmes LB. Chapter 1. What are common birth defects in humans, and how are they diagnosed? Teratology Primer available at http://teratology.org, 2005.
3. Jeans PC, Smith MB, Stearns G. Incidence of prematurity in relation to maternal nutrition. J Am Diet Assoc 1955; 31:576–581.

4. Burke BS, Beal VA, Kirkwood SB, Stuart HC. The influence of nutrition occurring pregnancy upon the condition of the infant at birth. J Nutr 1943; 26:569–583.
5. Ebbs JH, Tisdall FF, Scott WA. The influence of prenatal diet on the mother and child. J Nutr 1941; 22:515–521.
6. Daston GP, Uriu-Adams J. Chapter 24. How does nutrition influence development? Teratology Primer available at http://teratology.org, 2005.
7. Singh J, Hood RD. Maternal protein deficiency enhances the teratogenicity of ochratoxin A in mice. Teratology 1985; 32:381–388.
8. Snoeck A, Remacle C, Reusens B, et al. Effect of low protein diet during pregnancy on fetal rat endocrine pancreas. Biol Neonate 1990; 57107–118.
9. Pond WG, Yen JT, Merrsmann HJ, et al. Reduced mature size in pregnancy of swine severely restricted in protein intake during pregnancy. Growth Dev Aging 1990; 54:77–84.
10. Keen CL, Clegg MS, Hanna LA, et al. The plausibility of micronutrient deficiencies being a significant contributing factor to the occurrence of pregnancy complications. J Nutr 2003; 133:1597S–1605S.
11. King JC. Determinants of maternal zinc status during pregnancy. Am J Clin Nutr 2000; 71(suppl):1334S–1343S.
12. Solomons NW, Pineda O, Viteri F, et al. Studies on the bioavailability of zinc in humans: Mechanism of the intestinal interaction of nonheme iron and zinc. J Nutr 1983; 113:337–349.
13. Hambidge KM, Krebs NF, Jacobs MA, et al. Zinc nutritional status during pregnancy: A longitudinal study. Am J Clin Nutr 1983; 37:429–442.
14. Simmer K, Iles CA, James C, et al. Are iron-folate supplements harmful? Am J Clin Nutr 1987; 45:122–125.
15. Tamura T, Goldenberg RL, Freeberg LE, et al. Maternal serum folate and zinc concentrations and their relationships to pregnancy outcome. Am J Clin Nutr 1992; 56:365–370.
16. Hansen DK. Alterations in folate metabolism as a possible mechanism of embryotoxicity. In: Handbook of Experimental Pharmacology, Vol. 124/I. Drug Toxicity in Embryonic Development, Vol. I. R. J. Kavlock, Daston GP, eds. Springer-Verlag, Berlin, 1997;407–432.
17. Hansen DK. Developmental toxicants potentially acting via folate perturbation. In: Folate and Human Development, Massaro EJ, Rogers JM, eds. Humana Press, New Jersey, NJ, 2002;183–203.
18. Stover PJ. Influence of human genetic variation on nutritional requirements. Am J Clin Nutr 2006; 83(2):436S–442S.
19. Wang K, Zhou B, Kuo YM, et al. A novel member of a zinc transpoter family is defective in acrodermatitis enteropathica. Am J Hum Genet 2002; 71:66–73.
20. Rozen R. Genetic variation in folate metabolism. Impact on development. In: Folate and Human Development, Massaro EJ, Rogers JM, eds. Humana Press, New Jersey, NJ, 2002;27–42.
21. Oommen AM, Griffin JB, Sarath G, et al. Roles for nutrients in epigenetic events. J Nutr Biochem 2005; 16:74–77.
22. Stover PJ, Garza C. Nutrition and developmental biology—Implications for public health. Nutr Rev 2006; 64(5):S60–S71.

23. Lau C, Rogers JM. Embryonic and fetal programming of physiological disorders in adulthood. Birth Defects Res C 2004; 72:300–312.

24. Keen CL, Hanna LA, Lanoue L, et al. Developmental consequences of trace mineral deficiencies in rodents: Acute and long-term effects. J Nutr 2003; 133: 1477S–1480S.

25. Schardein JL. Foods, Nutrients, and Dietary Factors. In: Chemically Induced Birth Defects. 3rd edition. Marcel Dekker, Inc., New York, NY, 2000;653–691.

26. Hassan YI, Zempleni J. Epigenetic regulation of chromatin structure and gene function by biotin. J Nutr 2006; 136:1763–1765.

27. Katan MB, Boekschoten MV, Connor WE, et al. Which are the greates recent discoveries and the greatest future challenges in nutrition? Eur J Clin Nutr Advance online publication, October, 2007.

28. MRC Vitamin Study Research Group. Prevention of neural tube defects: Results of the Medical Research Council Vitamin Study. Lancet 1991; 338:131–137.

29. Czeizel AE, Dudas I. Prevention of the first occurrence of neural-tube defects by periconceptional vitamin supplementation. Lancet 1992; 327:1832–1835.

30. Berry RJ, Li Z, Erickson JD, et al. Prevention of neural-tube defects with folic acid in China. New Eng J Med 1999; 341:1485–1490.

31. Pitkin RM. Folate and neural tube defects. Am J Clin Nutr 2007; 85(suppl):285S–288S.

32. Honein MA, Paulozzi LJ, Mathews TJ, et al. Impact of folic acid fortification of the US food supply on the occurrence of neural tube defects. J Am Med Assoc 2001; 285:2981–2986.

33. Yang Q-H, Carter HK, Mulinare J, et al. Race-ethnicity differences in folic acid intake in women of childbearing age in the United States after folic acid fortification: Findings from the National Health and Nutrition Examination Survey, 2001–2002. Am J Clin Nutr 2007; 85:1409–1416.

34. Centers for Disease Control and Prevention. Knowledge and use of folic acid by women of childbearing age—United States, 1995 and 1998. MMWR 1999; 48(16):325–327.

35. Centers for Disease Control and Prevention. Use of vitamins containing folic acid among women of childbearing age—United States, 2004. MMWR 2004; 53(36):847–850.

36. Centers for Disease Control and Prevention. Use of supplements containing folic acid among women of childbearing age—United States, 2007. MMWR 2008; 57(1):5–8.

37. Centers for Disease Control and Prevention. Folate status in women of childbearing age—United States, 1999. MMWR 2000; 49(42):962–965.

38. Pfeiffer CM, Caudill SP, Gunter EW, et al. Biochemical indicators of B vitamin status in the US population after folic acid fortification: Results from the National Health and Nutrition Examination Survey 1999–2000. Am J Clin Nutr 2005; 82:442–450.

39. Ganji V, Kafai MR. Trends in serum folate, RBC folate, and circulating total homocysteine concentrations in the United States: Analysis of data from National Health and Nutrition Examination Surveys, 1988–1994, 1999–2000, and 2001–2002. J Nutr 2006; 136:153–158.

40. Centers for Disease Control and Prevention (2007) Folate status in women of childbearing age, by race/ethnicity—United States, 1999–2000, 2001–2002, and 2003–2004. MMWR 2007; 55(51):1377–1380.

41. Rader JI, Weaver CM, Angyal G. Total folate in enriched cereal-grain products in the United States following fortification. Food Chem 2000; 70:275–289.
42. Johnston KE, Tamura T. Folate content in commercial white and whole wheat sandwich breads. J Agric Food Chem 2004; 52:6338–6340.
43. Williams LJ, Mai CT, Edmonds LD, et al. Prevalence of spina bifida and anencephaly during the transition to mandatory folic acid fortification in the United States. Teratology 2002; 66:33–39.
44. Williams LJ, Rasmussen SA, Flores A, et al. Decline in the prevalence of spina bifida and anencephaly by race/ethnicity: 1995–2002. Pediatrics 2005; 116:580–586.
45. Centers for Disease Control and Prevention. Spina bifida and anencephaly before and after folic acid mandate—United States, 1995–1996 and 1999–2000. MMWR 2004; 53(17):362–365.
46. Canfield MA, Collins JS, Botto LD, et al. Changes in the birth prevalence of selected birth defects after grain fortification with folic acid in the United States: Findings from a multi-state populatin-based study. Birth Defects Res A Clin Mol Teratol 2005; 73(10): 679–689.
47. Botto LD, Lisi A, Bower C, et al. Trends of selected malformations in relation to folic acid recommendations and fortification: An international assessment. Birth Defects Res A 2006; 76:693–705.
48. Centers for Disease Control and Prevention. Improved national prevalence estimates for 18 selected major birth defects—United States, 1999–2001. MMWR 2006; 54:1301–1305.
49. Tolarova M, Harris J. Reduced recurrence of orofacial clefts after periconceptional supplementation with high-dose folic acid and multivitamins. Teratology 1995; 51:71–78.
50. Shaw GM, Lammer EJ, Waserman CR, et al. Risks of orofacial clefts in children born to women using multivitamins containing folic acid periconceptionally. Lancet 1995; 346:393–396.
51. Czeizel AE, Toth M, Rockenbauer M. Population-based case control study of folic acid supplementation during pregnancy. Teratology 1996; 53:345–351.
52. Bailey LB, Berry RJ. Folic acid supplementation and the occurrence of congenital heart defects, orofacial clefts, multiple births, and miscarriage. Am J Clin Nutr 2005; 81(suppl):1213S–1217S.
53. Badovinac RL, Werler MM, Williams PL, et al. Folic acid-containing supplement consumption during pregnancy and risk for oral clefts: A meta-analysis. Birth Defects Res A 2007; 79:8–15.
54. Yazdy MM, Honein MA, Xing J. Reduction in orofacial clefts following folic acid fortification of the U. S. grain supply. Birth Defects Res A 2007; 79:16–23.
55. Czeizel AE. Nutritional supplementation and prevention of congenital abnormalities. Curr Opin Obstet Gynecol 1995; 7:88–94.
56. Czeizel AE. Reduction of urinary tract and cardiovascular defects by periconceptional multivitamin supplementation. Am J Med Genet 1996; 62:179–183.
57. Shaw GM, O'Malley CD, Wasserman CR, et al. Maternal periconceptional use of multivitamins and reduced risk for conotruncal heart defects and limb deficiencies among offspring. Am J Med Genet 1995; 59:536–545.
58. Botto LD, Khoury MJ, Mulinare J, et al. Periconceptional multivitamin use and the occurrence of conotruncal heart defects: Results from a population-based, case-control study. Pediatrics 1996; 98:911–917.

59. Botto LD, Mulinare J, Erickson JD. Occurrence of congenital heart defects in relation to maternal multivitamin use. Am J Epidemiol 2000; 151:878–884.
60. Czeizel AE. Limb-reduction defects and folic acid supplementation. Lancet 1995; 345(8954):932.
61. Yang Q, Khoury MJ, Olney RS, et al. Does periconceptional multivitamin use reduce the risk for limb deficiency in offspring? Epidemiology 1997; 8:157–161.
62. Werler MM, Hayes C, Louik C, et al. Multivitamin supplementation and risk of birth defects. Am J Epidemiol 1999; 150:675–682.
63. James SJ, Pogribna M, Pogribny IP, et al. Abnormal folate metabolism and mutation in the methylenetetrahydrofolate reductase gene may be maternal risk factors for Down syndrome. Am J Clin Nutr 1999; 70:495–501.
64. Hobbs CA, Sherman SL, Yi P, et al. Polymorphisms in genes involved in folate metabolism as maternal risk factors for Down syndrome. Am J Hum Genet 2000; 67(3):623–630.
65. O'Leary VB, Parle-McDermott A, Molloy AM, et al. MTRR and MTHFR polymorphism: Link to Down syndrome? Am J Med Genet 2002; 107(2):151–155.
66. Ray JG, Meier C, Vermeulen MJ, et al. Prevalence of trisomy 21 following folic acid food fortification. Am J Med Genet A 2003; 120(3):309–313.
67. Czeizel AE, Puho E. Maternal use of nutritional supplements during the first month of pregnancy and decreased risk of Down's syndrome: Case-control study. Nutrition 2005; 21(6):698–704.
68. Martinez-Frias ML, Perez B, Desviat LR, et al. Maternal polymorphisms 677 C-T and 1298 A-C of MTHFR, and 66 A-G MTRR genes: Is there any relationship between polymorphisms of the folate pathway, maternal homocysteine levels, and the risk for having a child with Down's syndrome? Am J Med Genet 2006; 140(9):987–997.
69. Scala I, Granese B, Sellitto M, et al. Analysis of seven maternal polymorphisms of genes involved in homocysteine/folate metabolisma nd risk of Down syndrome offspring. Genet Med 2007; 8(7):409–416.
70. Zintzaras E. Maternal gene polymorphisms involved in folate metabolism and risk of Down syndrome offspring: a meta-analysis. J Hum Genet 2007; 52(11):943–953.
71. Wald N. Folic acid and the prevention of neural tube defects. Ann NY Acad Sci 1993; 678:112–129.
72. Molloy AM, Kirke P, Hillary I, et al. Maternal serum folate and vitamin B_{12} concentrations in pregnancies associated with neural tube defects. Arch Dis Child 1985; 60:660–665.
73. Gardiki-Kouidou P, Seller MJ. Amniotic fluid folate, vitamin B_{12} and trancobalamins in neural tube defects. Clin Genet 1988; 33:441–448.
74. Economides DL, Ferguson J, Mackenzie IZ, et al. Folate and vitamin B_{12} concentrations in maternal and fetal blood, and amniotic fluid in second trimester pregnancies complicated by neural tube defects. Br J Obstet Gynaecol 1992; 99:23–25.
75. Mills JL, Tuomilehto J, Yu KL, et al. Maternal vitamin levels during pregnancies producing infants with neural tube defects. J Pediatr 1992; 120:863–881.
76. Lucock MD, Wild J, Schorah CJ, et al. The methylfolate axis in neural tube defects: *In vitro* characterization and clinical investigation. Biochem Med Metab Biol 1994; 52:101–114.
77. Wild J, Seller MJ, Schorah CJ, et al. Investigation of folate intake and metabolism in women who have had two pregnancies complicated by neural tube defects. Br J Obstet Gynaecol 1994; 101:197–202.

78. Yates JRW, Ferguson-Smith MA, Shenkin A, et al. Is disorderd folate metabolism the basis for the genetic predisposition to neural tube defects? Clin Genet 1987; 31:279–287.

79. Kirke PN, Molloy AM, Daly LE, et al. Maternal plasma folate and vitamin B_{12} are independent risk factors for neural tube defects. Q Rev Med 1993; 86:703–708.

80. van der Put NMJ, Thomas CMG, Eskes TKAB, et al. Altered folate and vitamin B_{12} metabolism in families with spina bifida offspring. Q J Med 1997; 90:505–510.

81. van der Put NMJ, Steegers-Theunissen RPM, Frosst P, et al. Mutated methylenetetrahydrofolate reductase as a risk factor for spina bifida. Lancet 1995; 346:1070–1071.

82. McNulty H, Dowey LRC, Strain JJ, et al. Riboflavin lowers homocysteine in individuals homozygous for the MTHFR 677C-T polymorphism. Circulation 2006; 113(1):74–80.

83. Wilcken B, Bamforth F, Li Z, et al. Geographical and ethnic variation of the 677 C>T allele of 5,10-methylenetetrahydrofolate reductase (MTHFR): Findings from over 7000 newborns from 16 areas world wide. J Med Genet 2003; 40:619–625.

84. Blom HJ, Shaw GM, den Heijer M, et al. Neural tube defects and folate: Case far from closed. Nat Neurosci 2006; 7:724–731.

85. Steegers-Theunissen RPM, Boers GHJ, Trijbels FJM, et al. Maternal hyperhomocysteinemia: A risk factor for neural-tube defects? Metabolism 1994; 43:1475–1480.

86. Lucock MD, Wild J, Lumb CH, et al. Risk of neural tube defect-affected pregnancy is associated with a block in maternal one-carbon metabolism at the level of N-5-methyltetrahydrofolate: Homocysteine methyltransferase. Biochem Mol Med 2007; 61:28–40.

87. Bjorke-Monsen AL, Ueland PM, Schneede J, et al. Elevated plasma total homocysteine and C677T mutation of the methylenetetrahydrofolate reductase gene in patients with spina bifida. Q J Med 1997; 90:593–596.

88. Graf WD, Oleinik OE, Jack RM, et al. Plasma homocysteine and methionine concentrations in children with neural tube defects. Eur J Pediatr Surg 1996; 6(Suppl I):7–9.

89. Ubbink JB, Christianson A, Beter MJ, et al. Folate status, homocysteine metabolism, and methylenetetrahydrofolate reductase genotype in rural South African blacks with a history of pregnancy complicated by neural tube defects. Metabolism 1999; 48:269–274.

90. Rosenquist TH, Ratashak SA, Selhub J. Homocysteine induces congenital defects of the heart and neural tube: Effect of folic acid. Proc Natl Acad Sci U S A 1996; 93:15227–15232.

91. Hansen DK, Grafton TF, Melnyk S, et al. Lack of embryotoxicity of homocysteine thiolactone in mouse embryos in vitro. Reprod Toxicol 2001; 15:239–244.

92. Greene ND, Dunlevy LE, Copp AJ. Homocysteine is embryotoxic but does not cause neural tube defects in mouse embryos. Anat Embryol 2003; 206(3):185–191.

93. Bennett GD, Vanwaes J, Moser K, et al. Failure of homocysteine to induce neural tube defects in a mouse model. Birth Defect Res B 2006; 77(2):89–94.

94. Da Costa M, Sequeira JM, Rothenberg SP, et al. Antibodies to folate receptors impair embryogenesis and fetal development in the rat. Birth Defect Res A 2003; 67:837–847.

95. Rothenberg SP, Da Costa MP, Sequeira JM, et al. Autoantibodies against folate receptors in women with a pregnancy complicated by a neural tube defect. New Eng J Med 2004; 350:134–142.

96. Bliek JB, Rothenberg SP, Steegers-Theunissen RPM. Maternal folate receptor autoantibodies and cleft lip and/or palate. Int J Gynecol Obstet 2006; 93:142–143.

97. Pacheco-Alvarez D, Solorzano-Vargas RS, Leon Del Rio A. Biotin in metabolism and its relationship to human disease. Arch Med Res 2002; 33:439–447.

98. Hymes J, Wolf B. Human biotinidase isn't just for recycling biotin. J Nutr 1999; 129:485S–489S, 1999.

99. Wolf B. Biotinidase: Its role in biotinidase deficiency and biotin metabolism. J Nutr Biochem 2005; 16:441–445.

100. Zempleni J, Mock DM. Biotin biochemistry and human requirements. J Nutr Biochem 1999; 10:128–138.

101. Mock DM. Biotin status: Which are valid indicators and how do we know? J Nutr 1999; 129:498S–503S.

102. Bhagavan HN. Biotin content of blood during gestation. Int J Vitam Res 1969; 39:235–237.

103. Dostalova L. Vitamin status during puerperium and lactation. Ann Nutr Metab 1984; 28:385–408.

104. Mock DM, Stadler D. Conflicting indicators of biotin status from a cross-sectional study of normal pregnancy. J Am Coll Nutr 1997; 16:252–257.

105. Mock DM, Stadler D, Stratton S, et al. Biotin status assessed longitudinally in pregnant women. J Nutr 1997; 127:710–716.

106. Zempleni J, Mock DM. Marginal biotin deficiency is teratogenic. Proc Soc Exp Biol Med 2000; 223:14–21.

107. Mock DM. Marginal biotin deficiency is teratogenic in mice and perhaps humans: A review of biotin deficiency during human pregnancy and effects of biotin deficiency on gene expression and enzyme activities in mouse dam and fetus. J Nutr Biochem 2005; 16:435–437.

108. Karl P, Fisher SE. Biotin transport in microvillous membrane vesicles, cultured trophoblasts, and the isolated perfused cotyledon of the human placenta. Am J Physiol 1992; 262:C302–C308.

109. Hu Z-Q, Henderson GI, Mock DM, et al. Biotin uptake by basolateral membrane of human placenta: Normal characteristics and role of ethanol. Proc Soc Exp Biol Med 1994; 206:404–408.

110. Mantagos S, Malamitsi-Puchner A, Antsaklis A, et al. Biotin plasma levels of the human fetus. Biol Neonate 1998; 74:72–74.

111. Watanabe T. Morphological and biochemical effects of excessive amounts of biotin on embryonic development in mice. Experientia 1996; 52:149–154.

112. Said HM, Redha R, Nylander W. Biotin transport and anticonvulsant drugs. Am J Clin Nutr 1989; 49:127–131.

113. Krause K-H, Kochen W, Berlit P, et al. Excretion of organic acids associated with biotin deficiency in chronic anticonvulsant therapy. Int J Vitam Nutr Res 1984; 54:217–222.

114. Mock DM, Dyken ME. Biotin catabolism is accelerated in adults receiving long-term therapy with anticonvulsants. Neurology 1997; 49:1444–1447.

115. Rathman SC, Blanchard RK, Badinga L, et al. Dietary carbamazepine administration decreases liver pyruvate carboxylase activity and biotinylation by decreasing protein and mRNA expression in rats. J Nutr 2003; 13:2119–2124.

116. Watanabe T. Teratogenic effects of biotin deficiency in mice. J Nutr 1983; 113:574–581.

117. Watanabe T, Endo A. Teratogenic effects of avidin-induced biotin deficiency in mice. Teratology 1984; 30:91–94.

118. Mock DM, Mock NI, Stewart C, et al. Marginal biotin deficiency is teratogenic in CD-1 mice. J Nutr 2003; 133:2519–2525.

119. Levin SW, Roecklein BA, Mukherjee AB. Intrauterine growth retardation caused by dietary biotin and thiamine deficiency in the rat. Res Exp Med 1985; 185:375–381.

120. Watanabe T, Endo A. Species and strain differences in teratogenic effects of biotin deficiency in rodents. J Nutr 1989; 119:255–261.

121. Watanabe T. Dietary biotin deficiency affects reproductive function and prenatal development in hamsters. J Nutr 1993; 123:2101–2108.

122. Heard GS, Blevins TL. Is biotin deficiency teratogenic in mice? J Nutr 1989; 119:1348–1349.

123. Watanbe T, Endo A. Reply to the leter of G. S. Heard and T. L. Blevins. J Nutr 1989; 119:1350.

124. Watanabe T, Endo A. Biotin deficiency per se is teratogenic in mice. J Nutr 1991; 121:101–104.

125. Valenciano AI, Mayordomo R, de la Rosa EJ, et al. Biotin decreases retinal apoptosis and induces eye malformations in the early chick embryo. NeuroReport 2002; 13:297–299.

126. Suchy SF, Rizzo WB, Wolf B. Effect of biotin deficiency and supplementation on lipid metabolism in rats: Saturated fatty acids. Am J Clin Nutr 1986; 44:475–480.

127. Mock DM, Mock NI, Johnson SB, et al. Effects of biotin deficiency on plasma and tissue fatty acid composition: Evidence for abnormalities in rats. Pediatr Res 1988; 24:396–403.

128. Proud VK, Rizzo WB, Patterson JW, et al. Fatty alterations and carboxylase deficiencies in the skin of biotin-deficient rats. Am J Clin Nutr 1990; 51:853–858.

129. Mock DM, Johnson SB, Holman RT. Effects of biotin deficiency on serum fatty acid composition: Evidence for abnormalities in humans. J Nutr 1988; 118:342–348.

130. Watanabe T, Dakshinamurti K, Persaud TVN. Biotin influences palatal development of mouse embryos in organ culture. J Nutr 1995; 125:2114–2121.

131. Dakshinamurti K. Biotin—a regulator of gene expression. J Nutr Biochem 2005; 16;419–423.

132. Dakshinamurti K, Litvak S. Biotin and protein synthesis in rat liver. J Biol Chem 1970; 245:5600–5605.

133. Rodriguez-Melendez R, Cano S, Mendez ST, et al. Biotin regulates the genetic expression of holocarboxylase synthetase and mitochondrial carboxylases in rats. J Nutr 2001; 131:1909–1913.

134. Solorzano-Vargas RS, Pacheco-Alvarez D, Leon-Del-Rio A. Holocarboxylase synthethase is an obligate participant in bioin-mediated regulation of its own expression and of biotin-dependent carboxylases mRNA levels in human cells. Proc Natl Acad Sci U S A 2002; 99:5325–5330.

135. Sealey WM, Stratton SL, Mock DM, et al. Marginal maternal biotin deficiency in CD-1 mice reduces fetal mass of biotin-dependent carboxylases. J Nutr 2005; 135:973–977.

136. Wiedmann S, Eudy JD, Zempleni J. Biotin supplementation causes increased expression of genes encoding interferon-γ, interleukin-1ß, and 3-methylcrotonyl-CoA carboxylase, and causes decreased expression of the gene encoding interleukin-4 in human peripheral blood mononuclear cells. J Nutr 2003; 133:716–719.

137. Prasad PD, Wang H, Kekuda T, et al. Cloning and functional expression of a cDNA encoding a mammalian sodium-dependent vitamin transporter mediating the uptake of pantothenate, biotin, and lipoate. J Biol Chem 1998; 273:7501–7506.

138. Manthey KC, Griffin JB, Zempleni J. Biotin supply affects expression of biotin transporters, biotinylation of carboxylases, and metabolism of interleukin-2 in Jurkat cells. J Nutr 2002; 132:887–892.

139. Pacheco-Alvarez D, Solorzano-Vargas RS, Gonzalez-Noriega A, et al. Biotin availability regulates expression of the sodium-dependent multivitamin transporter and the rate of biotin uptake in HepG2 cells. Mol Genet Metab 2005; 85:301–307.

140. Zempleni J. Uptake, localization, and noncarboxylase roles of biotin. Ann Rev Nutr 2005; 25:175–196.

141. Wiedmann S, Rodriguez-Melendez R, Ortega-Cuellar D, et al. Clusters of biotin-responsive genes in human peripheral blood mononuclear cells. J Nutr Biochem 2004; 15:433–439.

142. Rodriguez-Melendez R, Griffin JB, Zempleni J. Biotin supplementation increases expression of the cytochrome P450 1B1 gene in Jurkat cells, increasing the occurrence of single-stranded DNA breaks. J Nutr 2004; 134:2222–2228.

143. Griffin JB, Rodriguez-Melendez R, Zempleni J. The nuclear abundance of transcription factors Sp1 and Sp3 depends on biotin in Jurkat cells. J Nutr 2003; 133:3409–3415.

144. Rodriguez-Melendez R, Griffin JB, Sarath G, et al. High-throughput immunoblotting identifies biotin-dependent signaling proteins in HepG2 hepatocarcinoma cells. J Nutr 2005; 135:1659–1666.

145. Hymes J, Fleischhauer K, Wolf B. Biotinylation of histones by human serum biotinidase: assessment of biotinyltransferase activity in sera from normal individuals and children with biotinidase deficiency. Biochem Mol Med 1995; 56:76–83.

146. Narang MA, Dumas R, Ayer LM, et al. Reduced histone biotinylation in multiple carboxylase deficiency patients: A nuclear role for holocarboxylase synthetase. Hum Molec Genet 2004; 13(1):15–23.

147. Sarath G, Kobza K, Rueckert B, et al. Biotinylation of human histone H3 and interactions with biotinidase. FASEB J 2004; 18:A103.

148. Camporeale G, Shubert EE, Sarath G, et al. K8 and K12 are biotinylated in human histone H4. Eur J Biochem 2004; 271:2257–2263.

149. Chew YC, Camporeale G, Kothapalli N, et al. Lysine residues in N-terminal and C-terminal regions of human histone H2A are targets for biotinylation by biotinidase. J Nutr Biochem 2006; 17:225–233.

150. Stanley JS, Griffin JB, Zempleni J. Biotinylation of histones in human cells. Effect of cell proliferation. Eur J Biochem 2001; 268:5424–5429.

151. Stanley JS, Mock DM, Griffin JB, et al. Biotin uptake into human peripheral blood mononuclear cells increases early in the cell cycle, increasing carboxylase activities. J Nutr 2002; 132:1854–1859.

152. Kothapalli N, Zempleni J. Biotinylation of histones depends on the cell cycle in NCI-H69 small cell lung cancer cells. FASEB J 2005; 19:A55.
153. Kothapalli N, Camporeale G, Kueh A, et al. Biological functions of biotinylated histones. *J. Nutr. Biochem.* 2005; 16:446–448.
154. Rodriguez-Melendez R, Schwab LD, Zempleni J. Jurkat cells respond to biotin deficiency with increased nuclear translocation of NF-κB, mediating cell survival. Int J Vitam Nutr Res 2004; 74:209–216.
155. Kothapalli N, Sarath G, Zempleni J. Biotinylation of K12 inhistone H4 decreases in response to DNA double-strand breaks in human Jar choriocarcinoma cells. J Nutr 2005; 135:2337–2342.
156. Camporeale G, Oommen AM, Griffin JB, Sarath G and Zempleni J. K12-biotinylated histone H4 marks heterochromatin in human lymphoblastoma cells. J Nutr Biochem 2007; 18:760–768.
157. Afman LA, Trijbels FJM, Blom HJ. The H475Y polymorphism in the glutamate carboxypeptidase II gene increases plasma folate without affecting the risk for neural tube defects in humans. J Nutr 2003; 133:75–77.
158. Morin I, Devlin AM, Leclerc D, et al. Evaluation of genetic variants in the reduced folate carrier and in glutamate carboxypeptidase II for spina bifida risk. Mol Genet Metab 2003; 79:197–200.
159. Relton CL, Wilding CS, Pearce MS, et al. Gene-gene interaction in folate-related genes and risk of neural tube defects in a UK population. J Med Genet 2004; 41:256–260.
160. O'Leary VB, Pangilinan F, Cox C, et al. Reduced folate carrier polymorphisms and neural tube defect risk. Mol Genet Metab 2006; 87:364–369.
161. De Marco P, Calevo MG, Moroni A, et al. Polymorphisms in genes involved in folate metabolism as risk factors for NTDs. Eur J Pediatr Surg 2001; 11(Suppl):S14–S17.
162. De Marco P, Calevo MG, Moroni A, et al. Reduced folate carrier polymorphism (80 A→G) and neural tube defects. Eur J Hum Genet 2003; 11:245–252.
163. Relton CL, Wilding CS, Laffling AJ, et al. Low erthrocyte folate status and polymorphic variation in folate-related genes are associated with risk of neural tube defect pregnancy. Mol Genet Metab 2004; 81:273–281.
164. Shang Y, Zhao H, Niu B, et al. Correlation of polymorphism of MTHFRs and RFC-1 genes with neural tube defects in China. Birth Defects Res A 2008; 82(1):3–7.
165. Afman LA, Van der Put NMJ, Thomas CMG, Trijbels JMF and Blom HJ. Reduced vitamin B12 binding by transcobalamin II increases the risk of neural tube defects. Q J Med 2001; 94:159–166.
166. Swanson DA, Pangilinan F, Mills JL, et al. Evaluation of transcobalamin II polymorphisms as neural tube defect risk factors in an Irish population. Birth Defects Res A 2005; 73(4):239–244.
167. Barber R, Shalat S, Hendricks K, et al. Investigation of folate pathway gene polymorphisms and the incidence of neural tube defects in a Texas Hispanic population. Mol Genet Metab 2000; 70:45–52.
168. De Marco P, Calevo MG, Moroni A, et al. Study of the MTHFR and MS polymorphisms as risk factors for NTD in the Italian population. J Hum Genet 2002; 47(6):319–324.
169. Morrison K, Papepetrou C, Hol FA, et al. Susceptibility to spina bifida; an association study of five candidate genes. Ann Hum Genet 1998; 62:379–396.

170. Zhu H, Wicker NJ, Shaw GM, et al. Homocysteine remethylation enzyme polymorphisms and increased risks for neural tube defects. Mol Genet Metab 2003; 78:216–221.

171. Doolin M-T, Barbaux S, McDonnell M, et al. Maternal genetic effects, exerted by genes involved in homocysteine remethylation, influence the risk of spina bifida. Am J Hum Genet 2002; 71:1222–1226.

172. Gros M, Sliwerska E, Szpecht-Potocka A. Mutation incidence in folate metabolism genes and regulatory genes in Polish families with neural tube defects. J App Genet 2004; 45(3):363–368.

173. Van der Linden IJM, den Heijer M, Afman LA, et al. The methionine synthase reductase 66 A>G polymorphism is a maternal risk factor for spina bifida. J Mol Med 2006; 84:1047–1054.

174. Volcik KA, Shaw GM, Zhu H, et al. Associations between polymorphisms within the thymidylate synthase gene and spina bifida. Birth Defects Res A 2003; 67:924–928.

175. Grandone E, Corrao AM, Colaizzo D, Vecchione G, Di Girgenti C, Paladini D, Sardella L, Pellegrino M, Zelante L, Martinelli P and Margaglione M. Homocysteine metabolism in families from southern Italy with neural tube defects: role of genetic and nutritional determinants. Prenat Diagn 2006; 26:1–5.

176. Afman LA, Lievers KJA, Kluijtmans LAJ, et al. Gene-gene interaction between the cystathionine β-synthase 31 base pair variable number of tandem repeats and the methylenetetrahydrofolate reductase 677 C>T polymorphism on homocysteine levels and risk for neural tube defects. Mol Genet Metab 2003; 78:211–215.

177. Ramsbottom D, Scott JM, Molloy A, et al. Are common mutations of cystathionine β-synthase involved in the aetiology of neural tube defects? Clin Genet 1997; 51:39–42.

178. Zhu H, Curry S, Wen S, et al. Are the betaine-homocysteine methyltrasferase (BHMT and BHMT2) genes risk factors for spina bifida and orofacial clefts? Am J Med Genet A 2005; 135(3):274–277.

179. Morin I, Platt R, Weisberg I, et al. Common variant in betaine-homocysteine methyltransferase (BHMT) and risk for spina bifida. Am J Med Genet A 2003; 119(2):172–176.

180. Heil SG, Van Der Put NMJ, Waas ET, et al. Is mutated serine hydroxymethyltransferase (SHMT) involved in the etiology of neural tube defects? Mol Genet Metab 2001; 73:164–172.

181. Brody LC, Conley M, Cox C, et al. A polymorphism, R653Q, in the trifunctional enzyme methylenetetrahydrofolate dehydrogenase/methenyltetrahydrofolate cyclohydrolase/formyltetrahydrofolate synthetase is a maternal genetic risk factor for neural tube defects: Report of the birth defects research group. Am J Hum Genet 2002; 71:1207–1215.

182. Parle-McDermott A, Kirke PN, Mills JL, et al. Confirmation of the R653Q polymorphism f the trifunctional C1-synthase enzyme as a maternal risk for neural tube defects in the Irish population. Eur J Hum Genet 2006; 14:768–772.

4

Epigenetic Mechanisms: Role of DNA Methylation, Histone Modifications, and Imprinting

Robert G. Ellis-Hutchings and John M. Rogers

Reproductive Toxicology Division, National Health and Environmental Effects Research Laboratory, U.S. Environmental Protection Agency, Research Triangle Park, North Carolina, U.S.A.

INTRODUCTION TO EPIGENETICS

The primary DNA sequence is only a foundation for understanding how the genetic program is read. Superimposed upon the DNA sequence is a layer of heritable "epigenetic" information, a facet of the genetic code that we have only just begun to discover and appreciate. Epigenetics is defined as "mitotically and/or meiotically heritable changes in gene function that cannot be explained by changes in DNA sequence" (1). This epigenetic information is stored as chemical modifications falling into two main categories: (1) DNA methylation and (2) changes to the histone proteins that package the genome (2). By regulating DNA accessibility and chromatin structure, these chemical changes influence how the genome is translated across a diverse array of developmental stages, tissue types, and disease states (3–5).

In this chapter, we will first discuss key features of the DNA methylation and histone protein modification landscapes, including structural and chemical characteristics of DNA methylation and histone protein changes, the enzymes involved in such changes, and the effects of such modifications on gene transcription. We will also discuss the interplay of these dynamic modifications and the emerging role of small RNAs in epigenetic gene regulation. Epigenetic modifications are particularly dynamic within the complex regulatory environment controlling proper development; therefore, a substantive component of this chapter will focus on epigenetic regulation during development. Alterations in epigenetic programming during development may be environmentally induced, and have the potential to adversely affect the development of the offspring as well as health during adulthood. This concept will be discussed in the final portion of this chapter, devoted to the emerging concept of developmental origins of health and disease (DOHaD), in which the effects of environmental factors on epigenetic programming and adult health will be explored.

MOLECULAR MECHANISMS OF EPIGENETIC REGULATION

DNA Methylation

In vertebrates, DNA methylation occurs almost exclusively in the context of methylation at the 5-position of the cytosine residue within cytosine-guanine dinucleotides (CpG). This results in the formation of 5-methylcytosine (m^5C), which has been designated as the fifth base of DNA (6, 7) (Fig. 1).

It has been estimated that 60 to 90% of cytosine residues located within CpG dinucleotides are methylated (4,8). Interrupting this relatively featureless sea of genomic methylation are the CpG islands-short sequence domains characterized by high (G+C) and CpG content that generally remain unmethylated at all times, regardless of gene expression (3,9). CpG islands cover about 0.7% of the human genome, but contain 70% of the CpG dinucleotides (10). About 60% of human gene promoters are associated with CpG islands. Although it has been suggested that most CpG islands are always unmethylated, a subset has been shown to be subject to tissue-specific methylation during development (3,11), leading to long-term shutdown of the associated genes (12–14). Also, many promoters that lack strictly defined CpG islands have nonetheless been shown to have tissue-specific

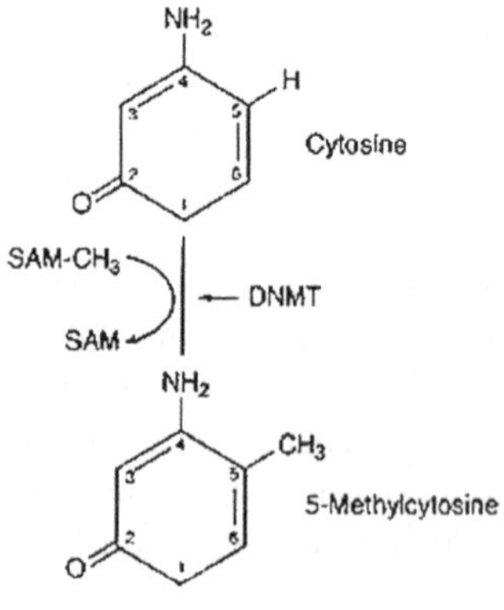

Figure 1 Formation of 5-methylcytosine on DNA.

methylation patterns that strongly correlate with transcriptional activity (e.g., *Oct4* and *Nanog* promoters) (2).

Methylation of genomic DNA can also occur at the N6 position of adenine or the N4 position of cytosine residues of prokaryotic or eukaryotic genomic DNA (15). Although non-CpG methylation has established functional roles in plants (16), there is only limited evidence that it might also act in mammals. Non-CpG methylation has been observed at a low frequency in the early mouse embryo (17) and embryonic stem (ES) cells (18), but is significantly decreased in somatic tissues.

The methylation of mammalian genomic DNA is catalyzed by DNA methyltransferases (DNMTs) that use S-adenosyl-l-methionine (SAM or AdoMet) as a donor of methyl groups (19) (Fig. 1). DNMTs can be divided into de novo and maintenance methyltransferases (20,21). De novo DNMTs can effectively methylate C to m^5C postreplicatively in unmethylated DNA, whereas maintenance DNMT preferentially attaches a methyl group to hemimethylated DNA during replication (15). The eukaryotic DNMT family has five members: DNMT2, DNMT3 A, DNMT3B, DNMT3 L, and DNMT1. DNMT3 A and DNMT3B are de novo methyltransferases (particularly during embryogenesis), whereas DNMT1 is involved in the maintenance of DNA methylation (20,21). DNMT3A and 3B methylate CpG dinucleotides without preference for hemimethylated DNA. DNMT3 A and 3B activity is reduced upon differentiation of ES cells and remains low in adult somatic tissues. During the replication of eukaryotic genomic DNA, approximately 40 million m^5CpG dinucleotides are converted into the hemimethylated state in the newly synthesized DNA strand. These hemimethylated CpG sites must be methylated precisely to maintain the original DNA methylation pattern. DNMT1 is the major enzyme responsible during replication for maintenance of the DNA methylation pattern (15). DNMT1 displays a 5- to 40-fold higher activity in vitro for hemimethylated DNA than for unmethylated DNA (20,22). However, this enzyme also exhibits very weak de novo methylation activity, which is stimulated by DNMT3A (23).

DNA methylation is largely associated with repression of gene transcription, which occurs through several different mechanisms. A general route is through exclusion of proteins that affect transcription through their DNA-binding sites (24). The presence of m^5CpG dinucleotides in the first gene exon or promoter may have a direct effect on gene transcription through the interference of m^5CpG dinucleotides with transcription factors binding to a promoter. An example of such exclusion applies to the chromatin boundary element-binding protein, CTCF, which can block interactions between an enhancer and its promoter when placed between the two elements (25,26). CpG methylation blocks the binding of CTCF to DNA and thus allows an enhancer to stimulate promoter activity across the inert boundary site (24). This binary switching of CTCF binding through DNA methylation is clearly important in imprinting of the *IGF2* gene, which is expressed exclusively from the paternal allele during development (26,27).

Repression of transcription can also occur indirectly through DNA binding to m⁵CpG dinucleotide-specific proteins, which block the interaction of transcription factors with certain DNA sequences (28). These protein suppressors of promoters mainly include methyl-CpG-binding domain proteins and m⁵CpG-binding proteins. These proteins are able to form large complexes (e.g., NuRD and Sin3a) containing histone deacetylases (HDACs) and adenosine triphosphate (ATP)-dependent chromatin remodeling proteins (Mi-2, Sin3a), which are involved in the stabilization of heterochromatin structure (15) (Fig. 2). Biochemical evidence indicates that DNA methylation is just one component of a wider epigenetic

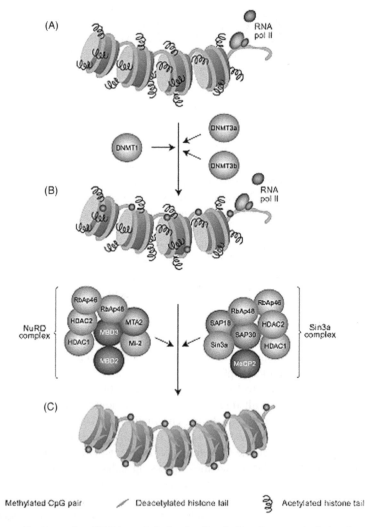

Figure 2 Example of DNA methylation leading indirectly to transcriptional repression. *Source*: From Ref. 172.

(A)

(B)

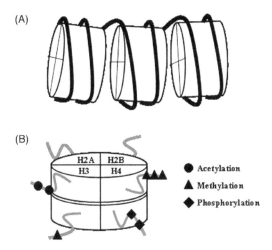

Figure 3 (**A**) The nucleosome showing the repeating units of histones (*cylinders*) around which 147 bp of DNA is wrapped (*solid lines*). (**B**) The core histone composed of H3, H4, H2A, and H2B with the unstructured N-terminal tails (*gray curves*) and their most common modifications.

program that includes posttranslational modification of chromatin proteins. This will be discussed next.

Histone Modifications

In order to properly understand histone modifications, one must first have a working knowledge of the molecular composition of chromatin. Chromatin functionally serves to compact DNA and is composed of repeating units of histones and short sections of DNA (29). The nucleosome is the fundamental unit of chromatin and is composed of an octamer of the four core histones (H) 3, 4, 2A, and 2B [Fig. 3(B)] around which 147 base pairs of DNA are wrapped [Fig. 3(A)] (30). There are 14 contact points between histones and DNA, making the nucleosome one of the most stable protein–DNA complexes under physiological conditions (29). The core histones are predominantly globular except for their N-terminal "tails," which are unstructured (30) [Fig. 3(B)]. The core histones that make up the nucleosome are subject to more than 100 different posttranslational modifications, including 8 distinct types of modifications (acetylation, methylation, phosphorylation, deimination, ubiquitylation, sumoylation, adenosine diphosphate (ADP) ribosylation, and proline isomeration) of which acetylation, methylation, and phosphorylation are best understood. These occur primarily at specific positions within the amino-terminal histone tails (2). Additional complexity comes partly from the fact that methylation at lysines or arginines may be in one of three different forms: mono-, di-, or trimethylated for lysines and mono- or dimethylated (asymmetric or symmetric) for arginines. There are numerous enzymes that mediate the modifications, each with their own mechanisms of action and biological function of the chromatin modification. The term "histone code" has been loosely used to describe the role of modifications to enable DNA functions (30). In this chapter, we will limit our discussion of histone modifications to methylation and acetylation, as these are the only two modifications that have been demonstrated to be transmitted in a heritable manner. This is not to say that other histone modifications are not involved in epigenetic processes, but there is little or no evidence that they are actually transmitted.

There are two characterized mechanisms for the function of histone modifications. The first is the disruption of contacts between the nucleosomes in order to "unravel" chromatin and the second is the recruitment of nonhistone proteins, with the latter best understood to date (30). Thus, depending on the composition of modifications on a given histone, a set of proteins is encouraged to bind to or is occluded from binding to chromatin. These proteins carry with them enzymatic activities (e.g., remodeling ATPases) that further modify chromatin (30). Modifications may affect higher-order chromatin structure by affecting the contact between different histones in adjacent nucleosomes or the interaction of histones with DNA. Of all the known modifications, acetylation has the most potential to unfold chromatin since it neutralizes the basic charge of the lysine. Phosphorylation is another modification that may, well, have important consequences for chromatin compaction via charge alterations (30). Simplistically, histone modifications serve to either establish global chromatin environments or orchestrate DNA-based biological tasks (transcription, DNA repair, replication, kinetochore, and centromere formation) (29,30). Most modifications are distributed in distinct localized patterns within the upstream promoter region: the core promoter, the 5' end of the open reading frame, and the 3' end of the open reading frame (29). The location of the modification is tightly regulated and is crucial for its effect on transcription. It is important to note that modifications on histones are dynamic and rapidly changing. Acetylation, methylation, phosphorylation, and deimination can appear and disappear on chromatin within minutes of a stimulus arriving at the cell surface (30).

Global Chromatin Environments

To establish the global chromatin environment, modifications help partition the genome into distinct domains such as euchromatin, where DNA is kept "accessible" for transcription, and heterochromatin, where chromatin is "inaccessible" for transcription (30). Each of these chromatin environments in the genome (eu- and heterochromatin) is associated with a distinct set of modifications. In mammals, demarcation between these different environments is set up by boundary elements, which recruit enzymes to modify the chromatin (30). Experiments in fission yeast have shown that the heterochromatin boundaries are maintained by the presence of methylation at H3K4 and H3K9 in adjacent euchromatic regions (30). There is also evidence from fission yeast that the nucleation of heterochromatin (rather than its spreading) involves the production of small interfering RNA (siRNAs) from transcripts emanating from centromeric repeats (30). Recruitment of HP1 (a "reader" or "effector" protein recognizing H3K9 methylation) then allows spreading and maintenance of the heterochromatic state (31).

Euchromatin Modifications

Euchromatin structure often contains unmethylated first gene exons. Histone modifications associated with active transcription include lysine acetylation and methylation. Typically, histone acetylation occurs at multiple lysine residues, most commonly on histones H3 and H4, and is usually carried out by a variety of histone acyltransferase complexes (29,32). Acetylation is enriched at specific sites in the promoter and 5′ end of the coding regions; within the promoter there are two nucleosomes flanking the initiation site that are hypoacetylated at certain lysines and are enriched in the H2 A variant Hzt1 (33,34).

Whereas lysine acetylation almost always correlates with chromatin accessibility and transcriptional activity, lysine methylation can have different effects depending on which residue is modified (2). Lysine methyltransferases have enormous specificity compared to acetyltransferases. They usually methylate one single lysine on a single histone and their output can be either activation or repression of transcription, depending on the number of methyl groups transferred (up to three) (35). Three methylation sites on histones are implicated in activation of transcription: H3K4 (di- or trimethylated), H3K36 (tri-me), and H3K9 (mono-me), all of which have been implicated in transcriptional elongation. H3K4me3 localizes to the 5′ end of active genes and is found associated with the initiated form of RNA pol II (phosphorylated at serine 5 of its C-terminal domain) (36). H3K36me is found to accumulate at the 3′ end of active genes and is found associated with the serine 2 phosphorylated elongating form of RNA pol II. One role for H3K36me is the suppression of inappropriate initiation from cryptic start sites within the coding region (37–39). Very little is known about the function of other histone lysine methylation sites associated with transcriptional activation, including H3K79 (mono- and tri-me), H4K20 (mono-me), H3K27 (mono-me), H4K20 (mono-me), H2BK5 (mono-me), and H3K79.

Heterochromatin Modifications

Transcriptionally inactive chromatin is associated with methylated DNA and several unique histone modifications. The formation of heterochromatin may start with the deacetylation of H3K9 by HDAC, which enables the methylation of H3K9. Trimethylation of histone H3K27 and H4K20 also generally correlate with repression (15).

Recently, bivalent domains have been found that possess both activating and repressive modifications, which somewhat shatters our simplistic view that activating versus silencing modifications dictate distinct types of chromatin environments (40). For example, methylation at H3K36 has a positive effect when it is found on the coding region and a negative effect when it is found in the promoter. Methylation at H3K9 may be the same: negative in the promoter and positive in the coding region (41). The abundance of modifications on the histone tail makes "cross talk" between modifications very likely due to the diversity of modifications occurring on lysine residues. This undoubtedly could result in antagonism, protein-binding disruptions, or alterations in enzymatic activities. The best-studied example is the case of ubiquitinilation of H2B being required for methylation of H3K4me3 (30).

Interactions Between Epigenetic Pathways: DNA Methylation, Histone Modifications, Complex Recruitment and Small RNAs

Much of our understanding regarding the interactions between DNA methylation and other epigenetic pathway components comes from research using DNA methyltransferase 1 and 3 (DNMT1 and 3) knockout cell lines. DNMT1 activity is crucial for the maintenance of DNA methylation and the appropriate histone H3 modification, which is important for organization of the chromatin domains (42,43). Altered nuclear organization of DNMT3 ($-/-$) cells suggests that there is some underlying change in chromatin structure when DNA methylation is absent. Indeed, such changes have been identified at two levels of primary chromatin structures: histone modifications and linker histone binding.

ES cells lacking DNA methylation have globally elevated levels of histone acetylation at H3K9, H4K5, and H4K16 (44). Increased acetylation at H4K5 has been earlier reported in these cells (45). Reduced levels of H3K9 di-methylation and a redistribution of histone methylation within satellite-containing heterochromatin compartments are also evident in DNMT3 ($-/-$) cells (44). Both DNMTs and methyl-CpG-binding proteins can physically associate with HDACs (46) and histone methyltransferases (47,48), so the altered histone modification in cells that completely lack CpG methylation may reflect the loss of these interactions (44). Histone modifications other than deacetylation are also strongly implicated in triggering the de novo methylation of DNA. Most markedly, H3K9 di- or trimethylation is clearly linked to gene silencing, and has been shown to be essential for DNA methylation in the fungus *Neurospora crassa* (49) and in the plant *Arabidopsis thaliana* (50).

Loss of DNA methylation also leads to the altered binding of linker histones. $T_{1/2}$-binding times for the linker histones H1 and H5 have been shown to be increased in DNMT3-cells compared with wild type cells. Alternatively, reduction of linker histone levels in vivo can give rise to altered DNA methylation at specific genomic sites (44). There seems, however, to be conflicting evidence about whether DNA methylation influences linker histone binding in chromatin. Unmethylated CpG islands appear to be depleted of H1 (51), and H1-containing nucleosomes contain 80% of the 5'-methylcytosine (52), suggesting that linker histones may prefer to bind to methylated DNA (44).

Interactions between chromatin-recruited protein complexes are also evident. m^{5}CpG-binding proteins 2 interacts with a corepressor complex containing HDACs. Methyl-CpG-binding domain proteins 2 is also associated with HDACs, extending a relationship between histone modification and DNA methylation that is likely to be better understood in the future (Fig. 2) (24). Acetylated lysines are recognized by bromodomains within nucleosome remodeling complexes. An interaction between methylated H3K4 and Chd1 chromodomain appears to recruit activating complexes to chromatin (53). In contrast, methylated H3K9 and H3K27 are bound by HP1 and Polycomb, respectively, which mediate chromatin compaction (5). The relative "methylation-state" of a given lysine can influence chromodomain binding. Polycomb preferentially interacts with trimethylated H3K27, while HP1 shows preference for both di- and trimethylated H3K9 (54).

Lastly, we are just beginning to understand how small RNAs can act in concert with various components of the cell's chromatin and DNA methylation machinery to achieve stable silencing. Double-stranded RNA-induced posttranscriptional gene silencing (PTGS), also known as RNA interference (RNAi) in animals, involves small interfering double-stranded RNAs (siRNAs) inducing homology-dependent degradation of messenger RNA (55). RNAi can also suppress gene expression through a transcription-mediated pathway, transcriptional gene silencing (TGS) (56), which has been shown to be due to RNA-dependent DNA methylation (57). RNAi-mediated TGS has been implicated in regulating H3K9 methylation-induced heterochromatic silencing (58). Small RNAs have also been shown to be involved in silencing of the inactive X chromosome where *Xist* RNA, DNA methylation, histone modification, and their writers and readers all play a role (59). Although one might not consider PTGS-inducing RNAs (e.g., microRNAs, siRNAs, etc) to be epigenetic in nature, TGS-evoking RNAs (e.g., Repeat-associated siRNAs, *Xist* RNA, and small RNAs in *S. pombe*) are more clearly epigenetic, as they can induce long-term silencing effects that can be inherited through cell division (60).

Epigenetic Inheritance

How epigenetic marks are propagated is currently a very active area of research. Cytosine methylation patterns are clearly propagated through cell division. Their preservation involves the "maintenance" methyltransferase DNMT1, which has specificity for hemimethylated CpG dinucleotides; the enzyme can thus methylate CpGs in a newly synthesized DNA strand based upon the presence of methylation in the CpG dinucleotides in the complementary template strand (3,4).

A subset of histone modifications also appears to show epigenetic inheritance. If epigenetic memory is mediated by one or more of the histone modifications, then there should be a mechanism for the transmission of such modifications onto the chromatin of the replicating DNA. Such a mechanism has been proposed for H3K9 methylation in the transmission of the heterochromatin: recruitment of HP1 brings in further H3K9-methylating activity that modifies nucleosomes on the daughter strand, thus ensuring the transmission of the H3K9me mark through cell divisions. This mechanism of transmission, along with the observation that H3K4 tri-me and H3K27 patterns persist, has given lysine methylation an epigenetic status. The argument that histone methylation is a permanent mark is now on shaky ground, given the discovery of demethylases (30).

Another determinant likely to transmit information for the assembly of a correct local chromatin structure is RNA. Small RNAs may emanate from many loci in the genome and once transmitted to the next generation, these RNAs may deliver chromatin-modifying complexes to specific genes or to specific locations, thus generating the pattern of chromatin that we observe (61). One appealing aspect of this model is that small RNAs are likely to be highly precise in their delivery since their guiding system is nucleic acid. However, identification of such an RNA-mediated mechanism does not imply that histone modifications are unnecessary for epigenetic events. It merely points out that histone modifications may be the executors of the epigenetic phenomenon rather than the carriers of the memory (30). Models of inheritance are further obscured by replication-independent histone deposition and by the potentially significant role of histone variants (62).

Age-associated Changes in Epigenetic Marks

Aging has been demonstrated to be associated with both hypo- and hypermethylation of DNA. A progressive loss of overall methylation is seen during the in vitro culture of fibroblasts (63), and in aging animals (64,65). Hypermethylation of specific genes has been observed in tissues of aging individuals. For example, methylation of the CpG islands associated with many genes, including those encoding the estrogen receptor (66), *IGF2*, and *MYOD*, is undetectable in young individuals, but becomes progressively detectable with age in normal tissues (67). The functional significance of age-related changes in DNA methylation, which may primarily affect repeated sequences such as transposable elements (68), remains to be determined.

EPIGENETIC REGULATION DURING DEVELOPMENT

Of the approximately 30,000 genes in the human genome, different subsets will be expressed at different stages of development and in different cells, tissues, and organs of the offspring. It has long been known that differential gene expression controlled by transcription factors is critical for normal development. The pluripotent cells of the cleavage-stage conceptus progressively differentiate along specific lineages to give rise to the embryo and fetus. While regulation of differential gene expression by transcription factors is a key feature of development, it is now understood that gene expression patterns during development (as well as in the adult) can also be dependent on epigenetic modifications (69,70), including those described earlier: methylation of DNA at CpG sequences (8,69), posttranslational modification of histone protein tails (30,71), and non-nucleosomal proteins complexed with DNA (72). These epigenetic "marks" may be transient, such as the histone modifications that, during cleavage, repress genes needed for later development, or long-lived, such as the DNA methylation and other chromatin modifications that result in X-chromosome inactivation or the silencing of imprinted genes and transposons (see later).

X-Chromosome Inactivation

Female eutherian mammals undergo inactivation of one X chromosome in every cell so that only a single X chromosome remains active, as is the situation in the male (73). This occurs early in development through a process of heterochromatinization. The gene *Xist* (X-inactivation specific transcript, which is noncoding) and its antisense partner *Tsix* control inactivation, and X-chromosome inactivation is maintained through DNA methylation and histone acetylation and methylation. Interestingly, although inactivation affects most of the X chromosome, several X-linked genes are known to escape silencing, despite being embedded in heterochromatin (73).

Imprinted Genes

Imprinted genes are those genes or genomic domains that are marked with information about their parental (maternal or paternal) origin (74,75). Expression of such genes is restricted to one parental allele. For some imprinted genes only the paternal allele is expressed, while for other imprinted genes, it is the maternal allele that is expressed. Monoallelic expression of imprinted genes results from epigenetic marks, including DNA methylation. Imprinted genes usually occur in clusters on a chromosome, forming an imprinted domain that may be under the control of a single differentially methylated region (DMR). Abnormalities at imprinted loci are involved in a number of human developmental disorders including Angelman syndrome, Prader-Willi syndrome, and Beckwith Wiedemann syndrome, as well as some cancers (76). It is not known how many imprinted alleles exist. Early estimates suggested the existence of 100 to 200 imprinted genes in mice (77,78). Using new techniques, more recent studies have identified 600 candidate imprinted mouse genes (79) and perhaps half as many in humans (80). Imprinted genes in the mouse are not randomly distributed, and about half are located in five imprinted domains on chromosome 7 (Fig. 4). An interesting finding that makes it difficult to estimate the total number of imprinted genes is that some imprinted genes exhibit monoallelic expression only in a limited number of cell lineages (81).

Transposable Elements and Metastable Epialleles

Approximately 40% of the human genome consists of transposable elements and other repetitive DNA. This DNA is usually highly methylated in somatic tissues (8), which is critical for protecting against expression or translocation. However, some transposons exhibit variable methylation and can affect the expression of other genes. One result can be the formation of metastable epialleles, which are alleles that are identical in sequence but variably expressed because of epigenetic

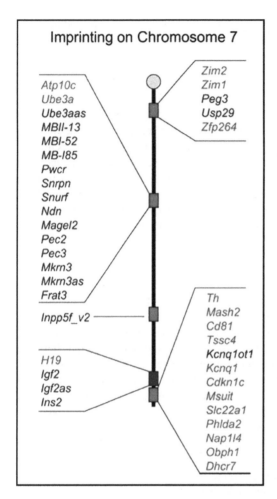

Figure 4 (*See color insert*) The five imprinted domains on chromosome 7 in the mouse. This figure demonstrates the lack of random distribution in imprinted genes, which occur in clusters on a chromosome. *Source*: From Ref. 81.

modifications established during development. One example that will be discussed further below is the murine A^{vy} allele that resulted from insertion of an IAP retrotransposon upstream of the transcription start site of the *Agouti* gene. A cryptic promoter in the IAP promotes constitutive *Agouti* expression, and the degree of CpG methylation in the A^{vy} IAP correlates with ectopic *Agouti* expression (Fig. 5).

Critical Periods for Epigenetic Regulation

There are specific stages of the life cycle during which long-term epigenetic marks may be erased and reestablished (Fig. 6) (82). There are two periods during which

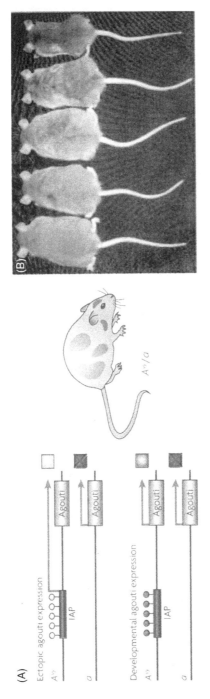

Figure 5 (*See color insert*) (**A**) The murine A^{vy} metastable epiallele in which a cryptic promoter in the IAP promotes constitutive *Agouti* expression. (**B**) The degree of CpG methylation in the A^{vy} IAP correlates with ectopic *Agouti* expression. *Source*: From Ref. 95.

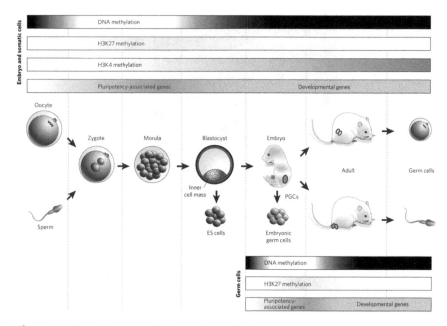

Figure 6   (*See color insert*) Critical periods of epigenetic regulation during development. *Source*: From Ref. 82.

large-scale, but not complete, demethylations of the genome are known to occur. One is during migration and proliferation of the primordial germ cells (PGCs). This takes place between days E10.5 and E12.5 in the mouse embryo, and during this time imprinted genes are demethylated (69,83), primarily at CpG island in imprinted gene DMRs. Methylation is reestablished later, in a parental gender-specific manner, during gametogenesis, by the de novo methyltransferase DNMT3 A and its cofactor DNMT3 L.

Demethylation of the DNA in the PGCs also serves to reactivate pluripotency-related genes needed in the early conceptus. It is not known whether the demethylation of DNA in the PGCs is via an active (i.e., by the action of a demethylase) or a passive mechanism. Not all genomic methylation is lost in the PGCs and many transposons remain highly methylated (84). Transposons may be both methylated themselves and marked by repressive histone modifications such as H3K9 methylation. The other period of widespread epigenetic reprogramming occurs early after fertilization. Following removal of protamines (sperm proteins) from, and acquisition of histones by, the paternal genome, many paternal alleles become demethylated. While the specific paternal alleles being demethylated have not been enumerated, sequences known not to be affected include IAPs and paternally methylated DMRs in imprinted domains. This active demethylation of the paternal genome is followed by passive demethylation (i.e., lack of methylation of corresponding bases in the nascent strand during DNA replication) of both paternal and maternal genes. During this period of widespread demethylation, the methylation patterns of imprinted genes are maintained first by DNMT1o (the oocyte form of DNMT1) followed by DNMTs (the somatic form) in the embryo, fetus, and adult. Despite maintenance of methylation in imprinted genes by DNMT1 s, total genome methylation in the early embryo decreases, reaching a nadir at the blastocyst stage. Generalized demethylation in the embryo at this stage may play a role in returning cells to pluripotency by activating cells such as *Nanog* and *Oct4*, necessary for the establishment of the inner cell mass (85).

It is likely that the patterns and extent of epigenetic marks on the genome may be specified or altered in part by the developmental environment. Since these epigenetic marks can last a lifetime, it is plausible that epigenetic programming could result in permanent changes in the physiology and, therefore, adult disease risks of the offspring. Indeed, the refractory nature of IAPs to reprogramming, and their effect on the expression of neighboring genes, suggests that the state of expression of such genes may be inherited across several generations. The potential role of epigenetic programming in long-term health of offspring will be the topic of the remainder of this chapter.

EPIGENETIC PROGRAMMING AND THE DOHaD

There is now compelling epidemiological and laboratory experimental evidence that the in utero environment in which a conceptus develops, as well its early postnatal environment, affects the lifelong health and disease susceptibility of the offspring. The framework in which this occurs is called the DOHaD concept. Recent reviews cover exhaustively the origins of the concept, the evolutionary perspective, the extant human and animal evidence, and possible underlying mechanisms (86–90). The implications of this concept for toxicology have been considered (91–95). Lifelong metabolic programming could occur through a number of mechanisms involving growth trajectories, cell proliferation and differentiation, organ maturation, and paracrine and endocrine effects. For the purposes of this review, discussion will be limited to those examples demonstrating epigenetic chromatin changes underlying or associated with long-term effects of the developmental environment. Examples will include effects of the nutritional, toxicological, and behavioral environments during development and the effects of cloning and assisted reproduction technologies (ART) on the epigenome.

Nutrition

Some of the first indications that the developmental environment could influence risk of adult diseases came from studies by Barker and colleagues that showed inverse relationships between birth weight and incidence of adult coronary heart disease mortality (96), elevated blood pressure (97), non-insulin-dependent diabetes (98), and risk of the metabolic syndrome (99). The associations between birth weight and adult disease risks have been confirmed in a number of studies around the world, and have been attributed to inadequate maternal nutrition during key periods before and during pregnancy. The basis for this includes studies of famines associated with World War II, such as the siege of Leningrad (100) and the Dutch Hunger Winter (101). Studies of the Dutch famine demonstrate relationships between prenatal undernutrition and increased risk of adult coronary heart disease (101), obesity (102), kidney disease (103), and non-insulin-dependent diabetes (104).

Limitations inherent to epidemiological studies led to the development of animal models of the effects of maternal nutrition on the adult disease risk of offspring (105). Species used in these studies include the rat, mouse, guinea pig, sheep, and pig (106). Many of these studies have involved either maternal protein deprivation or global undernutrition during pregnancy. In sum, the results of these studies recapitulate the results of the human epidemiology studies, demonstrating elevated blood pressure, insulin resistance, renal insufficiency, and obesity in offspring following developmental malnutrition or undernutrition.

Use of animal models in studies of the DOHaD concept has a number of advantages, including the ability to perform experiments to elucidate mechanisms of action. Using microarray gene expression analysis, it has been demonstrated that prenatal protein restriction results in changes in the expression of a large number of genes in organs and tissues of the offspring (107). However, it is difficult to determine cause and affect relationships in such studies, since there were effects on the histology of the tissues and organs as well. As the results of additional studies become available, attention can be focused on those genes which are most central to fetal programming.

Several studies have now been carried out to determine whether there are effects of prenatal undernutrition on the epigenome of offspring. Offspring of rats fed a low-protein diet during pregnancy had lower PPAR-a gene methylation and higher gene expression than controls (108). Glucocorticoid receptor (GR) gene methylation was also lower and expression higher than controls. Interestingly, supplementing the low-protein diet with 1 mg/kg folic acid prevented these changes in gene methylation and expression. In a subsequent study (109), GR promoter methylation in offspring of protein-restricted mothers was 33% lower than controls, and GR gene expression was 84% higher. Reverse transcription-polymerase chain reaction (RT-PCR) showed that DNA methyltransferase-1 (DNMT1) expression was 17% lower in protein-deprived offspring, while DNMT3a/b and methyl-binding domain protein-2 expression was not affected. The observed reduction in expression of DNMT1 may underlie the decreased methylation and increased expression of the GR gene. Gluckman and colleagues (110) studied 170-day old female offspring of rats undernourished (fed 30% of ad libitum control intake) during pregnancy. This offspring had been injected with leptin or saline on postnatal days 3 to 10, and then fed a normal or high-fat diet from weaning. Leptin exposure affected the expression of 11β-hydroxysteroid dehydrogenase type 2 (11β-HSD2) in a manner dependent on maternal nutritional status. In offspring of well-nourished mothers, neonatal leptin exposure increased 11β-HSD2 expression in adulthood, while in offspring of undernourished dams, 11β-HSD2 expression was decreased by neonatal leptin exposure. PPAR-a expression was suppressed by leptin in offspring of well-nourished mothers and increased by leptin in offspring of undernourished mothers. Effects of leptin on methylation of PPAR-a and GR were also observed. Methylation of the PPAR-a promoter was elevated by leptin in offspring of well-nourished dams and decreased by leptin in offspring of undernourished dams. GR promoter methylation was elevated by leptin in offspring of well-nourished dams but unchanged by leptin in those of undernourished dams. These results indicate that the response of offspring to leptin was programmed by their developmental nutritional status, and that this occurred, at least in part, by changes in the epigenome.

Maternal dietary folate supplementation has been shown to reduce the incidence of birth defects including neural tube defects. The mechanism of this protective effect is poorly understood, but is not likely a simple correction of a folate deficiency. Work with a variety of mouse genetic models of neural tube defects and other birth defects suggest a connection between folate metabolism and developmental control genes (111). While at present there is no evidence that such a connection involves epigenetic changes in gene expression, DNA methylation can be affected by dietary levels of methyl-donor constituents, including folic acid (112). It has been demonstrated that maternal dietary methyl-donor supplementation can increase methylation at CpG sites in the upstream IAP transposable element of the A^{vy} metastable epiallele. Expression of this gene results in ectopic agouti protein, which causes yellowing of the coat, obesity, diabetes, and tumorigenesis. Increased methylation of the IAP by dietary methyl-donor supplementation decreases the expression of the gene and the severity of the phenotype (113–115) (Fig. 5). The A^{vy} mouse model has been used to study the interactive effects of exposure to environmental chemicals and maternal folate status on epigenetic programming. These studies will be discussed in the next section.

Toxicology

Environmental toxicity to the pregnant mother and her conceptus may occur through multiple pathways and mechanisms, and the processes underlying epigenetic programming are certainly plausible targets. As mentioned earlier, errors in imprinted chromosomal regions can result in severe developmental effects in humans. Yet, to date, there are few examples of chemical or physical agents inducing epigenetic alterations in the developing embryo and fetus. There is broad interest in applying new genomic technologies to the problem of environmentally induced birth defects, and several studies have emerged linking changes in patterns of embryonal gene expression to the etiology of chemically induced abnormal development (116–118). Applying this approach to elucidate gene expression changes associated with valproic acid teratogenicity in mice has produced evidence that valproic acid may exert its teratogenicity, at least in part, by inhibition of histone deacetylase (116,119). However, whether or not valproic acid actually results in changes in histone acetylation or other epigenetic alterations in the embryo has yet to be determined.

Cyclophosphamide is a chemotherapeutic and immunosuppressant agent that causes DNA alkylation at the N7 position of guanine, DNA–DNA and DNA–protein cross-links, and single-strand DNA breaks (120). In rats, paternal exposure to cyclophosphamide for 4 to 5 weeks prior to mating results in embryo loss, malformations, and behavioral effects in offspring (121), and these effects are transmissible to subsequent generations (122). In a study of the effects of paternal cyclophosphamide treatment on histone acetylation and DNA methylation in preimplantation rat embryos, Barton and colleagues (123) found that zygotes sired by cyclophosphamide-treated males exhibited disruption of epigenetic programming of both parental genomes. Early zygotic pronuclei were hyperacetylated, while mid-zygotic conceptuses showed hypomethylated male pronuclei and hypermethylated female pronuclei. Localization of histone H4 acetylation at lysine 5 at the nuclear periphery was disrupted in two-cell embryos. These authors hypothesized that these epigenetic changes might contribute to the transgenerational transmission of the developmental toxicity of cyclophosphamide.

The mouse A^{vy} mutant described earlier has been used to assess the effects of the endocrine disrupter chemicals genistein (124) and bisphenol A (80) on embryonal DNA methylation. Genistein is the major isoflavone phytoestrogen in soy. Maternal dietary exposure to genistein at levels similar to those in humans consuming a high-soy diet shifted the coat color of heterozygous viable yellow agouti (A^{vy}/a) offspring toward pseudoagouti. This shift to a less severe phenotype was accompanied by increased methylation of six CpG sites in the IAP upstream of the transcription start site of the *Agouti* gene. The genistein-induced hypermethylation appeared to be permanent, persisting into adulthood and protecting offspring from the deleterious health effects of ectopic *Agouti* expression. Bisphenol A, a high production volume chemical used in many products, is weakly estrogenic. In contrast to genistein, dietary exposure to bisphenol A in A^{vy}/a mice results in CpG hypomethylation of the IAP and shifting of coat color toward the yellow (more severe phenotype). Dietary supplementation with methyl donors or genistein ameliorated the DNA hypomethylation and coat-color shift induced by bisphenol A. The studies with genistein and bisphenol A clearly demonstrate that environmental chemicals can alter the epigenetic programming of embryos and thereby have permanent phenotypic effects on offspring.

Transgenerational effects of toxicants are those that are transmitted to successive generations without continued exposure to the agent in question. Exposure of a pregnant rodent to a chemical can result in some degree of exposure to three generations: the pregnant mother (F_0), the fetus(es) (F_1), and the germ cells (F_2) residing in the fetal gonads. Therefore, it is only at the F_3 generation that one can be reasonably sure that direct exposure to the agent is not responsible for phenotypic changes observed. Recent studies have demonstrated effects of xenobiotics on the F_3 generation through germline alterations in epigenetic programming (125–127). Vinclozolin, an environmental antiandrogen, induces transgenerational toxicity in rats, resulting in impaired spermatogenesis and male infertility, breast cancer, kidney and prostate disease, and immune abnormalities. These phenotypes have been followed and are transmitted for at least four generations (125,126). Methylation-sensitive endonucleases were used to identify DNA sequences putatively epigenetically reprogrammed in the male germline due to vinclozolin exposure in utero (125). It remains to be determined if these epigenetic changes are causal in the pathologies observed across generations.

Behavior

The adult offspring of female rats that exhibit high levels of licking/grooming (LG) during the first week of life have increased GR expression in the hippocampus, enhanced glucocorticoid feedback sensitivity, decreased hypothalamic corticotrophin releasing factor expression, and muted HPA stress responses compared to animals reared by low LG mothers (128,129). These investigators hypothesized that level of maternal care determined the epigenetic programming of offspring behavior (130). Subsequent studies provided evidence that increased maternal LG is associated with demethylation of the nerve growth factor–inducible protein, a transcription factor response element located in the GR promoter (131). Changes in methylation status of CpG sites in this sequence emerge over the first week of life, can be reversed with cross-fostering, persist into adulthood, and are associated with altered histone acetylation. These investigators have also shown that early postnatal maternal care affects the expression of hundreds of genes in the hippocampus of adult offspring (132). Effects on offspring behavior are permanent, but can be reversed by intervention with the histone deacetylase inhibitor trichostatin A (TSA). Treating adult offspring with TSA reversed the epigenetic changes in low LG offspring and ameliorated their stress response behavior such that it was indistinguishable from that of high LG offspring (133). The relationships between maternal care, epigenetic programming, and behavior have been reviewed (93). Among the striking implications of this work are that some epigenetic programming is still labile after birth and that interventions in adulthood may be able to rectify adverse epigenetic programming.

Cloning and ART

Cloning or somatic cell nuclear transfer (SCNT) involves the transfer of a diploid somatic cell nucleus into an enucleated oocyte to produce a "zygote" that has the appropriate chromosomal makeup and potentially will be able to activate all necessary genes in the proper sequence for normal development after implantation into a host mother. In vitro fertilization (IVF) involves mixing sperm and eggs in vitro, where fertilization occurs, and then transferring embryos to host mothers. Intracytoplasmic sperm injection (ICSI), typically used to counter male factor infertility, involves microinjecting sperm directly into the oocyte cytoplasm. The first human baby produced by IVF was in 1978 (134). ICSI was demonstrated in rabbits in 1989 (135) and in mice in 1995 (136). The birth of an apparently healthy baby by ICSI was reported in 1992 (137). Currently, at least 10 mammalian species have been successfully cloned (138). While application of cloning to humans has been claimed (139), there is no evidence that this has been achieved or is even possible (140). To achieve successful development following SCNT, the donor nucleus must dedifferentiate to a totipotent embryonic state and then redifferentiate to form the different somatic cell types later in development. For this to happen, the epigenetic marks of the differentiated somatic cell must be erased, and those needed for the totipotent and pluripotent cells of the early embryo must be established. Then the appropriate transitions to allow lineage-specific differentiation must occur.

Reproductive cloning in animals is plagued by errors (141). The efficiency of development of cloned embryos to the blastocyst stage is about 30 to 50%. Embryo production by IVF results in a similar viability to the blastocyst stage in species such as cows and pigs (142–144). Then, most cloned embryos that survive to the blastocyst stage die during postimplantation development or are born with malformations. Overall, the survival rate to birth from all nuclear transfer eggs is 2 to 3% at best (138) compared to 30 to 60% for IVF blastocysts (142). "Large offspring syndrome" is common in cloned cattle, sheep, and mice, and includes large size at birth and severe placental deficiency (145), prolonged gestation, dystocia, fetal and placental edema, hydramnios, respiratory problems, and perinatal death (145,146). In cattle, postnatal death rates are about 30% in the first six months (147). Reports on reprogramming of various candidate genes in cloned embryos are inconsistent. Expression of *Oct4*, a marker of pluripotency, has been reported to be normal (148,149) in cloned embryos but has also been reported to be abnormal after SCNT (150,151). Reprogramming of DNA methylation (152–154), expression of imprinted and nonimprinted genes (150,155–157), X-chromosome inactivation (158,159), and telomerase activity (160) are often abnormal in cloned early embryos.

IVF and other forms of ART are widely used in humans, and their use is increasing steadily (161). Approximately 100,000 IVF cycles are performed annually in the United States, and ART babies account for 0.6% of all births (162). A small but important proportion of these infants suffer from congenital problems. They have twice the rate of birth defects as babies born naturally (163). A number of reports indicate that IVF increases the risk of diseases caused by errors in genomic imprinting, including Angelman syndrome and Beckwith-Weidemann syndrome (164–168). Li and coworkers (169) examined the allelic expression of four imprinted genes: *Igf2*, *H19*, *Cdkn1 c*, and *Slc221 L* in morulas and blastocysts from C57BL/6J X *Mus spretus* F1 mice conceived in vivo and in vitro. Their results suggest that de novo DNA methylation on the maternal allele led to aberrant *Igf2/H19* imprinting in IVF-derived ES cells.

It is clear from the extant literature that the epigenetic effects of various ART procedures are not well understood (170,171). Continued elucidation of the mechanisms of imprinting and other forms of epigenetic programming and the effects of the developmental environment, especially in the very early embryo, will be important for understanding the risks of current ART approaches and allow for their continued improvement.

CONCLUSIONS

The interaction among DNA methylation, numerous posttranslational histone protein modifications (including acetylation and methylation), and the recruited protein complexes and small RNAs comprises the epigenetic regulation of gene transcription. These heritable changes in gene function are highly dynamic during specific developmental periods, are maintained throughout an individual's life span, and contribute to the onset of various disease states. The nonstatic nature of these epigenetic marks and potential alteration by environmental factors is only now beginning to be examined to a significant extent. This chapter has served to provide an overview of the molecular mechanisms of this regulation, the epigenetic regulation which occurs particularly during development, and the epigenetic programming associated with DOHaD.

REFERENCES

1. Russo VEA, Martienssen RA, Riggs AD. Epigenetic mechanisms of gene regulation. Cold Spring Harbor Laboratory Press, Plainview, NY, 1996.
2. Bernstein BE, Meissner A, Lander ES. The mammalian epigenome. Cell 2007; 128(4):669–681.
3. Bird A. DNA methylation patterns and epigenetic memory. Genes Dev 2002; 16(1):6–21.
4. Goll MG, Bestor TH. Eukaryotic cytosine methyltransferases. Annu Rev Biochem 2005; 74:481–514.
5. Margueron R, Trojer P, Reinberg D. The key to development: Interpreting the histone code? Curr Opin Genet Dev 2005; 15(2):163–176.
6. Bonfils C, Beaulieu N, Chan E, et al. Characterization of the human DNA methyltransferase splice variant Dnmt1b. J Biol Chem 2000; 275(15):10754–10760.
7. Delaval K, Feil R. Epigenetic regulation of mammalian genomic imprinting. Curr Opin Genet Dev 2004; 14(2):188–195.
8. Bird A. DNA methylation patterns and epigenetic memory. Genes Dev 2002; 16(1):6–21.
9. Larsen F, Gundersen G, Lopez R, et al. CpG islands as gene markers in the human genome. Genomics 1992; 13(4):1095–1107.
10. Fazzari MJ, Greally JM. Epigenomics: beyond CpG islands. Nat Rev Genet 2004; 5(6):446–455.
11. Strichman-Almashanu LZ, Lee RS, Onyango PO, et al. A genome-wide screen for normally methylated human CpG islands that can identify novel imprinted genes. Genome Res 2002; 12(4):543–554.
12. Mohandas T, Sparkes RS, Shapiro LJ. Reactivation of an inactive human X chromosome: Evidence for X inactivation by DNA methylation. Science 1981; 211(4480):393–396.

13. Wolf SF, Jolly DJ, Lunnen KD, et al. Methylation of the hypoxanthine phosphoribo-syltransferase locus on the human X chromosome: Implications for X-chromosome inactivation. Proc Natl Acad Sci U S A 1984; 81(9):2806–2810.

14. Riggs AD, Xiong Z, Wang L, et al. Methylation dynamics, epigenetic fidelity and X chromosome structure. Novartis Found Symp 1998; 214:214–225; discussion 225–32.

15. Kouzarides T. Chromatin modifications and their function. Cell 2007; 128(4):693–705.

16. Chan SW, Henderson IR, Jacobsen SE. Gardening the genome: DNA methylation in Arabidopsis thaliana. Nat Rev Genet 2005; 6(5):351–360.

17. Haines TR, Rodenhiser DI, Ainsworth PJ. Allele-specific non-CpG methylation of the Nf1 gene during early mouse development. Dev Biol 2001; 240(2):585–598.

18. Ramsahoye BH, Biniszkiewicz D, Lyko F, et al. Non-CpG methylation is prevalent in embryonic stem cells and may be mediated by DNA methyltransferase 3 a. Proc Natl Acad Sci U S A 2000; 97(10):5237–5242.

19. Pradhan S, Bacolla A, Wells RD, et al. Recombinant human DNA (cytosine-5) methyltransferase. I. Expression, purification, and comparison of de novo and main-tenance methylation. J Biol Chem 1999; 274(46):33002–33010.

20. Margot JB, Ehrenhofer-Murray AE, Leonhardt H. Interactions within the mammalian DNA methyltransferase family. BMC Mol Biol 2003; 4:7.

21. Bestor TH. The DNA methyltransferases of mammals. Hum Mol Genet 2000; 9(16):2395–2402.

22. Hermann A, Goyal R, Jeltsch A. The Dnmt1 DNA-(cytosine-C5)-methyltransferase methylates DNA processively with high preference for hemimethylated target sites. J Biol Chem 2004; 279(46):48350–48359.

23. Fatemi M, Hermann A, Gowher H, et al. Dnmt3 a and Dnmt1 functionally cooperate during de novo methylation of DNA. Eur J Biochem 2002; 269(20):4981–4984.

24. Jaenisch R, Bird A. Epigenetic regulation of gene expression: how the genome integrates intrinsic and environmental signals. Nat Genet 2003; 33(Suppl):245–254.

25. Bell AC, West AG, Felsenfeld G. The protein CTCF is required for the enhancer blocking activity of vertebrate insulators. Cell 1999; 98(3):387–396.

26. Ohlsson R, Renkawitz R, Lobanenkov V. CTCF is a uniquely versatile transcription regulator linked to epigenetics and disease. Trends Genet 2001; 17(9):520–527.

27. Hark AT, Schoenherr CJ, Katz DJ, et al. CTCF mediates methylation-sensitive enhancer-blocking activity at the H19/Igf2 locus. Nature 2000; 405(6785):486–489.

28. Okano M, Bell DW, Haber DA, et al. DNA methyltransferases Dnmt3 a and Dnmt3b are essential for de novo methylation and mammalian development. Cell 1999; 99(3):247–257.

29. Schuettengruber B, Chourrout D, Vervoort M, et al. Genome regulation by polycomb and trithorax proteins. Cell 2007; 128(4):735–745.

30. Li B, Carey M, Workman JL. The role of chromatin during transcription. Cell 2007; 128(4):707–719.

31. Reinberg D, Sims RJ III. de FACTo nucleosome dynamics. J Biol Chem 2006; 281(33):23297–23301.

32. Brown CE, Lechner T, Howe L, et al. The many HATs of transcription coactivators. Trends Biochem Sci 2000; 25(1):15–19.

33. Zhang H, Roberts DN, Cairns BR. Genome-wide dynamics of Htz1, a histone H2 A variant that poises repressed/basal promoters for activation through histone loss. Cell 2005; 123(2):219–231.

34. Raisner RM, Hartley PD, Meneghini MD, et al. Histone variant H2A.Z marks the 5′ ends of both active and inactive genes in euchromatin. Cell 2005; 123(2)233–248.

35. Bannister AJ, Kouzarides T. Reversing histone methylation. Nature 2005; 436(7054):1103–1106.

36. Eissenberg JC, Shilatifard A. Leaving a mark: the many footprints of the elongating RNA polymerase II. Curr Opin Genet Dev 2006; 16(2):184–190.

37. Carrozza MJ, Li B, Florens L, et al. Histone H3 methylation by Set2 directs deacetylation of coding regions by Rpd3S to suppress spurious intragenic transcription. Cell 2005; 123(4):581–592.

38. Joshi AA, Struhl K. Eaf3 chromodomain interaction with methylated H3-K36 links histone deacetylation to Pol II elongation. Mol Cell 2005; 20(6):971–978.

39. Keogh MC, Kurdistani SK, Morris SA, et al. Cotranscriptional set2 methylation of histone H3 lysine 36 recruits a repressive Rpd3 complex. Cell 2005; 123(4):593–605.

40. Bernstein BE, Kamal M, Lindblad-Toh K, et al. Genomic maps and comparative analysis of histone modifications in human and mouse. Cell 2005; 120(2):169–181.

41. Vakoc CR, Mandat SA, Olenchock BA, et al. Histone H3 lysine 9 methylation and HP1gamma are associated with transcription elongation through mammalian chromatin. Mol Cell 2005; 19(3):381–391.

42. Espada J, Ballestar E, Fraga MF, et al. Human DNA methyltransferase 1 is required for maintenance of the histone H3 modification pattern. J Biol Chem 2004; 279(35):37175–37184.

43. Milutinovic S, Brown SE, Zhuang Q, et al. DNA methyltransferase 1 knock down induces gene expression by a mechanism independent of DNA methylation and histone deacetylation. J Biol Chem 2004; 279(27):27915–27927.

44. Gilbert N, Thomson I, Boyle S, et al. DNA methylation affects nuclear organization, histone modifications, and linker histone binding but not chromatin compaction. J Cell Biol 2007; 177(3):401–411.

45. Jackson M, Krassowska A, Gilbert N, et al. Severe global DNA hypomethylation blocks differentiation and induces histone hyperacetylation in embryonic stem cells. Mol Cell Biol 2004; 24(20):8862–8871.

46. Fuks F, Burgers WA, Godin N, et al. Dnmt3 a binds deacetylases and is recruited by a sequence-specific repressor to silence transcription. EMBO J 2001; 20(10):2536–2544.

47. Fuks F, Hurd PJ, Wolf D, et al. The methyl-CpG-binding protein MeCP2 links DNA methylation to histone methylation. J Biol Chem 2003; 278(6)4035–4040.

48. Sarraf SA, Stancheva I. Methyl-CpG binding protein MBD1 couples histone H3 methylation at lysine 9 by SETDB1 to DNA replication and chromatin assembly. Mol Cell 2004; 15(4):595–605.

49. Tamaru H, Selker EU. A histone H3 methyltransferase controls DNA methylation in Neurospora crassa. Nature 2001; 414(6861):277–283.

50. Jackson JP, Lindroth AM, Cao X, et al. Control of CpNpG DNA methylation by the KRYPTONITE histone H3 methyltransferase. Nature 2002; 416(6880):556–560.

51. Tazi J, Bird A. Alternative chromatin structure at CpG islands. Cell 1990; 60(6):909–920.

52. Ball DJ, Gross DS, Garrard WT. 5-methylcytosine is localized in nucleosomes that contain histone H1. Proc Natl Acad Sci U S A 1983; 80(18):5490–5494.
53. Pray-Grant MG, Daniel JA, Schieltz D, et al. Chd1 chromodomain links histone H3 methylation with SAGA- and SLIK-dependent acetylation. Nature 2005; 433(7024):434–438.
54. Fischle W, Wang Y, Jacobs SA, et al. Molecular basis for the discrimination of repressive methyl-lysine marks in histone H3 by Polycomb and HP1 chromodomains. Genes Dev 2003; 17(15):1870–1881.
55. Sharp PA. RNA interference–2001. Genes Dev 2001; 15(5):485–490.
56. Sijen T, Vijn I, Rebocho A, et al. Transcriptional and posttranscriptional gene silencing are mechanistically related. Curr Biol 2001; 11(6):436–440.
57. Mette MF, Aufsatz W, Van Der Winden J, et al. Transcriptional silencing and promoter methylation triggered by double-stranded RNA. EMBO J 2000; 19(19):5194–5201.
58. Volpe TA, Kidner C, Hall IM, et al. Regulation of heterochromatic silencing and histone H3 lysine-9 methylation by RNAi. Science 2002; 297(5588):1833–1837.
59. Heard E. Delving into the diversity of facultative heterochromatin: The epigenetics of the inactive X chromosome. Curr Opin Genet Dev 2005; 15(5):482–489.
60. Bernstein BE, Kellis M. Large-scale discovery and validation of functional elements in the human genome. Genome Biol 2005; 6(3):1–2.
61. Verdel A, Jia S, Gerber S, et al. RNAi-mediated targeting of heterochromatin by the RITS complex. Science 2004; 303(5658):672–676.
62. Henikoff S, Furuyama T, Ahmad K. Histone variants, nucleosome assembly and epigenetic inheritance. Trends Genet 2004; 20(7):320–326.
63. Wilson VL, Jones PA. DNA methylation decreases in aging but not in immortal cells. Science 1983; 220(4601):1055–1057.
64. Wilson VL, Smith RA, Ma S, et al. Genomic 5-methyldeoxycytidine decreases with age. J Biol Chem 1987; 262(21):9948–9951.
65. Mays-Hoopes L, Chao W, Butcher HC, et al. Decreased methylation of the major mouse long interspersed repeated DNA during aging and in myeloma cells. Dev Genet 1986; 7(2):65–73.
66. Issa JP, Ottaviano YL, Celano P, et al. Methylation of the oestrogen receptor CpG island links ageing and neoplasia in human colon. Nat Genet 1994; 7(4):536–540.
67. Issa JP. CpG-island methylation in aging and cancer. Curr Top Microbiol Immunol 2000; 249:101–118.
68. Barbot W, Dupressoir A, Lazar V, et al. Epigenetic regulation of an IAP retrotransposon in the aging mouse: progressive demethylation and de-silencing of the element by its repetitive induction. Nucleic Acids Res 2002; 30(11):2365–2373.
69. Li E. Chromatin modification and epigenetic reprogramming in mammalian development. Nat Rev Genet 2002; 3(9):662–673.
70. Morgan HD, Santos F, Green K, et al. Epigenetic reprogramming in mammals. Hum Mol Genet 2005; 14(Spec No 1):R47–R58.
71. Turner BM. Defining an epigenetic code. Nat Cell Biol 2007; 9(1):2–6.
72. Ringrose L, Paro R. Epigenetic regulation of cellular memory by the polycomb and trithorax group proteins. Annu Rev Genet 2004; 38:413–443.
73. Heard E, Disteche CM. Dosage compensation in mammals: Fine-tuning the expression of the X chromosome. Genes Dev 2006; 20(14):1848–1867.

74. Lewis A, Reik W. How imprinting centres work. Cytogenet Genome Res 2006; 113(1–4):81–89.
75. Reik W, Walter J. Genomic imprinting: parental influence on the genome. Nat Rev Genet 2001; 2(1):21–32.
76. Feinberg AP, Cui H, Ohlsson R. DNA methylation and genomic imprinting: Insights from cancer into epigenetic mechanisms. Semin Cancer Biol 2002; 12(5):389–398.
77. Barlow DP. Gametic imprinting in mammals. Science 1995; 270(5242):1610–1613.
78. Hayashizaki Y, Shibata H, Hirotsune S, et al. Identification of an imprinted U2af binding protein related sequence on mouse chromosome 11 using the RLGS method. Nat Genet 1994; 6(1):33–40.
79. Luedi PP, Hartemink AJ, Jirtle RL. Genome-wide prediction of imprinted murine genes. Genome Res 2005; 15(6):875–884.
80. Dolinoy DC, Huang D, Jirtle RL. Maternal nutrient supplementation counteracts bisphenol A-induced DNA hypomethylation in early development. Proc Natl Acad Sci U S A 2007; 104(32):13056–13061.
81. Wood AJ, Oakey RJ. Genomic imprinting in mammals: Emerging themes and established theories. PLoS Genet 2006; 2(11):1677–1685.
82. Reik W. Stability and flexibility of epigenetic gene regulation in mammalian development. Nature 2007; 447(7143):425–432.
83. Hajkova P, Erhardt S, Lane N, et al. Epigenetic reprogramming in mouse primordial germ cells. Mech Dev 2002; 117(1–2):15–23.
84. Lane N, Dean W, Erhardt S, et al. Resistance of IAPs to methylation reprogramming may provide a mechanism for epigenetic inheritance in the mouse. Genesis 2003; 35(2):88–93.
85. Smith AG. Embryo-derived stem cells: of mice and men. Annu Rev Cell Dev Biol 2001; 17:435–462.
86. McMillen IC, Robinson JS. Developmental origins of the metabolic syndrome: Prediction, plasticity, and programming. Physiol Rev 2005; 85(2):571–633.
87. Gluckman PD, Hanson MA, Beedle AS. Early life events and their consequences for later disease: A life history and evolutionary perspective. Am J Hum Biol 2007; 19(1):1–19.
88. Nathanielsz PW, Poston L, Taylor PD. In utero exposure to maternal obesity and diabetes: Animal models that identify and characterize implications for future health. Obstet Gynecol Clin North Am 2007; 34(2):201–212.
89. Ozanne SE, Constancia M. Mechanisms of disease: The developmental origins of disease and the role of the epigenotype. Nat Clin Pract Endocrinol Metab 2007; 3(7):539–546.
90. Godfrey KM, Lillycrop KA, Burdge GC, et al. Epigenetic mechanisms and the mismatch concept of the developmental origins of health and disease. Pediatr Res 2007; 61(5 Pt 2):5R–10R.
91. Lau C, Rogers JM. Embryonic and fetal programming of physiological disorders in adulthood. Birth Defects Res C Embryo Today 2004; 72(4):300–312.
92. Rogers JM. The Barker Hypothesis. Curr Opin Endocrinol Diabetes 2006; 13(6):536–540.
93. Szyf M, Weaver I, Meaney M. Maternal care, the epigenome and phenotypic differences in behavior. Reprod Toxicol 2007; 24(1):9–19.

94. Reamon-Buettner SM, Borlak J. A new paradigm in toxicology and teratology: Altering gene activity in the absence of DNA sequence variation. Reprod Toxicol 2007; 24(1):20–30.

95. Jirtle RL, Skinner MK. Environmental epigenomics and disease susceptibility. Nat Rev Genet 2007; 8(4):253–262.

96. Fall CH, Osmond C, Barker DJ, et al. Fetal and infant growth and cardiovascular risk factors in women. Bmj 1995; 310(6977):428–432.

97. Law CM, Barker DJ, Bull AR, et al. Maternal and fetal influences on blood pressure. Arch Dis Child 1991; 66(11):1291–1295.

98. Phillips DI, Barker DJ, Hales CN, et al. Thinness at birth and insulin resistance in adult life. Diabetologia 1994; 37(2):150–154.

99. Barker DJ, Hales CN, Fall CH, et al. Type 2 (non-insulin-dependent) diabetes mellitus, hypertension and hyperlipidaemia (syndrome X): Relation to reduced fetal growth. Diabetologia 1993; 36(1):62–67.

100. Stanner SA, Yudkin JS. Fetal programming and the Leningrad Siege study. Twin Res 2001; 4(5):287–292.

101. Roseboom TJ, Van Der Meulen JH, Ravelli AC, et al. Effects of prenatal exposure to the Dutch famine on adult disease in later life: An overview. Twin Res 2001; 4(5):293–298.

102. Ravelli AC, Van Der Meulen JH, Osmond C, et al. Obesity at the age of 50 y in men and women exposed to famine prenatally. Am J Clin Nutr 1999; 70(5):811–816.

103. Painter RC, Roseboom TJ, van Montfrans GA, et al. Microalbuminuria in adults after prenatal exposure to the Dutch famine. J Am Soc Nephrol 2005; 16(1):189–194.

104. Ravelli AC, Van Der Meulen JH, Osmond C, et al. Infant feeding and adult glucose tolerance, lipid profile, blood pressure, and obesity. Arch Dis Child 2000; 82(3):248–252.

105. Langley-Evans SC. Developmental programming of health and disease. Proc Nutr Soc 2006; 65(1):97–105.

106. Langley-Evans SC. Experimental models of hypertension and cardiovascular disease. In Fetal Nutrition and Adult Disease, Langley-Evans SC, ed. CAB International, Willingford, Oxfordshire 2004;. 129–156.

107. Langley-Evans SC, Bellinger L, McMullen S. Animal models of programming: Early life influences on appetite and feeding behaviour. Matern Child Nutr 2005; 1(3):142–148.

108. Lillycrop KA, Phillips ES, Jackson AA, et al. Dietary protein restriction of pregnant rats induces and folic acid supplementation prevents epigenetic modification of hepatic gene expression in the offspring. J Nutr 2005; 135(6):1382–1386.

109. Lillycrop KA, Slater-Jefferies JL, Hanson MA, et al. Induction of altered epigenetic regulation of the hepatic glucocorticoid receptor in the offspring of rats fed a protein-restricted diet during pregnancy suggests that reduced DNA methyltransferase-1 expression is involved in impaired DNA methylation and changes in histone modifications. Br J Nutr 2007; 97(6):1064–73.

110. Gluckman PD, Lillycrop KA, Vickers MH, et al. Metabolic plasticity during mammalian development is directionally dependent on early nutritional status. Proc Natl Acad Sci U S A 2007; 104(31)12796–12800.

111. Kappen C. Folate supplementation in three genetic models: Implications for understanding folate-dependent developmental pathways. Am J Med Genet C Semin Med Genet 2005; 135(1)24–30.

112. Van Den Veyver IB. Genetic effects of methylation diets. Annu Rev Nutr 2002; 22:255–282.

113. Wolff GL, Kodell RL, Moore SR, et al. Maternal epigenetics and methyl supplements affect agouti gene expression in Avy/a mice. Faseb J 1998; 12(11):949–957.

114. Waterland RA, Jirtle RL. Transposable elements: Targets for early nutritional effects on epigenetic gene regulation. Mol Cell Biol 2003; 23(15):5293–5300.

115. Cooney CA, Dave AA, Wolff GL. Maternal methyl supplements in mice affect epigenetic variation and DNA methylation of offspring. J Nutr 2002; 132(8 Suppl):2393 S–2400 S.

116. Kultima K, Nystrom AM, Scholz B, et al. Valproic acid teratogenicity: A toxicogenomics approach. Environ Health Perspect 2004; 112(12):1225–1235.

117. Ali-Khan SE, Hales BF. Novel retinoid targets in the mouse limb during organogenesis. Toxicol Sci 2006; 94(1):139–152.

118. Green ML, Singh AV, Zhang Y, et al. Reprogramming of genetic networks during initiation of the Fetal Alcohol Syndrome. Dev Dyn 2007; 236(2):613–631.

119. Gurvich N, Berman MG, Wittner BS, et al. Association of valproate-induced teratogenesis with histone deacetylase inhibition in vivo. FASEB J 2005; 19(9):1166–1168.

120. Colvin ME, Sasaki JC, Tran NL. Chemical factors in the action of phosphoramidic mustard alkylating anticancer drugs: roles for computational chemistry. Curr Pharm Des 1999; 5(8):645–663.

121. Trasler JM, Hales BF, Robaire B. A time-course study of chronic paternal cyclophosphamide treatment in rats: Effects on pregnancy outcome and the male reproductive and hematologic systems. Biol Reprod 1987; 37(2):317–326.

122. Auroux M, Dulioust E, Selva J, et al. Cyclophosphamide in the F0 male rat: physical and behavioral changes in three successive adult generations. Mutat Res 1990; 229(2):189–200.

123. Barton TS, Robaire B, Hales BF. Epigenetic programming in the preimplantation rat embryo is disrupted by chronic paternal cyclophosphamide exposure. Proc Natl Acad Sci U S A 2005; 102(22)7865–7870.

124. Dolinoy DC, Weidman JR, Waterland RA, et al. Maternal genistein alters coat color and protects Avy mouse offspring from obesity by modifying the fetal epigenome. Environ Health Perspect 2006; 114(4):567–572.

125. Anway MD, Leathers C, Skinner MK. Endocrine disruptor vinclozolin induced epigenetic transgenerational adult-onset disease. Endocrinology 2006; 147(12):5515–5523.

126. Chang HS, Anway MD, Rekow SS, et al. Transgenerational epigenetic imprinting of the male germline by endocrine disruptor exposure during gonadal sex determination. Endocrinology 2006; 147(12):5524–5541.

127. Turusov VS, Nikonova TV, Parfenov Yu D. Increased multiplicity of lung adenomas in five generations of mice treated with benz(a)pyrene when pregnant. Cancer Lett 1990; 55(3):227–231.

128. Francis D, Diorio J, Liu D, et al. Nongenomic transmission across generations of maternal behavior and stress responses in the rat. Science 1999; 286(5442):1155–1158.

129. Liu D, Diorio J, Tannenbaum B, et al. Maternal care, hippocampal glucocorticoid receptors, and hypothalamic-pituitary-adrenal responses to stress. Science 1997; 277(5332):1659–1662.

130. Meaney MJ, Szyf M. Maternal care as a model for experience-dependent chromatin plasticity? Trends Neurosci 2005; 28(9):456–463.

131. Weaver IC, Cervoni N, Champagne FA, et al. Epigenetic programming by maternal behavior. Nat Neurosci 2004; 7(8):847–854.

132. Weaver IC, Meaney MJ, Szyf M. Maternal care effects on the hippocampal transcriptome and anxiety-mediated behaviors in the offspring that are reversible in adulthood. Proc Natl Acad Sci U S A 2006; 103(9):3480–3485.

133. Weaver IC, Diorio J, Seckl JR, et al. Early environmental regulation of hippocampal glucocorticoid receptor gene expression: Characterization of intracellular mediators and potential genomic target sites. Ann N Y Acad Sci 2004; 1024:182–212.

134. Steptoe PC, Edwards RG. Birth after the reimplantation of a human embryo. Lancet 1978; 2(8085):366

135. Iritani A, Hosoi Y. Microfertilization by various methods in mammalian species. Prog Clin Biol Res 1989; 294:145–149.

136. Kimura Y, Yanagimachi R. Intracytoplasmic sperm injection in the mouse. Biol Reprod 1995; 52(4):709–720.

137. Palermo G, Joris H, Devroey P, et al. Pregnancies after intracytoplasmic injection of single spermatozoon into an oocyte. Lancet 1992; 340(8810):17–18.

138. Sakai RR, Tamashiro KL, Yamazaki Y, et al. Cloning and assisted reproductive techniques: Influence on early development and adult phenotype. Birth Defects Res C Embryo Today 2005; 75(2):151–162.

139. Vogel G. Reproductive biology. Cloning: Could humans be next? Science 2001; 291(5505):808–809.

140. Schatten G, Prather R, Wilmut I. Cloning claim is science fiction, not science. Science 2003; 299(5605):344

141. Yang D, Zhang ZJ, Oldenburg M, et al. Human embryonic stem cell-derived dopaminergic neurons reverse functional deficit in parkinsonian rats. Stem Cells 2008; 26:55–63.

142. Cibelli JB, Campbell KH, Seidel GE, et al. The health profile of cloned animals. Nat Biotechnol 2002; 20(1):13–14.

143. Kruip TAM, den Daas JHG. In vitro pruduced and cloned embryos: Effects on pregnancy, parturition and offspring. Theriogenology 1997; 47:43–52.

144. Zhu J, Telfer EE, Fletcher J, et al. Improvement of an electrical activation protocol for porcine oocytes. Biol Reprod 2002; 66(3):635–641.

145. Young LE, Sinclair KD, Wilmut I. Large offspring syndrome in cattle and sheep. Rev Reprod 1998; 3(3):155–163.

146. Farin PW, Piedrahita JA, Farin CE. Errors in development of fetuses and placentas from in vitro-produced bovine embryos. Theriogenology 2006; 65(1):178–191.

147. Chavatte-Palmer P, Remy D, Cordonnier N, et al. Health status of cloned cattle at different ages. Cloning Stem Cells 2004; 6(2):94–100.

148. Daniels R, Hall V, Trounson AO. Analysis of gene transcription in bovine nuclear transfer embryos reconstructed with granulosa cell nuclei. Biol Reprod 2000; 63(4):1034–1040.

149. Jouneau A, Zhou Q, Camus A, et al. Developmental abnormalities of NT mouse embryos appear early after implantation. Development 2006; 133(8):1597–1607.
150. Boiani M, Eckardt S, Scholer HR, et al. Oct4 distribution and level in mouse clones: Consequences for pluripotency. Genes Dev 2002; 16(10):1209–1219.
151. Bortvin A, Eggan K, Skaletsky H, et al. Incomplete reactivation of Oct4-related genes in mouse embryos cloned from somatic nuclei. Development 2003; 130(8): 1673–1680.
152. Kang YK, Koo DB, Park JS, et al. Aberrant methylation of donor genome in cloned bovine embryos. Nat Genet 2001; 28(2):173–177.
153. Kang YK, Park JS, Koo DB, et al. Limited demethylation leaves mosaic-type methylation states in cloned bovine pre-implantation embryos. EMBO J 2002; 21(5): 1092–1100.
154. Ohgane J, Wakayama T, Kogo Y, et al. DNA methylation variation in cloned mice. Genesis 2001; 30(2):45–50.
155. Humpherys D, Eggan K, Akutsu H, et al. Abnormal gene expression in cloned mice derived from embryonic stem cell and cumulus cell nuclei. Proc Natl Acad Sci U S A 2002; 99(20):12889–12894.
156. Inoue K, Kohda T, Lee J, et al. Faithful expression of imprinted genes in cloned mice. Science 2002; 295(5553):297.
157. Wrenzycki C, Wells D, Herrmann D, et al. Nuclear transfer protocol affects messenger RNA expression patterns in cloned bovine blastocysts. Biol Reprod 2001; 65(1)309–317.
158. Eggan K, Akutsu H, Hochedlinger K, et al. X-Chromosome inactivation in cloned mouse embryos. Science 2000; 290(5496):1578–1581.
159. Xue F, Tian XC, Du F, et al. Aberrant patterns of X chromosome inactivation in bovine clones. Nat Genet 2002; 31(2):216–220.
160. Shiels PG, Kind AJ, Campbell KH, et al. Analysis of telomere lengths in cloned sheep. Nature 1999; 399(6734):316–317.
161. From the Centers of Disease Control and Prevention. Use of assisted reproductive technology–United States, 1996 and 1998. MMWR Morb Mortal Wkly Rep 2002; 51(5):97–101.
162. Schultz RM, Williams CJ. The science of ART. Science 2002; 296(5576): 2188–2190.
163. Hansen JE, Embryonic stem cell production through therapeutic cloning has fewer ethical problems than stem cell harvest from surplus IVF embryos. J Med Ethics 2002; 28(2):86–88.
164. Cox GF, Burger J, Lip V, et al. Intracytoplasmic sperm injection may increase the risk of imprinting defects. Am J Hum Genet 2002; 71(1):162–164.
165. DeBaun MR, Niemitz EL, Feinberg AP. Association of in vitro fertilization with Beckwith-Wiedemann syndrome and epigenetic alterations of LIT1 and H19. Am J Hum Genet 2003; 72(1):156–160.
166. Gicquel C, Gaston V, Mandelbaum J, et al. In vitro fertilization may increase the risk of Beckwith-Wiedemann syndrome related to the abnormal imprinting of the KCN1OT gene. Am J Hum Genet 2003; 72(5):1338–1341.
167. Maher ER, Afnan M, Barratt CL. Epigenetic risks related to assisted reproductive technologies: Epigenetics, imprinting, ART and icebergs? Hum Reprod 2003; 18(12):2508–2511.

168. Orstavik KH. Intracytoplasmic sperm injection and congenital syndromes because of imprinting defects. Tidsskr Nor Laegeforen 2003; 123(2):177.

169. Li T, Vu TH, Ulaner GA, et al. IVF results in de novo DNA methylation and histone methylation at an Igf2-H19 imprinting epigenetic switch. Mol Hum Reprod 2005; 11(9):631–640.

170. Horsthemke B, Ludwig M. Assisted reproduction: The epigenetic perspective. Hum Reprod Update 2005; 11(5):473–482.

171. Swales AK, Spears N. Genomic imprinting and reproduction. Reproduction 2005; 130(4): 389–399.

172. Strathdee G, Brown R. Aberrant DNA methylation in cancer: Potential clinical interventions. Expert Rev Mol Med 2002; 4(4):1–17.

5

Personalized Nutrition and Medicine in Perinatal Development

Jim Kaput and James J. Chen

*Division of Personalized Nutrition and Medicine,
FDA/National Center for Toxicological Research,
Jefferson, Arkansas, U.S.A.*

William Slikker, Jr.

*Office of the Director, FDA/National Center for Toxicological Research,
Jefferson, Arkansas, U.S.A.*

INTRODUCTION

The goal of personalized nutrition and medicine is to get the right nutrient or therapeutic to the right individual at the right time for the right outcome. The need for this informed approach is greatest in pregnant women and infants because lack of key nutrients in critical developmental windows prevent full mental and physical development and may alter health outcomes in adults (1,2). Modern evaluation tools will give clinicians the best available information about how to use a product to maximize its benefit and minimize its side effects. Many of these new technologies may allow individualization of treatment by identifying the individual who is likely to respond well to a treatment and protecting those that may respond adversely. Because food represents more than calories and fiber and is known to contain signaling molecules and morphogens, the "omics" approach (toxicogenomics, epigenomics, pharmacogenomics, proteomics, and metabolomics) applied to pharmaceuticals may be equally applied to nutrients. "Omics" refers to the simultaneous measurement of large numbers of a type of analyte (e.g., protein), or the simultaneous measurement of large numbers of many different analytes (nutrigenomics—which may measure many chemicals in foods, RNA, and gene variants).

Vitamin A and folic acid are prime examples of food-based chemicals that can be analyzed through "omic" technologies and thereby consumed in a manner promoting normal perinatal development. Recent findings indicate that variations in specific genes can aid in defining disease susceptibility and predicting drug-induced adverse events. While genetic and lifestyle data are necessary, they are not sufficient for developing personalized recommendations because of the high dimensionality (number of data points) and complexities of genotype–environment interactions. Classification algorithms can combine diverse, high-dimensional biomarkers consisting of gene variants, expression profiles, proteomic and metabolomic measurements, and their interactions to achieve increased accuracy in predicting which patients will benefit most from medical products and which are most likely to experience toxicities that outweigh benefits. The anticipated benefit of personalized nutrition and medicine is the identification of subgroups of people that respond favorably to specific nutrients or drug therapies and avoid toxicity, especially during development.

PERSONALIZED HEALTH CARE

The driving force behind personalized medicine and nutrition includes expectations for safe and effective treatments for specific populations of individuals. Even though clinicians provide personalized health care by definition, the opportunity to address personalized medicine at the molecular level has only come about since the sequencing of the human genome (3). The medical community is now poised to take a quantum leap toward evidence-based personalized health care based on the new molecular understanding of the individual. It is expected that new drugs may be developed that will exhibit greater safety and efficacy for specific populations and individuals. Therapeutic doses may be adjusted to the individual's genetic makeup, diminishing the need for the one-size fits all medical paradigm. Genomic information is also being used to identify disease targets, which will speed clinical trials by presorting high- versus low-responder populations. Together, these advances may result in more cost–effective health care by reducing costs due to avoidance of futile treatments and improved clinical outcomes.

PERINATAL DEVELOPMENT

A systems biology approach is required to analyze and understand the complexities and temporal manifestations resulting from exposures to toxicants during development. Developmental toxicity may manifest as structural or functional effects of a biological, chemical, or physical agent that diminishes the ability of an organism to survive, reproduce, or adapt to its environment. These effects can often be measured by a variety of functional, anatomical, biochemical, or molecular techniques. Extrapolation across species is feasible, but must take into account the relative ontogeny of the organ systems among species. Insults to the developing organism may take various forms and may be quite subtle. Although its manifestations may change with age, developmental toxicity may occur at any time in the life cycle from gestation through senescence. The developing organism may be more or less susceptible to toxic insult depending on the stage of development (4) and the genetic susceptibility to the toxicant.

PERSONALIZED NUTRITION AND MEDICINE

Personalized nutrition has typically been aimed toward optimizing health of adults by providing the right nutrient to the right individual at the right time while avoiding those environmental chemicals capable of disrupting biological processes. The application of this knowledge during pregnancy and development is particularly important to ensure optimal health in individuals (1,2). Developing the knowledge base for personalized dietary advice is the goal of nutrigenomics, the study and application of gene–nutrient interactions that account for differences in response to diet in different genetic backgrounds. Individualizing health care through pharmacogenomics, the analyses of drug–genotype interactions, will give clinicians the best available information and treatments to reduce the personal and public health impact and cost of disease, and improve the length and quality of life for patients with chronic diseases. Toxicogenomics will help identify the environmental chemicals that may produce developmental abnormalities and influence the probability of late onset chronic diseases. Nutrients, toxins, and drugs are xenobiotics with different concentrations and intended affects, but nevertheless interact with individual genetic makeups; the approaches described by nutrigenomics, pharmacogenomics, and toxicogenomics are conceptually similar. These chemical entities are a continuum of xenobiotics metabolized by genetic makeups, which are unique to the individual and affect human health through changes in gene expression or genome structure (5). As such, the challenges of analyzing gene–chemical interactions, their effect on health in individuals, and application of knowledge for these three broad disciplines are similar and all depend upon a systems approach to biology (6). These approaches have to address three overarching challenges: nutrient (i.e., environmental) complexity, genetic heterogeneity of humans, and the resulting intricacies of health and disease processes caused by interactions between a person's unique genetic makeup and environmental factors (5). While these challenges are confronted in studies of gene–environment interactions that influence health of adults, they also will affect the study, understanding, and long-term consequences of fetal or postnatal programming of metabolism. This emerging concept, that environmental influences act in utero and during early childhood to reprogram the metabolic outcome (1,2,7–10) encoded in the genome, will also add complexity to understanding individual health and disease processes (chap. 5, this volume).

DIVERSITY CHALLENGES

Complexity of Foods and Cultures

Epidemiological studies show associations between food intake and health and the incidence and severity of chronic diseases (11,12). Food is not only chemically complex, but the composition varies because of genotypic variation, climate, season, and processing methods (13). Processing also introduces additional variability because of physical matrices (14,15) used for palatability and stabilization. While 75% of worldwide retail food sales are processed (16), varying chemical ingredients from locally produced food sources used in global formulations may still yield different foods. Cultural, familial, and individual food preparation differences also affect nutrient bioavailability. For example, frying meat not only alters the food matrix, composition, and bioavailability of certain nutrients but also produces carcinogens (17) depending on the cooking time and conditions. Exposure during key developmental windows to natural chemicals or those produced during food preparation will produce different effects in each individual depending on their unique genetic makeup. Although the molecular pathways and reactions affected by individual dietary chemicals are being identified and characterized (18–20), accurately measuring the intake of dietary chemicals in complex mixtures has proven difficult (21,22) and few studies have examined exposure during development.

Genetic Heterogeneity

Microarray-based genotype analyses of DNA from 270 individuals from Africa (Yoruba), Japan, China (Han), and Europe (Centre d'Etude du Polymorphisme Humain—CEPH—samples) identified about 4 million single nucleotide polymorphisms (SNPs—http://www.hapmap.org/downloads/index.html.en) in each ancestral group (23,24). A total of 10 to 11 million SNPs are expected to be characterized in humans. The majority (85–90%) of SNPs identified thus far were shared among populations with the remainder unique to a given ancestral group (25,26). Although most SNPs are shared, marked differences occur in how frequently they occur among ancestral populations. Differences in allele frequencies produce the variations in visible and metabolic phenotypes of the world's population.

These differences arose because environments exert a selective pressure on genomes. Lactase persistence, the best-known example of gene–nutrient interactions, exemplifies this concept: In Europeans and their descendents, a polymorphism in promoter of the lactase gene (*LCT*) allows expression of the gene in adults. Adult consumption of large quantities of lactose is atypical for most mammals. The C-13910 T variant responsible for the phenotype arose as a mutation around 9000 years ago (27) in northern Europe. Milk represented another nutrient, calorie, and water source in the winter climate and the mutation provided a selective advantage. An independent mutation (a G-to-C mutation at -14010) occurred ~7000 years ago in central Africa (28) and was selected because the dry, hot environments lacked water, nutrients for energy, and calcium (28). Both mutations were fixed in their descendents because of the selective advantage of this nutrient source. This example of convergent evolution demonstrated the importance of analyzing genetic differences among ancestral groups: the same phenotype can result from different variations in the genome.

In addition to lactase persistence, genes involved in skin color (e.g., *SCL45A2, SCL24A5*) also vary among populations. Data from the HapMap project indicated that 22 chromosomal regions had 480 SNPs that occurred after the continents were populated (29). Eight of these polymorphisms were within coding regions and the remainder in regulatory sequences. In addition to *BLZF1* (= golgin 45, upregulated by retinoic acid—online mendelian inheritance in man ID number [OMIM] 608692), *SLC19A2* (thiamine transporter—OMIM 603941) and *CHST5* (carbohydrate sulfotransferase—OMIM 604817) are involved in nutrient metabolism or regulated by nutrients. Selective pressure on ancestral loci (those that arose prior to migration from Africa) is more difficult to detect, but has been shown. The most publicized polymorphisms result in the variable metabolism of certain drugs in different ancestral groups (30–33). Genes involved in nutrient metabolism, for example, β-glucosidase (*GBA*) and the homeobox genes, *NKX2.2* (34), vary in allele frequencies among populations. Comprehensive analyses of allele frequencies for genes involved in nutrient metabolism or regulated by nutrients have not been done, in part, because genotype or sequence data are not yet available for many populations.

Other environmental pressures also affected genes involved in nutrient metabolism (34,35). Malaria, for instance, exerts strong selective pressure on glucose 6 phosphate dehydrogenase (*G6PDH*), among other genes (e.g., *hemoglobin S*). G6PDH is solely responsible for reduced nicotinamide adenine dinucleotide phosphate (NADPH) production in RBCs. Reduced activities of this enzyme are advantageous because H_2O_2 produced by the parasite cannot be eliminated with inadequate levels of NADPH. The infection process is blunted because of peroxide-induced lysis of host RBCs. G6PDH deficiencies are highest in populations exposed to malaria, such as southern China (e.g., $\sim$17% of the Dai population) compared to northern regions (0% of those tested in Han Shandong) of the same country (36). Different polymorphisms resulting in G6PDH deficiency occur in other tropical areas where malaria is present (37). Since some of the G6PDH variants have altered Km for NADP+, increased dietary intake of niacin may restore enzymatic activity to normal levels [for a full discussion, see Ames (38)].

The human genome also has structural differences among individuals because of nucleotide insertions or deletions (indels), chromosomal segment rearrangements and variation, and increases or decreases in the number of genes [i.e., copy number variants (39)]. Copy number variation has been shown for at least one gene involved in nutrient metabolism: individuals whose ancestors were exposed to either high-starch or low-starch environments differed in the number of amylase genes (40,41). A telling aspect of this structural variation, which may apply to other types of genetic changes, was that different populations in the same geographical region (e.g., central Africa) had (on average) different numbers of amylase genes depending on the long-standing availability to specific components in foods, in this case, starch levels. Hence, differences in nutrient metabolism were not specific to a particular skin color or observable phenotype.

Allele frequencies are important not only because they affect one reaction step but also because many genes are affected by the activity of other genes through direct protein–protein interactions, sharing substrates or reaction products, or through other molecular interactions. Epistasis, or gene–gene interactions (42–44), can also influence the expression of a trait (45–48). Even monogenic diseases, as first demonstrated for phenylketonuria, have variable onsets and severities indicating that other loci modify the effect of the primary genetic defect (49,50). Polygenic diseases are also influenced by epistasis as demonstrated by a common polymorphism in ghrelin (*GHRL*) that abolished the association of an allele of its receptor (*GHSR*) with myocardial infarction and cardiovascular disease (51).

Epistatic interactions have also been shown to differ among ancestral populations (52) or in populations with recently mixed ancestral backgrounds [i.e., admixed populations (53)]. An allele (Haplotype K or HapK) of the leukotriene 4A hydrolase gene (*LTA4H*) increases the population risk for cardiovascular disease and myocardial infarction from 1.35 in European-Americans to almost 5 in African-Americans who carry the same allele. The HapK allele (consisting of 4 SNPs) arose in Europe and occurs in a subgroup of African-Americans only because of admixture occurring during the past 300-plus years. This epistatic interaction may be explained by (1) *LTA4H* interacting differently with one or more gene variants between African versus European chromosomal regions, resulting in increased effect of LTA4H in African-Americans, and/or (2) different environmental factors alter the influence of *LTA4H* on myocardial infarction (54) or (3) a combination of epistatic and gene–environment interactions (52). LTAH4 participates in leukotriene and prostaglandin metabolism, which are linked to dietary fatty acid intake (55). While epistasis may be apparent for recently admixed populations such as African-Americans or Latino-Americans, population structure exists even within seemingly homogenous groups such as in Iceland (56). Others data demonstrated that European populations are stratified north to southeast and east to west (57).

The effect of an allele (or SNP) on some measurable phenotype is therefore context specific and may vary depending upon alleles of other genes. In addition, environmental factors may influence not only the gene (or SNP) linked to a specific phenotype but also the genes that interact with it. Since many published studies have not accounted for stratification or environmental factors that influence expression of genetic information, existing results may be specific only to the specific population and environment tested (52,53,58–62).

Epigenesis

Additional variability in expression of genetic information occurs through epigenomic mechanisms. Alteration of chromosome structure through methylation or by altering chromatin structure [i.e., epigenetics/epigenomics (2,63–68)] changes regulation of gene expression without alterations in the DNA sequence. Chromatin remodeling is an ongoing and active process regulated in part by certain chemicals or metabolites derived from the diet (18,69,70). Hence, chromatin remodeling responds to short- and long-term nutrient availability (10), including energy availability (71).

The effects of imbalances of calories or nutrients for physiological needs and the effect on human health are exemplified by the Dutch Famine of late 1944 through Spring of 1945. Individuals born to women who were pregnant during the famine had increased incidence of late onset conditions (72,73), such as coronary heart disease, dyslipidemias, obesity, obstructive airways, and glucose intolerance (72). Recent studies have demonstrated an increased risk for overweight and obesity in children of mothers who smoked during pregnancy. These data and other clinical observations and animal model studies support the hypothesis that obesity in children of mothers who smoked during pregnancy is the result of long-lasting behavioral teratogenic effects of nicotine exposure in utero (74).

Experimental research in animal models has demonstrated alterations in gene expression and physiologies depending upon maternal diets (68,75–77). Dams fed diets supplemented with methyl group donors or cofactors involved in one carbon metabolism [choline, betaine, folic acid, vitamin B12, methionine, and zinc (76,77)] produced pups with altered phenotype through changes in expression of genes. Differences in expression were correlated with changes in DNA methylation (70,76,77). Differences in DNA methylation status of regulatory regions are often associated with changes in gene expression (78). The simple story that all epigenomic regulation occurs through methylation has been challenged by recent data showing that multiple molecular pathways (79), specific nutrients (68,75–77,80–83), and calorie levels (84) can influence and control epigenetic effects on gene expression. Many of these pathways converge on reversible modifications of chromatin proteins that influence DNA packaging and gene expression (85,86). Suggestive evidence exists that some of these effects may be transgenerational, that is, grandmothers' (87) and fathers' (88) diets may alter gene expression in later generations, although these findings have not been confirmed (89).

Regardless of the strength of individual studies, the sum of human association studies and animal experiments has led to a concept called developmental plasticity (90) or nutritional epigenomics (10,66). This concept is also referred to as polypheny (91), metabolic (9,92–94), or fetal (28) (8) programming. Plasticity is based on the response of a fetus to maternal environments or the individual's early nutritional environment through epigenetic mechanisms. Programming may be advantageous because the fetus/child would be prepared for the existing environment (90). When the environment no longer matches the programming, health outcomes are affected. Health outcomes are late onset, demonstrating the influence of long-term exposure to imbalanced (for one's genotype and epigenotype) nutrient intake. In the Dutch Famine example, fetuses were primed for nutritionally poor environments but were exposed to nutritionally rich environments after birth. Such mechanisms could explain the projected increase in diabetes incidence in developing countries ($\sim$170%) versus that in developed countries ($\sim$45%) by 2025 (95). Determining the optimum nutrient intake for individuals is a challenge and goal of nutrigenomic research.

The effects of environment on the fetus would necessarily depend upon the genotype of the mother, specifically, her metabolic capacity to maintain (or not) the appropriate sensing of the environment. Such sensing must come from the inherited genes and epigenetic modifications and their response to the environment. Hence, genetic and physiological complexities, in addition to nutritional factors, will influence fetal programming. Teratology also provides conceptual information about the influence of teratogen exposure during different developmental stages. Environmental, food, and manufactured chemicals that are teratogens frequently produce an all-or-none response in the first 2 weeks postconception, structural abnormalities if exposure occurs between $\sim$18 days and 60 days, or a variety of organ-specific defects at other times during pregnancy (96). A wide variety of clinical drugs [antiepileptics, angiotensin-converting enzyme inhibitors, indomethacin, retinoids, tetracycline derivatives, etc], agricultural chemicals (triazole fungicide, dioxins, chlorinated products, etc.), and petrochemicals (kerosene-based jet fuels or additives, etc) are the most frequently cited and studied teratogens (96–100). The molecular mechanisms of action of an increasing number of teratogens are being characterized, but many others are less well understood (96,98,101). Because of the similarity of nutrient and teratogen modes of action, at least some of the effects of teratogens could be on the epigenome (102).

Studies of epigenomic effects demonstrate universal concepts for human health care—exposure to teratogens, toxins, or absence of nutrients (103,104), during different developmental windows, may alter the risk of late onset diseases. These concepts also apply to the design of research programs since many, if not most, human studies (for drugs, genetic association, or other) assume similar, if not identical, developmental paths to adulthood. Strategies for dissecting early environmental influences and the corresponding causative genes and pathways will be essential for understanding how to maintain health and prevent disease in adulthood. The United States Food and Drug Administration (FDA) and its research and regulatory partners have a long and rich tradition of developing evidence-based guidelines for reproductive and developmental toxicology that can guide future efforts and applications of nutritional and teratolgoical epigenomics (105,106).

HETEROGENEITY OF HEALTH AND DISEASE PROCESSES: THE T2DM EXAMPLE

Health and disease are often considered as dichotomous states. However, the challenge of clinically managing a chronic disease in different individuals highlights the variability in molecular, genetic, and nutritional processes that produce the disease. We have used type 2 diabetes (T2DM) as an example of this complexity (5,107) and applied the concept of such phenotypic heterogeneity to health. That is, both phenotypes (health and disease) may arise by the action of multiple genes interacting with multiple environmental factors through multiple pathways.

Glycemic control (Hb1Ac, hypoglycemia, ketosis), cardiovascular risk factors, peripheral neuropathies, eye disease, renal disease, and autonomic neuropathies are used to create four levels (low, moderate, high, and very high) of T2DM severity (108). This and similar classification schemes are used for clinical management of the disease. The first option for early stage or less severe cases of T2DM is to modify diet and lifestyle to attain glycemic control, which is successful in only ~20% of patients (109). Six classes of drugs target different pathways and organs: insulin secretion by the pancreas (sulfonylurea, meglitinides, and exenatide), glucose absorption by the intestines (α-glucosidase inhibitors), glucose production in the liver (biguanide = metformin), and insulin sensitivity in adipose and peripheral tissues [e.g., rosiglitazone and pioglitazone (107,110,111)]. Approximately 50% of T2DM patients take oral medications only, about 11% take combinations of oral agents with insulin, and the remainder takes no medications (20%) or insulin alone (16.4%) (109).

Since abnormalities in different pathways in different organs produce the similar clinical manifestations (e.g., poor glucose control) or at least a designation of diabetes, this disease, like all other chronic diseases, consists of subtypes differing by alterations in underlying molecular pathways. Identifying the genes that cause abnormal responses in these pathways would contribute to the development of diagnostic tests for sorting individuals into treatment groups at the initial visit and perhaps personalization of health care. A recent study demonstrated that gene-based diagnostics could aid in drug selection: permanent neonatal diabetes caused either by L213R or by I1424V mutations in *ABCC8* (sulfonyl receptor [SUR1]) could be treated with glyburide (a sulfonylurea) rather than usual treatment of insulin (112).

The identification of causative variants contributing to T2DM has met with limited and uneven success. For example, with the recent genome-wide association studies [GWA (113–119)] identifying eight new loci and potential candidate genes contributing to T2DM, a total of eleven candidate genes are associated with increased population risk to T2DM (Table 1). Genes that had previously been considered excellent candidates based on reproducibility or significance (120), including the glucagon receptor (*GCGR*), glucokinase (*GCK*), glucose transporter 1 (*SLC2A1*), and the aforementioned *ABCC8*, were not identified in this large genome-wide screen. Previous association studies were criticized because they could not be replicated, did not appropriately match cases to controls, were underpowered—that is, the sample size was small. (121–127). The more recent GWA analyses attempted to address those limitations by increasing the number of cases in each of the studies (range: 1215–2938) and controls (range: 1258–3550). The Welcome Trust Case Control consortium analyzed 14,000 cases of six common diseases (bipolar, coronary artery disease, Crohn's, rheumatoid arthritis, Alzheimer's, Type 1 diabetes, and T2DM) with 3000 shared controls (119). Several of the studies replicated results in separate populations or shared samples. Collectively, over 18,000 individuals were analyzed for T2DM with about 14,000 controls. To reduce population stratification and concomitant epistatic interactions, samples for individual studies came from defined populations

Table 1 Type 2 Diabetes Candidate Genes[a]

Genes	Function	Chr Pos	OMIM
FTO	2-oxoglutarate-dependent nucleic acid demethylase	26q12.1	610966
CDKN2B	Cyclin-dependent kinase inhibitor 2b	9p21	600431
CDKAL1	CDK5 regulatory subunit-associated protein 1—like 1	6p22.3	611259
PPARG	Peroxisome proliferator-activated receptor	3p25	601487
IGF2BP2	IGF2 mRNA-binding protein	3q28	608289
TCF7L2	Transcription factor	10q25.3	702228
KCNJ11	K+ channel rectifier	11p15.1	600937
HHEX	Hematopoietically expressed homeobox	10q24	604420
SLC30A8	β cell zinc transporter	8q24.11	611145

[a]From Refs. 113,114,116,118,159.
Abbreviations: Chr Pos, chromosomal position; OMIM, Online Mendelian Inheritance in Man ID number (http://www.ncbi.nlm.nih.gov/sites/entrez?db= omim).

(Finland, Sweden, England/Ireland, France, and Ashkenazi Jews). A method developed to test for stratification was used in some studies (128) or the data were analyzed by a variety of algorithms for potential population substructures (116).

While these large studies had some success, the eight new candidates collectively "explain" between 0.5% and 2.4% of the population attributable fraction (PAF) for T2DM (116). Several limitations contributed to the number of genes identified and their small PAF. The various GWA studies used different criteria for including individuals in the case population. In some of the experimental designs, cases included newly diagnosed (i.e., not yet on medications) as well as those on any type of T2DM medication, regardless of the molecular pathway targeted by the drug (see "Heterogeneity of Health and Disease" section of this chapter). Combining all the patients as T2DM regardless of disease subtype (based on medication or some quantitative phenotype) reduced (or averaged) the contributions of causative genes in different pathways. In addition, regardless of matching the overall genomic architecture of individual cases to individual controls, epistatic interactions may still occur in seemingly homogeneous populations [as described earlier (57)] because distinct chromosomal regions, and not average total genomic structure, may differ within and between cases and controls. Support for this explanation has been shown by the lack of replication of the association of *FTO* (fat mass and obesity associated) gene variants to adiposity (and therefore T2DM) in Han Chinese from Shanghai and Beijing (129) caused perhaps by differences in gene–gene interactions or, alternatively, differences in disease development in different ancestral groups (see later). New mapping strategies that are designed specifically for analyzing epistasis in association studies may provide a means to account for confounding of gene–gene interactions (45,48,130).

A significant limitation of these GWA studies is that they did not assess nutrient intakes even though diet–gene interactions are major contributors to the control of gene expression and could influence associations among genes—phenotype and dietary intakes (20,131–133). Differences in intake of calories, dietary fat, high glycemic index carbohydrates, and certain micronutrients are linked to T2DM (134–138) and other chronic diseases (139). We (132,140) and others proposed that certain dietary chemicals regulate pathways and genes involved in maintaining health or producing disease, although the specific gene–nutrient associations have not yet been identified. The development of this concept was based on the facts that different drugs target different pathways directly or indirectly involved in nutrient metabolism. For example, the peroxisome proliferator activated receptor gamma 2, a target of thiazolidinediones, is activated by the dietary lipids linoleic, linolenic, arachidonic, eicosapentaenoic acids (141,142), and their metabolites (5). Hence, measures of nutrient intakes would improve gene–phenotype association studies, as demonstrated by associations between certain nutrients and individual genes involved in cardiovascular disease (143–146).

While this discussion is specific to T2DM, it illustrates conceptually the issues that challenge the design of experiments for nutrigenomics, pharmacogenomics, toxicogenomics, and epigenomics. The methodological challenges can be addressed by (1) developing and using uniform nutrient intake measurements that can be compared within populations and among populations, (2) standardizing measures of relevant physiological parameters, and (3) testing for genetic ancestry within chromosomal segments, rather than overall genetic makeup. The international Nutrigenomics Society is undertaking the development of proposed standards (http://www.nugo.org/international).

Regardless of the detailed strategies for improving analyses of gene–disease, –nutrient, and –phenotype associations, these approaches yield the PAF (147–149). While these estimates are the best available for guiding preventive and treatment strategies, population-based risk fractions have limited utility for individuals: data derived from one ancestral group (e.g., European, African, or Asian) may not apply to individuals from other ancestral populations or to other individuals within the studied ancestral group because of uncharacterized epistatic or gene–environment interactions (150). Increasing the size or diversity in the study groups—for example, including African-Americans, Latino-Americans, and European Americans—may reduce the size of the biological response by averaging high and low responders, by gene–gene interactions (epistasis), and by gene–environment interactions. Hence, developing new methods to "personalize" research for generating recommendations for individuals remains a major hurdle for nutrigenomics and personal health care (150).

PREDICTING RISK THROUGH CLASSIFICATION ALGORITHMS

"Omic" technologies and better experimental designs will result in more reliable data sets that describe health and disease processes. A key challenge facing high-throughput technologies is discovering the functional relationships among elements in genomics, transcriptomics, proteomics, and metabolomics data sets (151,152). How are the various levels of quantitative information (DNA, RNA, protein, chromosome/gene methylation status, and metabolite) that define biological processes related to each other and how does diet alter their levels, relationships, interactions, and effects on biological processes? A further challenge will be associating varying nutrient intakes with these data sets, since practically each gene or physiology has some level of change of expression relative to a nutrient intake or treatment. Each gene–level measurement may have direct, indirect, or inverse correlative relationships to the levels of other gene transcripts, and the sum of these relationships defines the transcriptional response to the nutrient or other perturbation, such as the presence or absence of disease or complications of disease.

High-throughput "omic" data are characterized by a large number of variables with a relative small number of samples. In a typical microarray gene expression data set, the number of genes is in the tens of thousands but the number of samples rarely exceeds a few hundred and often less than one hundred. Many multivariate statistical methods and data-mining techniques have been applied to the analysis of high-dimensional "omic" data. The approaches are generally grouped into two types based on study objectives: supervised and unsupervised methods. Supervised methods involve developing classification algorithms to assign samples to a priori defined classes. Supervised methods are generally used to discriminate different biological phenotypes or predict treatment responses of patients. Unsupervised methods develop analytic algorithms, such as grouping samples on the basis of some measures of similarity, without using class membership information. Unsupervised analyses are descriptive in nature; they are used for exploring underlying biological structures or mechanisms, identifying disease process, or discovering new sample classes. Hierarchical clustering and k-means cluster analysis are two widely used unsupervised methods. These methods divide the samples into clusters; however, the interpretation of the clusters (or determining if the clusters are meaningful) or newly discovered classes may require further validation.

One key challenge in the analysis of high-throughput data is to extract information from the large number of the variables. Dimensionality reduction (DR) is a traditional technique developed to analyze the high-dimensional data by transforming the original high-dimensional data set into a lower dimensional coordinate system (usually two or three). The concept of DR is to discover simpler low-dimensional structures for high-dimensional data to help in identifying underlying biological structures or identifying relatedness of variables. Since the underlying dimensionality is not known, DR techniques must search for the number of coordinates (components) that can account for much of data variation. Principal component analysis and multidimensional scaling are the two most recognized DR methods. Principal components are the orthogonal linear combinations of variables that show the greatest variability of the data. Thus, DR techniques define a smaller number of hybrid components that are a composite of the original variables. These hybrid variables are chosen to provide independent information about different samples. Other nonlinear DR methods such as isomap (153) and locally embedding (154) are also been developed. DR techniques provide a vast reduction of dimensionality for identifying data structures and trends. The high-dimensional data are typically displayed graphically in a two- or three-dimensional space for easy visualization.

One major problem with the DR is that the interpretation of the components is often difficult. Other limitations on the use of DR for prediction are (1) that it requires measurements of all variables for future samples and (2) the components are not necessarily good predictors since DR generally does not use the class membership in constructing the components. The Partial-least-squares method, a DR method similar to the principal components, uses the class membership in constructing components; the partial-least-squares method has been applied to tumor classification in microarray gene expression data analysis (155).

A classification algorithm is developed to predict likelihood of disease of a new sample from the available data, including genomics, transcriptomics, proteomics, and metabolomics. In classification, typically, the original sample data set is divided into two subsets: a training set and a test set. The classification algorithm is built from the training samples and then its prediction rule is applied to the test samples to assess its performance. Classification development generally consists of two steps: (1) model building, including determining a prediction algorithm, selecting the predictor variables, and fitting the prediction model to training data and (2) assessment the performance of the classification algorithm.

High-dimensional data sets impose a challenge for model building, which is the selection of a subset of variables to improve predictability. In machine learning context, this is known as feature selection. One common characteristic of high-dimensional "omic" data is that although the number of variables is large, but many of them are not relevant to either unsupervised clustering or supervised classification. Data can be removed if they do not provide significant incremental information. If the measurement of a particular variable is the same in all samples, it will not be useful for distinguishing these samples. Keeping these data may confuse the analysis and make it unnecessarily complex in computations. On the other hand, if the measurements are very different over all samples, these variables may contain useful information to distinguish samples. Selection of an appropriate number of discriminatory variables from the original data set is critical to the accuracy of the classification. Feature selection procedures can be independent of the classification algorithms using some predetermined criterion such as statistical significance tests.

Classification algorithms, such as the Classification Tree (156) and support vector machines with recursive feature elimination (157,158), have incorporated feature selection into classification algorithms. These algorithms find a subset of predictors and evaluate its relevance for the classification; their classification rules are built from an optimal predictor subset. The goal of feature selection is to identify a minimum number of variables that are useful. The optimal set of variables may depend on a classification algorithm, the number selected, and the classification method. The optimal set can vary from data to data. No theoretical estimation of the optimal number of variables exists for a given specific classification algorithm on a particular application. In the development of classification model, the most important question is the ability of the model to predict a future sample. To ensure an unbiased assessment of accuracy, the prediction model is developed in the training data set and is applied to the test data set to estimate the predictive accuracy. An important feature of these algorithms is that different numbers of variables selected as part of the training set will yield different classification results.

Regardless of specific mathematical models, algorithms provide the fundamental basis to make personalized medicine a reality by enabling the assignment of therapies to patients to maximize efficacy and minimize toxicity or optimize the intake of nutrients for maintaining health of the mother, fetus, child, and adult. The data produced for such analyses must come from a systems biology approach that recognizes different developmental windows (6). While the relationships among nutritional and environmental exposures, genetic makeups including epigenetic differences, and developmental plasticity seem to be insurmountable, "omics" technologies, improved experimental design, and informed statistical analyses provide the tools to decipher gene–nutrient interactions that maintain health and prevent or delay the onset of chronic diseases.

REFERENCES

1. Seki Y, Hayashi K, Itoh K, et al. Extensive and orderly reprogramming of genome-wide chromatin modifications associated with specification and early development of germ cells in mice. Dev Biol 2005; 278(2):440–58.
2. Morgan HD, Santos F, Green K, et al. Epigenetic reprogramming in mammals. Hum Mol Genet 2005; 14 (Spec No 1):R47–R58.
3. Schnackenberg LK. Global metabolic profiling and its role in systems biology to advance personalized medicine in the 21st century. Expert Rev Mol Diagn 2007; 7(3):247–259.
4. Slikker WJ, Chang LW. Handbook of developmental neurotoxicology. Academic Press, New York, 1998; 748.
5. Kaput J, Perlina A, Hatipoglu B, et al. Nutrigenomics: Concepts and applications to pharmacogenomics and clinical medicine. Pharmacogenomics 2007; 8(4):369–390.
6. Slikker W Jr, Paule MG, Wright LK, et al. Systems biology approaches for toxicology. J Appl Toxicol 2007; 27(3):201–217.
7. Mathers JC. Early nutrition: Impact on epigenetics. Forum Nutr 2007; 60:42–48.
8. Burdge GC, Hanson MA, Slater-Jefferies JL, et al. Epigenetic regulation of transcription: A mechanism for inducing variations in phenotype (fetal programming) by differences in nutrition during early life? Br J Nutr 2007; 97(6)1036–1046.
9. McMillen IC, Robinson JS. Developmental origins of the metabolic syndrome: Prediction, plasticity, and programming. Physiol Rev 2005; 85(2):571–633.

10. Gallou-Kabani C, Junien C. Nutritional epigenomics of metabolic syndrome: New perspective against the epidemic. Diabetes 2005; 54(7):1899–906.

11. Jenkins DJA, Kendall CWC, Ransom TPP. Dietary fiber, the evolution of the human diet and coronary heart disease. Nutr Res 1998; 18(4):633–652.

12. Willett W. Isocaloric diets are of primary interest in experimental and epidemiological studies. Int J Epidemiol 2002; 31(3):694–695.

13. Cheynier V. Polyphenols in foods are more complex than often thought. Am J Clin Nutr 2005; 81(1 Suppl):223S–229S.

14. de Vos WM, Castenmiller JJ, Hamer RJ, et al. Nutridynamics—studying the dynamics of food components in products and in the consumer. Curr Opin Biotechnol 2006; 17(2):217–225.

15. Mezzenga R, Schurtenberger P, Burbidge A, et al. Understanding foods as soft materials. Nat Mater 2005; 4(10):729–740.

16. Regmi A, Gehlhar M. New directions in global food markets. USDA: Washington, DC, 2005; 81; http://www.ers.usda.gov/publications/aib794.pdf.

17. Malfattti MA, Felton JS. Susceptibility to exposure to heterocyclic amines from cooked food: Role of udp-glucuronosyltransferases. In Nutritional Genomics. Discovering the Path to Personalized Nutrition, Kaput J, Rodriguez RL, eds. John Wiley and Sons, Hoboken, NJ, 2006; 331–352.

18. Fenech M. The Genome Health Clinic and Genome Health Nutrigenomics concepts: Diagnosis and nutritional treatment of genome and epigenome damage on an individual basis. Mutagenesis 2005; 20(4):255–269.

19. Ordovas JM. The quest for cardiovascular health in the genomic era: Nutrigenetics and plasma lipoproteins. Proc Nutr Soc 2004; 63(1):145–152.

20. Kaput J, Rodriguez RL. Nutritional genomics: The next frontier in the postgenomic era. Physiol Genomics 2004; 16(2):166–177.

21. Rutishauser IH. Dietary intake measurements. Public Health Nutr 2005; 8(7A):1100–1107.

22. Arab L. Individualized nutritional recommendations: Do we have the measurements needed to assess risk and make dietary recommendations? Proc Nutr Soc 2004; 63(1):167–172.

23. The International HapMap Consortium. A haplotype map of the human genome. Nature 2005; 437(7063):1299–1320.

24. Frazer KA, Ballinger DG, Cox DR, et al. A second generation human haplotype map of over 3.1 million SNPs. Nature 2007; 449(7164):851–861.

25. Jorde LB, Wooding SP. Genetic variation, classification and "race." Nat Genet 36(Suppl 1):S28–S33.

26. Hinds DA, Stuve LL, Nilsen GB, et al. Whole-genome patterns of common DNA variation in three human populations. Science 2005; 307(5712):1072–1079.

27. Enattah NS, Sahi T, Savilahti E, et al. Identification of a variant associated with adult-type hypolactasia. Nat Genet 2002; 30(2):233–237.

28. Tishkoff SA, Reed FA, Ranciaro A, et al. Convergent adaptation of human lactase persistence in Africa and Europe. Nat Genet 2007; 39(1):31–40.

29. Sabeti PC, Varilly P, Fry B, et al. Genome-wide detection and characterization of positive selection in human populations. Nature 2007; 449(7164):913–918.

30. Burchard EG, Avila PC, Nazario S, et al. Lower bronchodilator responsiveness in Puerto Rican than in Mexican subjects with asthma. Am J Respir Crit Care Med 2004; 169(3):386–392.

31. Rosskopf D, Manthey I, Siffert W. Identification and ethnic distribution of major haplotypes in the gene GNB3 encoding the G-protein beta3 subunit. Pharmacogenetics 2002; 12(3):209–220.

32. Xie HG, Kim RB, Wood AJ, et al. Molecular basis of ethnic differences in drug disposition and response. Annu Rev Pharmacol Toxicol 2001; 41:815–850.

33. Suarez-Kurtz G. Pharmacogenomics in admixed populations. Trends Pharmacol Sci 2005; 26(4):196–201.

34. Nielsen R, Hellmann I, Hubisz M, et al. Recent and ongoing selection in the human genome. Nat Rev Genet 2007; 8(11):857–868.

35. Mitchell-Olds T, Willis JH, Goldstein DB. Which evolutionary processes influence natural genetic variation for phenotypic traits? Nat Rev Genet 2007; 8(11):845–856.

36. Jiang W, Yu G, Liu P, et al. Structure and function of glucose-6-phosphate dehydrogenase-deficient variants in Chinese population. Hum Genet 2006; 119(5)463–478.

37. Kwiatkowski DP. How malaria has affected the human genome and what human genetics can teach us about malaria. Am J Hum Genet 2005; 77(2):171–192.

38. Ames BN, Suh JH, Liu J. Enzymes lose binding affinity (increased km) for coenzymes and substrates with age: A strategy for remediation. In Nutritional Genomics. Discovering the Path to Personalized Nutrition, Kaput J, Rodriguez RL, eds. John Wiley and Sons, Hoboken, NJ, 2006; 277–294.

39. Feuk L, Carson AR, Scherer SW. Structural variation in the human genome. Nat Rev Genet 2006; 7(2):85–97.

40. Perry GH, Dominy NJ, Claw KG, et al. Diet and the evolution of human amylase gene copy number variation. Nat Genet 2007; 39(10):1256–1260.

41. Novembre J, Pritchard JK, Coop G. Adaptive drool in the gene pool. Nat Genet 2007; 39(10)1188–1190.

42. Hartman JL, Garvik B, Hartwell L. Principles for the buffering of genetic variation. Science 2001; 291(5506):1001–1004.

43. Carlborg O, Haley CS. Epistasis: Too often neglected in complex trait studies? Nat Rev Genet 2004; 5(8):618–625.

44. Moore JH. The ubiquitous nature of epistasis in determining susceptibility to common human diseases. Hum Hered 2003; 56(1–3)73–82.

45. Togawa K, Moritani M, Yaguchi H, et al. Multidimensional genome scans identify the combinations of genetic loci linked to diabetes-related phenotypes in mice. Hum Mol Genet 2006; 15(1):113–128.

46. Yang RC. Epistasis of quantitative trait loci under different gene action models. Genetics 2004; 167(3):1493–1505.

47. Cheverud JM, Vaughn TT, Pletscher LS, et al. Genetic architecture of adiposity in the cross of LG/J and SM/J inbred mice. Mamm Genome 2001; 12(1):3–12.

48. Chiu S, Diament AL, Fisler JS, et al. Gene-gene epistasis and gene environment interactions influence diabetes and obesity. In Nutritional Genomics. Discovering the Path to Personalized Nutrition, Kaput J, Rodriguez RL, eds. John Wiley and Sons, Hoboken, NJ, 2006; 135–152.

49. Scriver CR. Nutrient-gene interactions: The gene is not the disease and vice versa. Am J Clin Nutr 1988; 48(6):1505–1509.

50. Scriver CR. The PAH gene, phenylketonuria, and a paradigm shift. Hum Mutat 2007; 28(9):831–845.

51. Baessler A, Fischer M, Mayer B, et al. Epistatic interaction between haplotypes of the ghrelin ligand and receptor genes influence susceptibility to myocardial infarction and coronary artery disease. Hum Mol Genet 2007; 16(8):887–899.

52. Klos KLE, Kardia SLR, Hixson JE, et al. Linkage analysis of plasma apoe in three ethnic groups: Multiple genes with context-dependent effects. Ann Hum Genet 2005; 69(2):157–167.

53. Suarez-Kurtz G, Pena SD. Pharmacogenomics in the Americas: The impact of genetic admixture. Curr Drug Targets 2006; 7(12):1649–1658.

54. Helgadottir A, Manolescu A, Helgason A, et al. A variant of the gene encoding leukotriene A4 hydrolase confers ethnicity-specific risk of myocardial infarction. Nat Genet 2006; 38(1):68–74.

55. Kelley DS. Modulation of human immune and inflammatory responses by dietary fatty acids. Nutrition 2001; 17(7–8):669–673.

56. Helgason A, Yngvadottir B, Hrafnkelsson B, et al. An icelandic example of the impact of population structure on association studies. Nat Genet 2005;37(1):90–95.

57. Bauchet M, McEvoy B, Pearson LN, et al. Measuring European population stratification with microarray genotype data. Am J Hum Genet 2007; 80(5):948–956.

58. Tsai HJ, Kho JY, Shaikh N, et al. Admixture-matched case-control study: A practical approach for genetic association studies in admixed populations. Hum Genet 2006; 118(5)626–639.

59. Tang H, Coram M, Wang P, et al. Reconstructing genetic ancestry blocks in admixed individuals. Am J Hum Genet 2006; 79(1):1–12.

60. Choudhry S, Coyle NE, Tang H, et al. Population stratification confounds genetic association studies among Latinos. Hum Genet 2006; 118(5):652–64.

61. Yang N, Li H, Criswell LA, et al. Examination of ancestry and ethnic affiliation using highly informative diallelic DNA markers: Application to diverse and admixed populations and implications for clinical epidemiology and forensic medicine. Hum Genet 2005; 118(3–4):382–392.

62. Tsai HJ, Choudhry S, Naqvi M, et al. Comparison of three methods to estimate genetic ancestry and control for stratification in genetic association studies among admixed populations. Hum Genet 2005; 118(3–4):424–433.

63. Fowler AM, Alarid ET. Dynamic control of nuclear receptor transcription. Sci STKE 2004, (256):pe51.

64. Jiang YH, Bressler J, Beaudet AL. Epigenetics and human disease. Annu Rev Genomics Hum Genet 2004; 5:479–510.

65. Delaval K, Feil R. Epigenetic regulation of mammalian genomic imprinting. Curr Opin Genet Dev 2004; 14(2):188–195.

66. Jirtle RL, Skinner MK. Environmental epigenomics and disease susceptibility. Nat Rev Genet 2007; 8(4):253–262.

67. Esteller M. Cancer epigenomics: DNA methylomes and histone-modification maps. Nat Rev Genet 2007;8(4):286–298.

68. Dolinoy DC, Das R, Weidman JR, et al. Metastable epialleles, imprinting, and the fetal origins of adult diseases. Pediatr Res 2007; 61(5 pt2):30R–37R.

69. Cooney CA. Maternal nutrition: Nutrients and control of expression. In Nutritional Genomics. Discovering the Path to Personalized Nutrition, Kaput J, Rodriguez RL, eds. John Wiley and Sons, Hoboken, NJ, 2006; 219–254.

70. Dolinoy DC, Weidman JR, Waterland RA, et al. Maternal genistein alters coat color and protects Avy mouse offspring from obesity by modifying the fetal epigenome. Environ Health Perspect 2006; 114(4):567–572.

71. Picard F, Kurtev M, Chung N, et al. Sirt1 promotes fat mobilization in white adipocytes by repressing PPAR-gamma. Nature 2004; 429(6993):771–776.

72. Painter RC, Roseboom TJ, Bleker OP. Prenatal exposure to the Dutch famine and disease in later life: An overview. Reprod Toxicol 2005; 20(3):345–352.

73. Stein AD, Lumey LH. The relationship between maternal and offspring birth weights after maternal prenatal famine exposure: The Dutch Famine Birth Cohort Study. Hum Biol 2000; 72(4):641–654.

74. Slikker W Jr, Schwetz B. Childhood obesity: The possible role of maternal smoking and impact on public health. J Children's Health 2003; 1:29–40.

75. Waterland RA, Jirtle RL. Transposable elements: Targets for early nutritional effects on epigenetic gene regulation. Mol Cell Biol 2003; 23(15):5293–5300.

76. Cooney CA, Dave AA, Wolff GL. Maternal methyl supplements in mice affect epigenetic variation and DNA methylation of offspring. J Nutr 2002; 132(8 Suppl):2393–2400S.

77. Wolff GL, Kodell RL, Moore SR, et al. Maternal epigenetics and methyl supplements affect agouti gene expression in A^{vy}/a mice. FASEB J 1998; 12(11):949–957.

78. Esteller M. CpG island hypermethylation and tumor suppressor genes: A booming present, a brighter future. Oncogene 2002; 21(35):5427–5440.

79. Bishop NA, Guarente L. Genetic links between diet and lifespan: Shared mechanisms from yeast to humans. Nat Rev Genet 2007; 8(11):835–844.

80. Russo GL. Ins and outs of dietary phytochemicals in cancer chemoprevention. Biochem Pharmacol. 2007; 74(4):533–544.

81. Lagouge M, Argmann C, Gerhart-Hines Z, et al. Resveratrol improves mitochondrial function and protects against metabolic disease by activating SIRT1 and PGC-1alpha. Cell 2006; 127(6):1109–1122.

82. Baur JA, Pearson KJ, Price NL, et al. Resveratrol improves health and survival of mice on a high-calorie diet. Nature 2006; 444(7117):337–342.

83. Young LF, Hantz HL, Martin KR. Resveratrol modulates gene expression associated with apoptosis, proliferation and cell cycle in cells with mutated human c-Ha-Ras, but does not alter c-Ha-Ras mRNA or protein expression. J Nutr Biochem 2005; 16(11):663–674.

84. Lin S-J. Molecular mechanisms of longevity regulation and calorie restriction. In Nutritional Genomics. Discovering the Path to Personalized Nutrition, Kaput J, Rodriguez RL, eds. John Wiley and Sons, Hoboken, NJ, 2006; 207–218.

85. Ito T. Role of histone modification in chromatin dynamics. J Biochem (Tokyo) 2007; 141(5):609–614.

86. Richards EJ. Inherited epigenetic variation—revisiting soft inheritance. Nat Rev Genet 2006; 7(5):395–401.

87. Cropley JE, Suter CM, Beckman KB, et al. Germ-line epigenetic modification of the murine A vy allele by nutritional supplementation. Proc Natl Acad Sci U S A 2006; 103(46):17308–17312.

88. Pembrey ME, Bygren LO, Kaati G, et al. Sex-specific, male-line transgenerational responses in humans. Eur J Hum Genet 2006; 14(2):159–166.

89. Waterland RA, Travisano M, Tahiliani KG. Diet-induced hypermethylation at agouti viable yellow is not inherited transgenerationally through the female. FASEB J 2007; 21(12):3380–3385.

90. Bateson P, Barker D, Clutton-Brock T, et al. Developmental plasticity and human health. Nature 2004; 430(6998):419–421.

91.   Gluckman PD, Lillycrop KA, Vickers MH, et al. Metabolic plasticity during mammalian development is directionally dependent on early nutritional status. Proc Natl Acad Sci U S A 2007; 104(31):12796–12800.

92.   Jackson AA. Nutrients, growth, and the development of programmed metabolic function. Adv Exp Med Biol 2000; 478:41–55.

93.   Lucas A. Programming by early nutrition: An experimental approach. J Nutr 1998; 128(2 Suppl):401S–406S.

94.   DeRisi JL, Iyer VR, Brown PO. Exploring the metabolic and genetic control of gene expression on a genomic scale. Science 1997; 278(5338):680–686.

95.   King H, Aubert RE, Herman WH. Global burden of diabetes, 1995–2025: Prevalence, numerical estimates, and projections. Diabetes Care 1998; 21(9):1414–1431.

96.   Polifka JE, Friedman JM. Medical genetics: 1. Clinical teratology in the age of genomics. CMAJ 2002; 167(3):265–273.

97.   Menegola E, Broccia ML, Di Renzo F, et al. Postulated pathogenic pathway in triazole fungicide induced dysmorphogenic effects. Reprod Toxicol 2006; 22(2):186–195.

98.   Dolk H, Vrijheid M. The impact of environmental pollution on congenital anomalies. Br Med Bull 2003; 68:25–45.

99.   Huwe JK. Dioxins in food: A modern agricultural perspective. J Agric Food Chem 2002; 50(7):1739–1750.

100. Ritchie G, Still K, Rossi J III, et al. Biological and health effects of exposure to kerosene-based jet fuels and performance additives. J Toxicol Environ Health B Crit Rev 2003; 6(4):357–451.

101. Jelinek R. The contribution of new findings and ideas to the old principles of teratology. Reprod Toxicol 2005; 20(3):295–300.

102. Szyf M. The dynamic epigenome and its implications in toxicology. Toxicol Sci 2007; 100(1):7–23.

103. Zeisel SH. Choline: Critical role during fetal development and dietary requirements in adults. Annu Rev Nutr 2006; 26:229–250.

104. McCann JC, Ames BN. Is docosahexaenoic acid, an n-3 long-chain polyunsaturated fatty acid, required for development of normal brain function? An overview of evidence from cognitive and behavioral tests in humans and animals. Am J Clin Nutr 2005; 82(2):281–295.

105. Seed J, Carney EW, Corley RA, et al. Overview: Using mode of action and life stage information to evaluate the human relevance of animal toxicity data. Crit Rev Toxicol 2005; 35(8–9):664–672.

106. Collins TF. History and evolution of reproductive and developmental toxicology guidelines. Curr Pharm Des 2006; 12(12):1449–1465.

107. Kaput J, Noble J, Hatipoglu B, et al. Application of nutrigenomic concepts to type 2 diabetes mellitus. Nutr Metab Cardiovasc Dis 2007; 17(2):89–103.

108. Rosenzweig JL, Weinger K, Poirier-Solomon L, et al. Use of a disease severity index for evaluation of healthcare costs and management of comorbidities of patients with diabetes mellitus. Am J Manag Care 2002; 8(11):950–958.

109. Koro CE, Bowlin SJ, Bourgeois N, et al. Glycemic control from 1988 to 2000 among U.S. adults diagnosed with type 2 diabetes: A preliminary report. Diabetes Care 2004; 27(1):17–20.

110. Krentz AJ, Bailey CJ. Oral antidiabetic agents: Current role in type 2 diabetes mellitus. Drugs 2005; 65(3):385–411.

111. McCormick M, Quinn L. Treatment of type 2 diabetes mellitus: Pharmacologic intervention. J Cardiovasc Nurs 2002; 16(2):55–67.

112. Babenko AP, Polak M, Cave H, et al. Activating mutations in the ABCC8 gene in neonatal diabetes mellitus. N Engl J Med 2006; 355(5):456–466.

113. Zeggini E, Weedon MN, Lindgren CM, et al. Replication of genome-wide association signals in UK samples reveals risk loci for type 2 diabetes. Science 2007; 316(5829):1336–1341.

114. Sladek R, Rocheleau G, Rung J, et al. A genome-wide association study identifies novel risk loci for type 2 diabetes. Nature 2007; 445(7130):881–885.

115. Scott LJ, Mohlke KL, Bonnycastle LL, et al. A genome-wide association study of type 2 diabetes in Finns detects multiple susceptibility variants. Science 2007; 316(5829):1341–1345.

116. Saxena R, Voight BF, Lyssenko V, et al. Genome-wide association analysis identifies loci for type 2 diabetes and triglyceride levels. Science 2007; 316(5829):1331–1336.

117. Sandhu MS, Weedon MN, Fawcett KA, et al. Common variants in WFS1 confer risk of type 2 diabetes. Nat Genet 2007; 39(8):951–953.

118. Frayling TM, Timpson NJ, Weedon MN, et al. A common variant in the FTO gene is associated with body mass index and predisposes to childhood and adult obesity. Science 2007; 316(5826):889–894.

119. The Welcome Trust Consortium. Genome-wide association study of 14,000 cases of seven common diseases and 3,000 shared controls. Nature 2007; 447(7145):661–678.

120. Florez JC, Hirschhorn J, Altshuler D. The inherited basis of diabetes mellitus: Implications for the genetic analysis of complex traits. Annu Rev Genomics Hum Genet 2003; 4:257–291.

121. Hirschhorn JN, Lohmueller K, Byrne E, et al. A comprehensive review of genetic association studies. Genet Med 2002; 4(2):45–61.

122. Newton-Cheh C, Hirschhorn JN. Genetic association studies of complex traits: Design and analysis issues. Mutat Res 2005; 573(1–2):54–69.

123. Ioannidis JP. Why most published research findings are false. PLoS Med 2005; 2(8):e124.

124. Cardon LR, Bell JI. Association study designs for complex diseases. Nat Rev Genet 2001; 2(2):91–99.

125. Lander E, Kruglyak L. Genetic dissection of complex traits: Guidelines for interpreting and reporting linkage results [see comments]. Nat Genet 1995; 11(3):241–247.

126. Tabor HK, Risch NJ, Myers RM. OPINION: Candidate-gene approaches for studying complex genetic traits: Practical considerations. Nat Rev Genet 2002; 3(5):391–397.

127. Risch N. Evolving methods in genetic epidemiology. II. Genetic linkage from an epidemiologic perspective. Epidemiol Rev 1997; 19(1):24–32.

128. Devlin B, Roeder K. Genomic control for association studies. Biometrics 1999; 55(4):997–1004.

129. Li H, Wu Y, Loos RJ, et al. Variants in FTO gene are not associated with obesity in a Chinese Han population. Diabetes 2008; 57(1):264–268.

130. Motsinger AA, Ritchie MD. Multifactor dimensionality reduction: An analysis strategy for modelling and detecting gene-gene interactions in human genetics and pharmacogenomics studies. Hum Genomics 2006; 2(5):318–328.

131. Ordovas JM, Corella D. Nutritional genomics. Annu Rev Genomics Hum Genet 2004; 5:71–118.

132. Kaput J. Diet-disease gene interactions. Nutrition 2004; 20(1):26–31.
133. Ordovas JM, Corella D. Gene-environment interactions: Defining the playfield. In Nutritional Genomics. Discovering the Path to Personalized Nutrition, Kaput J, Rodriguez RL, eds. John Wiley and Sons, Hoboken, NJ, 2006; 57–76.
134. Nestel P. Nutritional aspects in the causation and management of the metabolic syndrome. Endocrinol Metab Clin North Am 2004; 33(3):483–492.
135. Bonnefont-Rousselot D. The role of antioxidant micronutrients in the prevention of diabetic complications. Treat Endocrinol 2004; 3(1):41–52.
136. Biesalski HK. Diabetes preventive components in the Mediterranean diet. Eur J Nutr 2004; 43(Suppl 1):I/26–30.
137. Neff LM. Evidence-based dietary recommendations for patients with type 2 diabetes mellitus. Nutr Clin Care 2003; 6(2):51–61.
138. Hung T, Sievenpiper JL, Marchie A, et al. Fat versus carbohydrate in insulin resistance, obesity, diabetes and cardiovascular disease. Curr Opin Clin Nutr Metab Care 2003; 6(2):165–176.
139. Wahlqvist ML. Dietary fat and the prevention of chronic disease. Asia Pac J Clin Nutr 2005; 14(4):313–318.
140. Kaput J, Swartz D, Paisley E, et al. Diet-disease interactions at the molecular level: An experimental paradigm. J Nutr 1994; 124(8 Suppl):1296S–1305S.
141. Chambrier C, Bastard JP, Rieusset J, et al. Eicosapentaenoic acid induces mRNA expression of peroxisome proliferator-activated receptor gamma. Obes Res 2002; 10(6):518–525.
142. Nosjean O, Boutin JA. Natural ligands of PPAR gamma: Are prostaglandin J(2) derivatives really playing the part? Cell Signal 2002; 14(7)573–583.
143. Corella D, Lai CQ, Demissie S, et al. APOA5 gene variation modulates the effects of dietary fat intake on body mass index and obesity risk in the Framingham Heart Study. J Mol Med 2007; 85(2):119–128.
144. Lai CQ, Corella D, Demissie S, et al. Dietary intake of n-6 fatty acids modulates effect of apolipoprotein A5 gene on plasma fasting triglycerides, remnant lipoprotein concentrations, and lipoprotein particle size: The Framingham Heart Study. Circulation 2006; 113(17):2062–2070.
145. Corella D, Qi L, Tai ES, et al. Perilipin gene variation determines higher susceptibility to insulin resistance in Asian women when consuming a high-saturated fat, low-carbohydrate diet. Diabetes Care 2006; 29(6):1313–1319.
146. Tai ES, Corella D, Demissie S, et al. Polyunsaturated fatty acids interact with the pparα-1162v polymorphism to affect plasma triglyceride and apolipoprotein c-iii concentrations in the framingham heart study. J Nutr 2005; 135(3):397–403.
147. Rockhill B, Newman B, Weinberg C. Use and misuse of population attributable fractions. Am J Public Health 1998; 88(1):15–19.
148. Levine B. What does the population attributable fraction mean? Prev Chronic Dis 2007; 4(1):A14.
149. Karp I, Topol E, Pilote L. Population attributable fraction: Its implications for genetic epidemiology and illness prevention. Am Heart J 2007; 154(4):607–609.
150. Kaput J. Nutrigenomics and genetics research for personalized nutrition and medicine Curr Opin Biotechnol, 2008; 19(2): 110–120.
151. Dawson K, Rodrigueez RL, Hawkes WC, et al. Biocomputation and the analysis of complex data sets in nutritional genomics. In Nutritional Genomics. Discovering the

Path to Personalized Nutrition, Kaput J, Rodriguez RL, eds. John Wiley and Sons, Hoboken, NJ, 2006; 375–402.

152. Sebastiani P, Gussoni E, Kohan IS, et al. MF statistical challenges in functional genomics. Statistical Science 2003; 18:33–70.

153. Tenenbaum JB, de Silva V, Langford JC. A global geometric framework for nonlinear dimensionality reduction. Science 2000; 290(5500):2319–2323.

154. Roweis ST, Saul LK. Nonlinear dimensionality reduction by locally linear embedding. Science 2000; 290(5500):2323–2326.

155. Nguyen DV, Rocke DM. Multi-class cancer classification via partial least squares with gene expression profiles. Bioinformatics 2002; 18(9):1216–1226.

156. Breiman L, Friedman J, Olshen RA, et al. CART: Classification and regression trees, CRC Press, Boca Raton, FL. 1995.

157. Guyon I, Weston J, Barnhill S, et al. Gene selection for cancer classification using support vector machines. Mach Learn 2002; 46:389–422.

158. Vapnik V. Statistical learning theory. Wiley and Sons, New York, NY, 1998.

159. Salonen JT, Uimari P, Aalto JM, et al. Type 2 diabetes whole-genome association study in four populations: The DiaGen consortium. Am J Hum Genet 2007; 81(2):338–345.

6

Targeted Gene Changes Affecting Developmental Toxicity

Sid Hunter and Phillip Hartig

Reproductive Toxicology Division, National Health and Environmental Effects Research Laboratory, U.S. Environmental Protection Agency, Research Triangle Park, North Carolina, U.S.A.

Developmental biologists have used a variety of tools to modulate gene expression in order to evaluate gene function and phenotypic consequences of altered expression during embryogenesis. As with the large number of tools used in these studies, the models are similarly diverse and range from *C. elegans*, zebra fish, and mice to human cells. This chapter will not cover the full range of tools and models, but will focus on studies and approaches that use transgenic rodent models, antisense oligonucleotides, RNA interference, and viral vectors to better understand structural birth defects induced by xenobiotics during mammalian development.

TRANSGENIC MODELS

The pioneering work of researchers such as Smithies, Capecchi, and Evans resulted in the development of the techniques needed for the genesis of transgenic mammalian models. Their research established the tools used to produce selective and specific gene mutation and the creation of offspring expressing the desired molecular alteration. The techniques for creating transgenic animals have undergone an evolution from the use of homologous recombination to modify a specific DNA sequence to a current use of time- and tissue-selective gene deletions using cre recombinase-loxP constructs and selective promoter-driven gene and reporter/marker gene expression. Public commercial sources and University core facilities specialize in generating transgenic animals and many reviews of transgenic techniques and protocols are available (such as Cold Spring Harbor Protocols http://www.cshprotocols.org/Taxonomy/transgenic_technology_11.dtl). Therefore, these subjects will not be discussed here. With regards to this chapter, it is important to note that all knockout techniques alter the DNA sequence for a selected gene resulting in the formation of an aberrant mRNA and no or a nonfunctional protein produced during translation. Similarly, knock-in techniques insert a sequence into the DNA that uses a promoter to express a selected gene. The promoters used in knock-in animals may be constantly driving mRNA production, or may mediate time and tissue specificity and selectivity. Transgenic animal models have been established and can be used to eliminate or selectively express specific gene products.

The published literature on the role of genes during embryonic development includes thousands of publications using knockout models. Transgenic knockout models of development cover almost the entire animal kingdom and many plant models as well. However, for purposes of this chapter, we will focus on studies evaluating the effects of xenobiotic-induced birth defects in transgenic mice. Transgenic model studies range from those that employ reporter/marker constructs to evaluate gene expression following xenobiotic exposure, to those evaluating the genotype-dependent response to stressors to determine the role of a specific gene in the toxic/dysmorphogenic response.

A specific use of transgenic mice has been to evaluate gene expression in reporter construct mice. For example, several studies have used retinoic acid receptor (RAR) promotor–reporter gene knock-in mice to determine the spatial distribution of retinoid-dependent transcriptional activation in order to assess the distribution of retinoids (1–6). These studies have shown that at teratogenic dose levels of retinoic acid there are ectopic and aberrant expressions of the reporter/marker. These studies indicate that retinoic acid was present in incorrect regions of the embryo, and at concentrations capable of altering gene expression. Using a lacZ-p53-binding reporter construct, Komarova (7) reported a robust activation of p53 in the neural tube and visceral arches following irradiation on gestation day (GD) (8). Another example of using a reporter construct to address an issue of developmental toxicity is the research of Wells' laboratory (8) that evaluated NFkB activity following exposure to phenytoin. In embryos grown in whole embryo culture, phenytoin exposure produced a time- and concentration-dependent ectopic NFkB activation. In another series of studies, Willey created a transgenic mouse using two dioxin-response element promoters and a lacZ reporter construct (9). Administration of TCDD on GD 13 to the transgenic animals produced a transcriptional response in the genital tubercle, primary and secondary palates, and other regions of the embryos. Thus, using a knock-in approach to create marker/reporter transgenic animals produces models useful for evaluating effects of xenobiotic exposure.

The predominant use of transgenic mice in developmental toxicology has been to evaluate the role(s) of genes and pathways in mediating teratogenesis.

There are many examples of this type of study that cover a full range of genes and chemical/stressor exposures. One group of studies focused on understanding the role of the aryl hydrocarbon receptor (AhR), epidermal growth factor (EGF), and transforming growth factor alpha (TGF alpha) in the toxicity of TCDD (10–15). These studies show that the majority of TCDD-induced teratogenic effects are dependent upon activation of the AhR. However, following TCDD exposure of AhR nullizygous $(-/-)$ fetuses, there was a low incidence of cleft palate, a trend toward a decreased body weight and a higher percentage of resorptions than observed in wild-type animals. These observations raise important questions of AhR-independent effects contributing to the developmental toxicity of TCDD. Additional studies from Abbott's group focused on the growth factor contribution to dysmorphogenesis, specifically comparing the induction of cleft palate and hydronephrosis produced by TCDD in transgenic mice that lack EGF, TGF alpha or a double knockout of these genes. This group demonstrated that EGF was a critical component of TCDD-mediated cleft palate, that is, the knockout was less sensitive to TCDD. Using a palate culture model, palates from EGF knockout mice were also resistant to the effects of TCDD. However, when EGF was added to the defined culture medium, the adverse effects of TCDD were the same as in palates from wild-type animals. Thus, using knockout animals, the authors were able to identify AhR-dependent and -independent processes and the role of EGF signaling in the etiology of malformations.

In another series of studies, Wells's laboratory has evaluated the roles of p53, ataxia telangiectasia mutated (Atm), and inducible nitric oxide synthase (iNOS) in mediating the toxic effects of benzo[a]pyrene and phenytoin. An excellent review of these studies is available (16). Administration of benzo[a]pyrene (BaP) or phenytoin produces an increase in adverse developmental effects in p53 nullizygous genotype $(-/-)$ fetuses compared to the wild genotype $(+/+)$ (17–19). Bhuller (20,21) reported that embryos lacking the *Atm* gene $(-/-)$ are more sensitive to phenytoin than wild-type embryos. Additionally, iNOS-deficient mice were used to evaluate the role of nitric oxide induction in the toxicity of benzo[a]pyrene and phenytoin. Deficient mice were partially protected from the adverse effects of these toxicants (22). Together these studies describe a reactive oxygen species (ROS) induced Atm-/p53-dependent cellular process associated with the toxicity of these compounds.

Another examples of transgenic mice being used to understand the toxicity of xenobiotics are those studies that use RAR and retinoid X receptor (RXR) knockout mice to understand the mechanisms responsible for retinoid teratogenicity (23–25). The roles of specific RARs and RXRs in development have been extensively studied using knockout mice [see (26,27) for reviews]. In studies of the teratogenicity of retinoic acid, knockout and wild-type mice were treated with retinoids during gestation and the morphology of the fetuses evaluated in order to determine if a specific receptor mediated the developmental toxicity. These studies revealed a critical role for RXR-alpha in limb defects (25) and a reduction in palatal fusion defects in RXR-alpha heterozygous $(+/-)$ fetuses (28) following administration of retinoic acid on GD 11.5. In contrast, there was no difference in the incidence of defects induced by retinoic acid administration on GD 8.5 or 11.5 in fetuses lacking all of the RAR-beta receptors (29) compared to wild-type fetuses. However, when RAR-beta knockout embryos were exposed to a pan-RAR agonist in whole embryo culture (GD 8.0; 2–4 somite stage), there was a decrease in the incidence of branchial arch defects compared to wild-type mice, indicating that RAR-betas do play a role in mediating retinoid-induced defects (30). There was a dramatic reduction in the defects produced by retinoic acid in RAR-gamma-deficient embryos (31). Following administration on GD 8.5, the caudal effects typically produced by retinoic acid were not observed in the knock-out embryos. This included reductions in spina bifida, degenerate or fused ribs, disorganized vertebral centers, and malformed neural arches in the lower thoracic-lumbar-sacral region. However, there was no reduction in craniofacial defects with this dosing regime. Interestingly, when retinoic acid was administered on GD 7.3, RAR-gamma knockout embryos were resistant to the embryolethality, craniofacial malformations, and neural tube closure defects observed in wild-type embryos (32). These studies have been essential in identifying unique roles for RARs and RXRs in mediating the teratogenic effects of retinoids. They have also elucidated temporal patterns in the biological redundancies within this family of receptors as exemplified by the resistance to retinoid-induced craniofacial defects during gastrulation but not neurulation in the RAR-gamma knockout mice.

Transformation related protein 53 (p53) is a transcription factor (and tumor suppressor gene) that responds to many stressors, such as DNA damage and oxidative stress, in adult tissues. It has been proposed that xenobiotic/stress-induced p53 activation is a central mediator of alterations in cellular proliferation and induction of cell death resulting in birth defects. To test this hypothesis, loss-of-function p53 mutant mice have been used to determine if the teratogenic effects of xenobiotics are more or less severe in the absence of functional p53 than in wild-type mice. In these studies, if the effects are less severe in the nullizygous ($-/-$) animal than in the wild type ($+/+$), this is evidence that p53 is functionally and causally linked to teratogen-induced birth defects. The effects of xenobiotics, radiation, and heat have been evaluated in this model system.

To evaluate the role of p53 in 2-chloro-2'-deoxyadenosine (2CdA-) induced eye defects, Wubah and coworkers administered 2CdA on GD 8 and evaluated the induction of cell death and optic maldevelopment, especially lens agenesis (33). p53 induction was observed as early as 3 hours after exposure and continued to 4.5 hours. In this model, p53-dependent terminal deoxynucleotidyl transferase dUTP nick end labeling (TUNEL) positive and nuclear p53 positive and p53-independent apoptoses (TUNEL positive and no p53 nuclear staining) were observed in all genotypes of embryos. The incidence of malformations and extent of cell death in the cranial head folds were highest in wild-type ($+/+$), intermediate in heterozygous ($+/-$), and lowest in nullizygous ($-/-$) offspring. With regard to understanding 2CdA-induced malformations, these studies indicate that the genotoxic stress induced by 2CdA produced a p53-dependent-induction of cell death and malformations. Additionally, these studies document the simultaneous activity of both gene-dependent and gene-independent events that can only be discerned using the knockout animals.

The role of p53 was evaluated in the induction of limb defects by 4-hydroperoxycyclophosphamide (4HPC), an active form of cyclophosphamide (34). In this study, limb buds were exposed to 4HPC in culture and evaluated after 6 days. Although limb buds from each p53 genotype grew comparably in control medium, p53 $-/-$ limbs were more sensitive to 4HPC than p53 $+/+$ limbs. Additionally, at the highest concentration used (3 μg/ml), the pattern and type of cell death observed in p53 $+/+$ limbs was diagnostic of apoptosis and localized predominately in the interdigital region. In p53 $-/-$ limbs, cell death was observed throughout the limb and had the morphological characteristics of necrosis. These studies clearly showed that chemical-induced p53-independent pathways could result in necrotic cell death and produce maldevelopment. Since exposure to 4HPC could result in alkylation of a large number of cellular targets, it is difficult to know if the differential sensitivity of wild-type and nullizygous animals is specific to a mechanism of toxicity or the 4HPC molecule.

To better understand the mechanisms by which p53 modulates cyclophosphamide-induced developmental toxicity, Pekar et al. (35) exposed mice to cyclophosphamide on GD 12 and compared the p53 genotypic consequences. In $+/+$ mice, there was an activation of caspases 3, 8, and 9; an inhibition of NFkB DNA binding coupled with a higher incidence of apoptosis; and alterations in the cell cycle and malformations. In this model, apoptosis effectors were not activated in the absence of p53 and there was a lower incidence of malformations (eye, limb, and tail) compared to the wild-type genotype animals. These studies established a role for p53-mediated activation of the cell death cascade associated with cyclophosphamide-induced developmental effects. Differences in this study's results and those of 4HPC (34) may be due to differences in the concentration of active metabolite reaching the target tissues.

The role of p53 in the induction of tail malformations produced by hyperthermia has also been evaluated (36). In these experiments, heterozygous male and female mice were mated producing the full complement of wild-type, heterozygous, and nullizygous offspring. On GD 10 or 11, pregnant female mice were heated to a core body temperature of 40.5°C for 60 minutes. When evaluated on GD 18, fetuses from mothers heated on GD 10 had shorter tails than fetuses from those mice not treated. In contrast, when mice were heated on GD 11, the tails were longer following hyperthermia than in nonheated controls. On both treatment days, there was no influence of p53 genotype on the phenotypic outcome, indicating that a p53-dependent process did not mediate these effects.

There are many studies of the role of p53 in radiation-induced birth defects. Using a lacZ-p53-binding reporter construct, Komarova (7) reported a robust activation of p53 by irradiation (5 Gy) on GD 8–9. The neural tube and visceral arches showed high levels of activation 3 hours after irradiation, reaching a maximum at 6 hours. Massive cell death was observed at the 6-hour time point and a gradual loss of beta-galatosidase (the enzyme produced from the *lacZ* gene) staining followed. This study suggests that there is an association between p53 activation, gene transcription, and cell death in the embryo, although the dose used was very high possibly leading to confounders of response.

Gottlieb (37) used an mdm2 promoter-lacZ reporter to assess p53-dependent transcriptional activation produced by irradiation at GD 8.5. This study showed a robust response throughout the embryo, with the exception of the heart where p53-mediated transcription was not induced. At GD 10.5, the induction was more limited, but was observed as quickly as 1 hour after irradiation. Of particular note was the observation that p53 activation began to decrease 3 hours after irradiation in p53 heterozygous (+/−) animals suggesting that the p53 response may be limited/limiting by the amount of p53 protein present in the tissue.

Norimura (38) evaluated the role of p53 in radiation-induced birth defects. Using a 2-Gy exposure on GD 9.5, they observed an increase in malformations and a decrease in lethality in p53 −/− mice compared to wild-type embryos. The number of apoptotic cells increased dramatically in p53 +/+ embryos following irradiation, but was not changed in the p53 −/− with treatment. This pattern of increased survival and increased maldevelopment in p53 −/− embryos was also present when mice were irradiated on GD 3.5.

Nomoto (39) evaluated the effects of 2-Gy irradiation on p53 +/+ and +/− embryos. Mice were irradiated on GD 9.5 of gestation and +/− embryos had a higher incidence of malformations and a lower incidence of lethality on GD 18 than irradiated +/+ mice. The frequency of apoptotic cells in the neural tube 4 hours after irradiation was greater in the +/+ mice than in +/− embryos. In p53 +/+ mice, there was no increase in mdm2 or p53 mRNA induced 4 hours after irradiation. This suggests that early production of apoptosis was not the result of induction of gene expression. When the time course of 3-Gy irradiation-induced apoptosis was evaluated in +/+ embryos, the frequency of dead cells peaked at 4 hours following irradiation (80%) then dropped dramatically at 12 hours (~50% frequency). This period of rapid change was followed by a gradual decrease in the frequency of apoptotic cells from 12 to 48 hours suggesting a biphasic response to irradiation.

Kato exposed ICR strain p53 (+/+) mice to radiation on GD 8.5–10.5 at 2 Gy (40). A 1.06-Gy/min dose rate induced malformations in embryos and GD 9.5 and 10.5 were more sensitive to the adverse effects than the GD 8.5 embryo. In contrast, when a lower exposure rate (1.2 mGy/min) exposure was used on GD 9.5–10.5, there was an increase in malformations in p53 −/−, but not in p53 +/+ fetuses. This exposure also produced cell death in the p53 +/+ embryos, but not in the p53 −/−. Thus, although irradiation at the lower exposure rate was sufficient to induce cell death in the neural tube of wild-type embryos, it was not sufficient to induce malformations. This suggests that there must be a balance between DNA repair, replication, and cell death that supports normal development. However, in the absence of p53, this balance is lost and an increase in malformations was produced.

Related to p53 activation following DNA damage is the sensor function of the *Atm* gene. Using a 0.5-Gy dose of radiation at GD 10.5, Laposa (41) reported a high incidence of runting and kinked tails in Atm (−/−) fetuses but no similar effects were found in Atm +/− or wild-type fetuses. A large increase in postnatal lethality was also observed in irradiated Atm −/− neonates than for the other genotypes. When evaluated at 6 hours after irradiation, there was little or no apoptosis observed in Atm −/− mice in contrast to a large number of TUNEL positive cells in the wild-type animals. In contrast, 48 hours after irradiation, there is an increase in cell death in irradiated Atm −/− fetuses that are TUNEL positive, yet have morphological characteristics consistent with necrosis. Thus, Atm is necessary for radiation-induced apoptotic cell death, suggesting that DNA damage maybe a critical event for initiating the Atm-/p53-dependent cell death cascade and malformations.

Irradiation of mice during late organogenesis produces malformations including limb defects. To evaluate the role of p53 in induction of limb defects and cell death, +/+, +/−, and −/− mice were irradiated at 1 or 3 Gy on GD 12 (42). Embryos without a functional p53 (−/−) did not exhibit an increase in limb defects on GD 18, while this dose produced a mean digit loss of 1.5 in the p53 +/+ animals. Similarly, irradiation produced a large increase in cell death in the predigital area of p53 +/+ limbs (64.8%) 6 hours after a 3-Gy exposure. In p53 −/− mice, the level of apoptosis was 16.2%. These studies demonstrate that limb effects produced by irradiation on GD 12 are mediated by p53-dependent cell death.

In addition to the adverse morphological effects of radiation on the developing embryo, several studies have evaluated the role of p53 in the radioprotective effects of a low-dose exposure. In one series of experiments, Wang (43) evaluated the potential for a 5- or 30-cGy exposure on GD 11 to ameliorate the effects of a subsequent 3-Gy exposure on GD 12. This exposure decreased the incidence of defects in p53 +/+ mice, but not in p53 +/−. In a separate series of studies, Mitchel (44) reported that a 30-cGy exposure 24 hours before a 4-Gy exposure on GD 11 reduced the extent of limb defects and tail-shortening in p53 +/+. In contrast, preirradiation on GD 10 enhanced the tail-shortening effects of irradiation on GD 11 in p53 −/− fetuses. These studies clearly demonstrate that p53 has an important role in the process of radioprotection. It is tempting to speculate that the same mechanisms responsible for the low-dose rate sensitivity of p53 (−/−) embryos may also contribute to the lack of radioprotection observed in p53-deficient embryos.

The roles of p53 in radiation-induced developmental effects have also been evaluated during gastrulation. Using an outbred mouse strain, Heyer (45) reported that the area of cell death in the embryo was highest (50–60%) in GD 6.5 and 7.5 compared to earlier- and later-staged conceptuses following a 0.5-Gy exposure. These studies demonstrate the exquisite sensitivity of gastrulation-staged embryos to radiation-induced damage. When p53 −/− embryos were exposed to this same dose of radiation, at the same stage there was no increase in cell death compared to the large area of cell death in the p53+/+ embryos. When mice lacking p19ARF (cyclin-dependent kinase inhibitor 2A) or DNA-PKscid/scid were exposed to 0.5 Gy, the apoptotic response in each knockout was the same as in wild-type animals. Activation of p53 was dependent upon signaling and interaction with other proteins. One of the proteins associated with p53 activation is ATM, which functions as a DNA damage sensor. In the absence of Atm, the p53 response to DNA damage is impaired. In Atm −/− mice, exposure to 0.5 Gy on GD 6.5 does not induce an increase in apoptosis, indicating a critical role of Atm signaling in mediating the response to irradiation and activation of p53. Unlike the results reported for later stages, both Atm and p53 were upregulated at 1 hour after irradiation in both embryonic and extraembryonic tissues. Thus, the molecular response to radiation in gastrulation-staged conceptuses is different than that of later stages and may be associated with the unique sensitivity at this stage.

In considering the studies that use p53-knockout mice to understand the toxicity of xenobiotics, it is tempting to propose the following. In consideration of radiation-induced damage, it has been shown that in A1–5 cells the induction of p53 (nuclear accumulation) is biphasic with an early response 1 to 2 hours after irradiation and a late response 12 to 24 hours after initial damage (46). Interestingly, the early response can be ameliorated by a free radical scavenger suggesting that ionizing radiation produces p53 activation and possibly damage via two distinct mechanisms; generation of free radicals that occurs immediately after irradiation and a delayed, long-lived DNA (and/or other macromolecular) damaging effect or other differences in DNA repair for different DNA damage types. Therefore, the effects of radiation on the embryo could also be explained as a biphasic response. Immediately, following irradiation, there is a large induction of free radicals resulting in a rapid and robust p53-induction and p53-dependent cell death. One of the critical events in mediating the p53 induction is DNA damage since Atm is required for the response. For many embryos, this wave of cell death is so massive that the embryo cannot survive because of the excessive loss of tissue. However, for a certain subset of embryos, there is an unknown mechanism (possibly antioxidant status) that blunts the effects of the ROS, p53-induction, and extent of cell death. These embryos survive the initial wave of damage, but are still subject to the longer-lived DNA damage (and other macromolecular damage). In these embryos, the long-lived damage results in malformations. Based on this hypothesis, the loss of functional p53 would result in a decrease in lethality following irradiation and ROS-generating exposures. However, it does not explain the increase in malformations seen in the animal model. One explanation for the increase in malformations is the observation that p53 is required for the balance between DNA (macromolecular) repair, cell proliferation, and cell death seen in low-level irradiation-induced effects. When p53 was present, low-level radiation produced cell death but did not result in malformations. Thus the elimination of damaged cells is essential in attaining normal development. In this biphasic response model, the second wave of p53 induction would be associated with surveillance for DNA damage and the elimination of those cells that have not or cannot be repaired. Based on this model, another consequence of a nonfunctional p53 would be the proliferation of damaged cells. The long-term consequences of this accumulation of damaged cells would be the death of those cells by necrotic or hybrid apoptotic/necrotic pathways that are p53 independent. This loss of cells would be associated with the induction of malformations. This is consistent with the observation that following irradiation, phenytoin, or benzo[a]pyrene exposure there is an increase in malformations among the p53-deficient fetuses compared to a wild-type p53 genotype. This suggests that a common mechanism (possibly ROS) is responsible for dysmorphogenesis

Although this model is consistent with the observed data for radiation and ROS-mediated malformations, it does not explain why there is a decrease in malformations in p53 $(-/-)$ embryos following 2CdA or cyclophosphamide exposure compared to wild-type embryos. Clearly, for these chemicals and the cellular perturbations they produce, induction of cell death is a critical event leading to dysmorphology. It will be interesting to see future experimental results that discern those chemicals and cellular perturbations where cell death is responsible for dysmorphology and where cell death is critical in supporting normal development.

In summary, developmental toxicology studies have made use of transgenic animals to provide important information about the mechanisms responsible for xenobiotic-induced birth defects. Both reporter/marker construct mice and single gene mutant (knockout) models have been used. Using reporter constructs, the temporal and spatial distributions of xenobiotic or activation of a signaling pathway have been evaluated. Knockout mice have made it possible to evaluate the roles of specific genes in the induction of malformations and the associated cell death. One cautionary note should be added about using transgenic mice. It is critical to know the genetic strain background for the animals used in developmental toxicity studies. Since many transgenic mice are derived on a mixed background (e.g., C57/129), it is essential that the effects of the xenobiotic are evaluated in that strain background. It is well established in the literature that there are profound strain differences in response to developmental toxicants. Thus, in future studies, it may be beneficial to backcross some knockout mice onto different strains of known sensitivity to the xenobiotic being evaluated.

Antisense Oligonucleotides

Antisense oligonucleotides have been used to study the functions of genes in development for many years. As early as 1988, antisense oligonucleotides were used to disrupt gene function in preimplantation mouse embryos (47,48). This technique continues to be extensively used in models of development such as preimplantation-staged rodents, Xenopus, and zebra fish. Antisense oligonucleotides have also been used to evaluate gene function and their role in morphogenesis in selected cells and tissues. For example, limb (49–54), kidney (55–68), and mandible (69–75) models have been used with antisense oligonucleotides.

Because of the extensive work of Karen Augustine and Tom Sadler, antisense oligonucleotide techniques were developed and adapted for use in rodent whole embryo culture (76–80). The technique of delivering antisense oligomers to neurulation-staged embryos requires a trans–yolk sac injection with intra-amniotic deposition. Using lipid micelle or ethoxylated polyethylenimine vehicles, oligomers injected into the amniotic cavity are taken up by the cells of the neural ectoderm and distributed to the embryo. These researchers set the stage for this important approach to knock down gene expression during the period of neurulation and early heart and craniofacial development.

The molecular structures of antisense oligonucleotides have undergone a radical change since the early days of using phosphate-linked nucleotide molecules, to phosphothioate, phosphorodiamidate morpholino (morpholino) oligomers, and peptide nucleic acids. There are many reviews describing the use, synthesis, and comparison of these molecules (81–88), and these topics will not be described in detail here. This approach to knocking down expression continues to thrive and antisense molecule technology continues to be developed. One recent development in antisense technology is the use of an UV-activated cage construct to regulate availability of the oligonucleotide (89,90). The intracellular delivery of antisense molecules has also undergone a revolution and ranges from electroporation, formation of neutral lipid micelles to cationic cell-penetrating peptides (91). Viral delivery systems will be described in a later section of the chapter.

One of the reasons for the advances in antisense oligonucleotide molecules structure was the improvement in stability and intracellular half-life of the molecule. This stability is critical in assessing gene function and for the success of any experiment using an antisense oligonucleotide approach. The transition from phosphate to phosphothioate and phosphorodiamidate morpholino oligomers rendered the molecule increasingly resistant to RNaseH degradation and increased the half-life from minutes to almost permanent. However, the increase in stability does not solve all of the issues in antisense experiments, because with each cell division the concentration of oligomer in the cell is decreased with cytokinesis and subsequent growth of the daughter cells.

Another issue to consider in all antisense experiments is the relationship between DNA, mRNA, and protein. If we consider a theoretical gene whose protein product has a long half-life (ca. 8 hours), then decreasing the abundance of the mRNA or translation of the mRNA for 8 hours produces a 50% decrease in the protein content and a 16-hour perturbation results in a 75% decrease in the gene's protein content. Also, if we consider that many proteins are produced in a form that needs to be modified in order to have biological activity (e.g., preproproteins), then understanding the relationship between mRNA abundance and content of biologically active peptide becomes that much more complex. Therefore, in antisense experiments, it is possible that a 90% decrease in mRNA will not produce any significant change in the protein content and thus the biological function of the gene being evaluated. Additionally, in the context of a 24-hour rodent embryo culture experiment, many of the morphological events that are studied occur during relatively small windows of morphogenesis, such as neural tube closure or lens induction. Thus, careful attention to detail is required in order to correctly interpret the results of antisense experiments. With this as background, it is easy to understand that an antisense approach to understanding gene function is most amenable to genes that are activated during the window of morphogenesis (i.e., expression goes from low to high levels), when the protein has a short biological half-life or when the antisense oligonucleotide can be delivered to the cell/tissue/embryo with sufficient time to decrease the protein content prior to the morphological period/event under study. This last point of using antisense oligomers with sufficient time prior to morphogenesis may be why antisense approaches continue to be extensively used in chick, Xenopus, and zebra fish models, but only in relatively few rodent whole embryo studies.

Despite the challenges of using this approach, several research groups have creatively and successfully used antisense oligonucleotide techniques to better understand the toxicity of xenobiotics. In one study, Hunter and Dix (92) evaluated the ability of antisense oligonucleotides to block the induction of heat shock protein (HSP) 70–1/3, a component of the heat shock response, and modulate the embryonic morphological response to an arsenical. Early somite-staged mouse conceptuses were prepared for whole embryo culture and phosphothioate oligomers were injected into the amniotic cavity as described by Sadler's laboratory (80). Conceptuses were cultured for 2 hours and then exposed to arsenite at a concentration that produced a 20% incidence of defects in the untreated conceptus. The incidence of maldevelopment was the same in the lipofectamine-only and untreated controls. However, the antisense oligomer produced a significant increase in arsenite-induced dysmorphology and prevented the induction of the HSP70–1/3 protein. The authors concluded that induction of the heat shock/stress response, specifically HSP70–1/3, was embryo protective.

Augustine-Rauch (93) used morpholino-antisense oligomers to knock down expression of a selected gene in order to compare the phenotypes produced by that perturbation to that produced by a xenobiotic. These studies were designed to determine the molecular basis for a serotonin receptor 1B agonist (SB-236,057-) induced teratogenesis. The authors proposed that the mechanism for SB-236,057-induced developmental effects was not associated with its agonistic effect. Using phage display, it was shown that SB-236,057 binds to a protein sequence found in r-esp1, a downstream component of Notch-1 signaling. To determine if perturbation of r-esp1 function would result in maldevelopment comparable to that produced by SB-230,657, morpholino antisense oligomers to r-esp1 were injected into the amniotic cavity of rat conceptuses prepared for whole embryo culture. Antisense-exposed embryos showed a strikingly similar morphological perturbation to that of SB-236,057-treated embryos. Additional gene expression changes further confirmed the similarity in molecular response of embryos to the antisense and SB-236,057. These studies provide strong evidence for a disruption of r-esp1 and Notch-1 signaling as responsible for SB-236,057-induced malformations.

Another example of using antisense oligonucleotides to address issues of developmental toxicity is the research of Wells' laboratory (8) to modulate NFkB and evaluate its role in mediating ROS-induced developmental effects. Wells' laboratory has established a critical role of ROS and oxidative stress as mediating phenytoin-induced birth defects. However, the proximal events that are altered by oxidative stress have not been determined. It has been clearly demonstrated that activation of NFkB can occur as a result of ROS. Therefore, following phenytoin exposure, ROS-mediated activation of NFkB may lead to abnormal development. To test this hypothesis, an NFkB promoter-lacZ reporter K1 mouse was used to evaluate activation of this signaling pathway and antisense oligomers used to decrease signaling. Phenytoin exposure produced a time- and concentration-dependent ectopic NFkB activation in embryos grown in whole embryo culture. The antisense oligomer (directed against the pgs subunit) blocked ectopic NFkB signaling and dramatically improved embryonic morphology during phenytoin exposure. Thus, the K1 mouse conceptus confirmed the localization and degree of aberrant NFkB activation and the antisense study confirmed that improper activation of NFkB signaling was part of the mechanism responsible for phenytoin-induced birth defects.

In summary, these studies are three examples of how antisense oligonucleotides can be used to further understand the mechanisms responsible for xenobiotic-induced birth defects. Using antisense oligomers to block activation of a stress pathway or aberrant induction of NFkB demonstrates the importance of these signals to birth defects. Additionally, antisense oligomers were used to knock down expression of a putative molecular target and confirm the similarities of phenotypic and signaling transactivational changes produced by the chemical and those when this specific molecular target was altered.

Gain-of-function Models

In addition to the loss-of-function models, discussed in the earlier sections of this chapter, there are several studies using gain-of-function models to better understand xenobiotic-induced developmental toxicity.

Many cell biology experiments use a gain-of-function paradigm to study the function and role of a selected gene in a disease state or to deliver a fully functional gene to replace an endogenous malfunctioning protein in gene therapy. Thus, gene delivery and overexpression has been the subject of intense research in the field of gene therapy. There are many reviews that evaluate the state of the science, including nonviral gene delivery (94–102) and viral-vector gene delivery (103–114).

For the study of developmental biology, although many vectors have been used for gene therapy, plasmids, adenoviruses, and retroviruses (e.g., lentivirus) have been most extensively used. Plasmid DNA has been used to express genes in many systems, but requires aggressive techniques to get the vector into the cells. Electroporation of plasmids has been extensively used in stem cells and in whole organisms (115–117). An alternative technique for getting plasmids into cells is to use high calcium phosphate levels or other treatments to make the membrane amenable to passage of the large plasmid molecule. In neurulation-staged whole mouse embryos, Hartig and Hunter (118) were partially successful in using plasmids to express marker genes by injecting naked plasmids or plasmids with lipid micelle into the amniotic cavity. Thus, plasmids have been successfully used to overexpress genes in many systems. However, using this vector to overexpress genes in neurulation-staged embryos would require further methods development.

Viral vectors are efficient and effective tools to deliver genes to model systems and have been successfully used in a large number of models to study developmental biology. This chapter focuses on two vectors: retroviruses and adenoviruses. Retroviruses are a highly effective vector for expressing genes in cells. One of the benefits of using retroviruses is that the viral DNA integrates into the host's DNA and expression of the exogenous gene(s) become part of the cell's normal transcriptional process. Using selected promoters and regulatory elements, the researcher can direct spatial and temporal expression. Additionally, as the host's DNA is replicated during cell division, the retroviral DNA is also replicated such that each daughter cell contains a copy of the exogenous gene under investigation. Despite these positive aspects, one important consideration in using a retroviral vector is the need for incorporation of the viral DNA into the host cell's genome before exogenous gene expression begins. Thus, in a cell that has a cell cycle time of 8 hours, there will be a significant delay between the time of exposure and exogenous gene expression. In experiments where the cells can be transfected prior to the morphogenic event in question (e.g., preimplantation mammalian embryos, Xenopus or zebra fish eggs, or early embryos), retroviruses offer a great tool for continuous expression of exogenous genes. However, in experiments that use whole embryo culture or organ cultures (e.g., limb bud or palate), the delay in exogenous gene expression may significantly impact the study's outcome.

Adenoviruses offer an alternative to retroviruses in the study of developmental biology and toxicology. Replication-deficient adenoviruses have been used to express exogenous genes in many developing systems. Unlike retroviruses, there is no requirement for incorporation of the viral DNA into the host's genome before exogenous gene expression begins. For example, expression of a marker gene (expressing lacZ under a cytomegalovirus (CMV) promoter) was seen in neurulating embryos as early as 4 to 6 hours after exposure (118). However, the lack of replication and integration leads to a decreasing intracellular concentration of vector with each round of cell replication in the test system. Thus, adenoviruses present a temporally limited expression vector.

In the neurulation-staged rodent embryo, adenovirus-mediated gene expression has been demonstrated in the heart and whole embryo (118–121). In a series of studies to evaluate the ability of adenovirus to transfect the heart and developing vascular system, Baldwin and coworkers used adenoviruses with either CMV or Rous sarcoma virus (RSV) promoters to drive expression of lacZ (119,121). Hartig and Hunter (118) injected an adenovirus expressing lacZ under a CMV promoter into the amniotic cavity to transfect the whole embryo. Both groups concluded that adenoviruses are effective tools for expressing exogenous genes in the postimplantation early organogenesis-staged embryo. Such an enthusiastic endorsement of adenoviruses must also be tempered with the observation that adenoviruses can also induce dysmorphology if too much virus is used for intra-amniotic injection. Baldwin's group has also shown that injection of adenovirus into the yolk sac vasculature can also result in embryonic expression of a marker gene when administered during late organogenesis or the early fetal period (122), GD-dependent and tissue-specific marker gene expression was observed. However, these studies clearly demonstrate fetal expression of exogenous genes and set the stage for future research of fetal gene therapy. Thus, adenoviruses are a useful tool to overexpress a gene in the post-implantation-staged embryo and fetus.

In addition to the many studies of developmental biology that use adenoviruses to evaluate gene function, several studies have used this approach to aid in understanding developmental toxicity. As described in section "Transgenic Models," a p53 loss-of-function model has been extensively used to evaluate the role of p53 as a mediator of developmental toxicity. Hunter et al. (in preparation) have used an adenovirus to overexpress human p53 in mouse embryos to determine if this overexpression induces maldevelopment and if this expression changes the response to chemical-induced dysmorphology. These studies demonstrate that overexpression of p53 was not dysmorphic in neurulation-staged rodent embryos in whole embryo culture. Using immunohistochemistry, it was shown that when conceptuses were cultured under control conditions, exogenous p53 was localized predominately in the cytoplasm and did not show nuclear accumulation typical of a stress response. Thus, it was proposed that overexpression of p53 did not induce dysmorphology because there was no activation of p53. In contrast, when the p53-transfected embryo was challenged with a chemical toxicant (3 μM arsenite), there was a dramatic increase in malformations compared to noninjected embryos, or those expressing a mutant (nonfunctional) p53 or green fluorescence protein marker genes. Additionally, the increase in maldevelopment was also dependent upon the level of virus injected into the amniotic cavity suggesting a gene-dose dependence. Immunohistochemistry showed that in the chemical-treated p53-overexpressing embryo, exogenous p53 was found primarily in the nucleus indicating activation and nuclear translocation induced by the toxicant. The authors proposed that the increased sensitivity of p53 overexpressing embryos was the result of p53-mediated gene expression changes, induction of cell cycle perturbation, and increased cell death. These studies further substantiated the critical role of p53 as a mediator of adverse developmental effects produced by xenobiotic exposure.

Adenoviruses have also been used to overexpress human superoxide dismutase (SOD) in mouse embryos in order to better understand the role of ROS-mediated dysmorphology (Smith et al., unpublished). Exposure of rodent conceptuses to ethanol in whole embryo culture produces dysmorphology and recapitulates many of the effects of ethanol-administration in vivo. It has also been shown that addition of Cu,Zn-superoxide dismutase protein from bovine erythrocytes will ameliorate the effects of ethanol exposure indicating that free radicals contribute to the genesis of the morphological effects (123). Using adenoviruses to overexpress the selected *hSOD* gene, Smith et al. (unpublished data) evaluated the protective effects of mitochondrial SOD2 (mitochondrial, Mn-superoxide dis-mutase) in mouse conceptuses exposed to high doses of ethanol in whole embryo culture. In embryos overexpressing SOD2, there was a significant reduction in ethanol-induced craniofacial and heart dysmorphologies compared to embryos transfected with a *lacZ*-marker gene or not treated with a virus. These results sug-gest that at a high concentration of ethanol, providing additional antioxidant protec-tion to mitochondria is effective at preventing some, but not all, dysmorphologies.

In a follow-up study to overexpression of SOD2 in whole embryos, Smith et al. (unpublished data) evaluated the protective effects of *SOD1* and *SOD2* genes in neural crest cell culture. Initial studies established that adenoviral-mediated gene expression was effective in the neural crest cell culture model and that marker gene expression was maintained throughout the culture period. In the neural crest cell culture model, exposure to ethanol induces concentration-dependent cell death. Overexpression of SOD1 was very effective at preventing ethanol-induced cell death, while overexpression of SOD2 provided limited protection. Since this result was different from that found by this same group in whole embryos, rt-PCR was used to characterize SOD expression in nontransfected migrating neural crest cells. Although a robust SOD2 expression was observed in the neural crest cells, there was no evidence for *SOD1* gene expression. The authors propose that the dramatic improvement in neural crest cell survival produced by overexpression of SOD1 was due to the apparent lack of endogenous SOD1 expression in that population. Additional experiments will be needed to determine if the protective effects of SOD1 in neural crest cells are mechanistically linked to ethanol-induced cell damages or if the protective effects of SOD1 are linked specifically to a lack of endogenous SOD1 expression in the crest cells.

In summary, experiments that use a gain-of-function paradigm will continue to be important in the future for developmental biology and toxicology studies. Viral vectors offer an effective and efficient mechanisms for gene delivery. Con-tinued development of the use of specific promoters and regulators will increase the ability of the investigator to control the spatial and temporal activation of the exogenous gene, thereby increasing the utility of these constructs. Recent advances in nanotechnology may improve delivery of constructs to the target tissues. One such advancement is the use of nanoparticulate polyelectrolyte complexes (PEC). Using an adenovirus PEC, it was shown that there was a dramatic increase in expression of a marker gene in cell culture compared to the adenovirus alone [see (94) for a recent review of PEC]. Because of the structure of PEC, it is also conceivable that they may be viable tools to facilitate uptake and distribution of expression vectors (such as shRNA, plasmid constructs, or viruses) across the placenta/yolk sac and to the postimplantation embryo in vivo.

Small Regulatory RNA and RNA Interference

It was recently estimated that ~2% of the transcriptional output of the human genome is RNA that codes protein. Within the remaining 98% of the noncoding

RNA (ncRNA) are RNA molecules required for cellular functions such as transfer RNA, ribosomal RNA, and small nuclear RNA. Additional small RNAs are also produced and include microRNA (miRNA) and small-interfering RNA (siRNA). These molecules are distinguished by their origin and not by cellular function. miRNA come from short hairpin precursors transcribed as pre-miRNA and processed in the nucleus before moving to the cytoplasm. Many are derived from introns of protein coding genes and the remainder from introns and exons of mRNA-like ncRNA. siRNA are longer double-stranded RNAs or long hairpin molecules. The siRNAs are produced from sense-antisense transcripts or mRNA-like ncRNA. miRNA and siRNA are processed to double-stranded RNA in the cytoplasm by the protein Dicer. The double-stranded RNA then forms complexes with proteins, such as Argonaut, and these complexes then perform the biological functions associated with miRNA and siRNA. Both miRNA and siRNA have been shown to suppress translation or cleave (slicing) target mRNAs using RNA interference (RNAi). There are many excellent reviews of ncRNA, miRNA, and siRNA (124–127).

Another class of small RNAs is the piwi-interacting RNA (piRNA). This group of RNAs was identified as binding to and forming complexes with rat protein homologs of the Piwi protein in *Drosophila*. The piRNA complex has DNA helicase activity and is implicated in transcriptional gene silencing. The piRNA are also unique in that they are found in gamete cells and can be found at very high levels [see (128) for an overview of piRNA].

RNA interference (RNAi) is a cellular process that regulates gene expression at transcriptional, posttranscriptional, and translational levels. Each miRNA or siRNA is processed in the cytoplasm by the dicer protein to form double-stranded RNA that then binds into a protein complex. siRNAs form complexes that are called RNA-induced silencing complexes (RISCs) that use the antisense strand as a guide to bind a target mRNA resulting in sequence-specific degradation (slicing) of the target. miRNAs also form protein complexes that bind to complementary sequence typically in the untranslated regions of the target mRNA, with a major effect of inhibiting translation. Certain mammalian miRNAs also facilitate target mRNA slicing. Additionally, RNAi has been linked to a decrease in gene expression produced by transcriptional gene silencing. It has been predicted that miRNA regulate up to one-third of all human genes (129). For reviews of RNAi see (124,130,131). Thus, RNAi uses small RNA to target selected sequences for modulation of gene expression from all aspects of transcription to translation.

Since the discovery that dicer can cleave synthetic exogenous siRNAs resulting in sequence targeting of specific mRNAs, the process of RNAi has been exploited to study gene function in nearly every conceivable model. As described in a review of RNAi (130) and their references, siRNAs have been delivered to cells as synthetic siRNA molecules or as short hairpin RNA (shRNA) in plasmids and viral vectors. Since plasmids and viral vectors produce a continuous supply of shRNA, these tools have been extensively used in "an almost bewildering number of shRNA expression systems" (130). A myriad of regulatory elements that direct spatial and temporal expression, selection tools to stimulate/repress expression, or select cells containing the vector or bicystronic expression of marker proteins have been used.

A large number of studies have used RNAi to study gene function during development and the morphological consequences of knocking down expression. The model systems include stem cells, chick, zebra fish, Xenopus, and rodent embryos. However, for studying mammalian developmental toxicity, RNAi has been underutilized. The reasons for this are complex, but may be related to the relative inaccessibility of the postimplantation embryo for genetic manipulation. Advances in transplacental/trans–yolk sac delivery of shRNAs or viral vectors may resolve this issue, but adequately tested and validated models are not available.

One well-tested and -proven model that uses RNAi is available and should be used in future developmental toxicity studies. This model is the creation of transgenic mice from shRNA-transfected mouse embryonic stem cells (mESCs) that would then be used to evaluate the developmental effects of changes in gene expression. In traditionally created knockout animals, a selected gene is mutated in mESCs and the knockout mESCs are injected into recipient blastocysts to create chimeric offspring. Chimeras with germ line transmission of the mutant gene are then used to create heterozygous and homozygous mutant offspring. In this traditional model, gene function is determined by evaluating the phenotype of the homozygous mutant animal. In contrast, in a seminal paper from Rossant's laboratory (132), a technique was described where mESCs are transfected with a shRNA expression vector to knock down gene expression. The shRNA-expressing mESCs are then used to create transgenic mice using the tetraploid aggregation method (133). As described by Kunath (132), this technique faithfully recapitulates the phenotype of traditional knockout techniques, but does so directly and quickly. The potential use of the shRNA technique takes advantage of the observation that RNAi is not 100% efficient and transfection of mESC results in a range of gene expression decreases. Thus, using the approach of creating knockout mice by the technique described by Rossant's laboratory, it is possible to evaluate the phenotypic consequences of reducing, but not eliminating, expression of any gene through selection of shRNA-expressing mESC with a reduced level of gene expression. This approach and assessment is critical in future studies of developmental toxicology, because in our studies (Karoly and Hunter unpublished observation) we rarely see a complete elimination of gene expression in our treated embryos compared to expression in control embryos. More typically, we see a 50% reduction in gene expression. It is important to determine if a decrease, but not elimination, of expression of a specific gene produces a phenotype in order to correctly begin to understand the effects of toxicants on embryogenesis. As an example, if we saw a twofold decrease in gene expression following exposure to a toxicant, we would select an mESC that had a 50% decrease in gene expression, compared to the wild-type cells, for creation of a transgenic animal. The techniques to create an shRNA knockdown of gene expression in mESC and the creation of tetraploid transgenic mice have been established. Using mESC with different levels of gene inactivation, we can then evaluate the relationship between gene expression level and dysmorphology. In the future, it may also be possible to design mismatch nucleotides into the antisense arm of the shRNA to decrease the binding efficiency of the target mRNA to the RISC, thereby selectively modulating gene expression in the mESC.

In summary, it is possible to take advantage of the cell's use of ncRNA to regulate gene expression, defend itself against viruses and other cellular functions by miRNA and siRNA. This regulation is termed RNAi and is associated with altered transcription, translation, and mRNA slicing. Exogenous synthetic shRNAs are processed by dicer and form RISCs similar to endogenous siRNA. A large number of expression vectors are available that allow the researcher to select many aspects of shRNA expression. This technique has been underutilized in the study of developmental toxicants in mammals, but is extensively used in other models such as zebra fish. One use of RNAi in mammalian developmental toxicity studies should be evaluations of reduced, but not eliminated, gene expression. These studies could take advantage of the rapid creation of transgenic animals by tetraploid aggregation with shRNA-expressing mESC at reduced, but not eliminated, levels of gene expression to better understand the biological consequences of reduced expression produced by xenobiotic exposure.

CONCLUSIONS

Studies of the modes and mechanisms responsible for xenobiotic-induced birth defects have used a large number of tools and techniques including those that modulate expression of selected genes. This chapter has focused on techniques and models, transgenic mice (including knockouts and promoter–reporter knock-in models), antisense oligonucleotides, and viral vectors to modulate expression of a selected gene. These models have significantly impacted the understanding of how chemicals produce birth defects and the role(s) of specific genes in mediating those effects. Lastly, siRNA and RNAi are briefly discussed with a proposal of how RNAi can be used to answer critical questions related to understanding the morphological consequences produced by reduced, but not eliminated, gene expression. Although this chapter focused on mammalian and rodent studies, there are important and exciting studies using these tools in nonrodent models, especially zebra fish.

REFERENCES

1. Tsou HC, Si SP, Lee X, et al. A beta 2RARE-LacZ transgene identifies retinoic acid-mediated transcriptional activation in distinct cutaneous sites. Exp Cell Res 1994; 214(1):27–34.
2. Zimmer A, Zimmer A. Induction of a RAR beta 2-lacZ transgene by retinoic acid reflects the neuromeric organization of the central nervous system. Development 1992; 116(4):977–983.

3. Balkan W, Colbert M, Bock C, et al. Transgenic indicator mice for studying activated retinoic acid receptors during development. Proc Natl Acad Sci U S A 1992; 89(8):3347–3351.

4. Reynolds K, Mezey E, Zimmer A. Activity of the beta-retinoic acid receptor promoter in transgenic mice. Mech Dev 1991; 36(1–2):15–29.

5. Mendelsohn C, Ruberte E, LeMeur M, et al. Developmental analysis of the retinoic acid-inducible RAR-beta 2 promoter in transgenic animals. Development 1991; 113(3):723–734.

6. Rossant J, Zirngibl R, Cado D, et al. Expression of a retinoic acid response element-hsplacZ transgene defines specific domains of transcriptional activity during mouse embryogenesis. Genes Dev 1991; 5(8):1333–1344.

7. Komarova EA, Chernov MV, Franks R, et al. Transgenic mice with p53-responsive lacZ: p53 activity varies dramatically during normal development and determines radiation and drug sensitivity in vivo. Embo J 1997; 16(6):1391–1400.

8. Kennedy JC, Memet S, Wells PG. Antisense evidence for nuclear factor-kappaB-dependent embryopathies initiated by phenytoin-enhanced oxidative stress. Mol Pharmacol 2004; 66(3):404–412.

9. Willey JJ, Stripp BR, Baggs RB, et al. Aryl hydrocarbon receptor activation in genital tubercle, palate, and other embryonic tissues in 2,3,7, 8-tetrachlorodibenzo-p-dioxin-responsive lacZ mice. Toxicol Appl Pharmacol 1998; 151(1):33–44.

10. Abbott BD, Buckalew AR, DeVito MJ, et al. EGF and TGF-alpha expression influence the developmental toxicity of TCDD: Dose response and AhR phenotype in EGF, TGF-alpha, and EGF + TGF-alpha knockout mice. Toxicol Sci 2003; 71(1):84–95.

11. Abbott BD, Buckalew AR, Leffler KE. Effects of epidermal growth factor (EGF), transforming growth factor-alpha (TGFalpha), and 2,3,7,8-tetrachlorodibenzo-p-dioxin on fusion of embryonic palates in serum-free organ culture using wild-type, EGF knockout, and TGFalpha knockout mouse strains. Birth Defects Res A Clin Mol Teratol 2005; 73(6):447–454.

12. Abbott BD, Lin TM, Rasmussen NT, et al. Lack of expression of EGF and TGF-alpha in the fetal mouse alters formation of prostatic epithelial buds and influences the response to TCDD. Toxicol Sci 2003; 76(2):427–436.

13. Bryant PL, Schmid JE, Fenton SE, et al. Teratogenicity of 2,3,7,8-tetrachlorodibenzo-p-dioxin (TCDD) in mice lacking the expression of EGF and/or TGF-alpha. Toxicol Sci 2001; 62(1):103–114.

14. Mimura J, Yamashita K, Nakamura K, et al. Loss of teratogenic response to 2,3,7,8-tetrachlorodibenzo-p-dioxin (TCDD) in mice lacking the Ah (dioxin) receptor. Genes Cells 1997; 2(10):645–654.

15. Peters JM, Narotsky MG, Elizondo G, et al. Amelioration of TCDD-induced teratogenesis in aryl hydrocarbon receptor (AhR)-null mice. Toxicol Sci 1999; 47(1):86–92.

16. Wells PG, Bhuller Y, Chen CS, et al. Molecular and biochemical mechanisms in teratogenesis involving reactive oxygen species. Toxicol Appl Pharmacol 2005; 207(2 Suppl):354–366.

17. Lapossa RR, Wiley MJ, Wells PG. Evidence for non-apoptotic paythways mediating phenytoin teratogenicity in mice. Fundam Appl Toxicol 1997; 36(S-1):303 (Abstract).

18. Nicol CJ, Harrison ML, Laposa RR, et al. A teratologic suppressor role for p53 in benzo[a]pyrene-treated transgenic p53-deficient mice. Nat Genet 1995; 10(2):181–187.

19. Wells PG, Winn LM. Biochemical toxicology of chemical teratogenesis. Crit Rev Biochem Mol Biol 1996; 31(1):1–40.

20. Bhuller Y, Jeng W, Wells PG. Variable in vivo embryoprotective role for ataxia-telangiectasia-mutated against constitutive and phenytoin-enhanced oxidative stress in atm knockout mice. Toxicol Sci 2006; 93(1):146–155.

21. Bhuller Y, Wells PG. A developmental role for ataxia-telangiectasia mutated in protecting the embryo from spontaneous and phenytoin-enhanced embryopathies in culture. Toxicol Sci 2006; 93(1):156–163.

22. Kasapinovic S, McCallum GP, Wiley MJ, et al. The peroxynitrite pathway in development: phenytoin and benzo[a]pyrene embryopathies in inducible nitric oxide synthase knockout mice. Free Radic Biol Med 2004; 37(11):1703–1711.

23. Look J, Landwehr J, Bauer F, et al. Marked resistance of RAR gamma-deficient mice to the toxic effects of retinoic acid. Am J Physiol 1995; 269(1 Pt 1):E91–E98.

24. Sucov HM, Dyson E, Gumeringer CL, et al. RXR alpha mutant mice establish a genetic basis for vitamin A signaling in heart morphogenesis. Genes Dev 1994; 8(9):1007–1018.

25. Sucov HM, Izpisua-Belmonte JC, Ganan Y, et al. Mouse embryos lacking RXR alpha are resistant to retinoic-acid-induced limb defects. Development 1995; 121(12):3997–4003.

26. Mark M, Ghyselinck NB, Chambon P. Function of retinoid nuclear receptors: Lessons from genetic and pharmacological dissections of the retinoic acid signaling pathway during mouse embryogenesis. Annu Rev Pharmacol Toxicol 2006; 46:451–480.

27. Ross SA, McCaffery PJ, Drager UC, et al. Retinoids in embryonal development. Physiol Rev 2000; 80(3):1021–1054.

28. Nugent P, Sucov HM, Pisano MM, et al. The role of RXR-alpha in retinoic acid-induced cleft palate as assessed with the RXR-alpha knockout mouse. Int J Dev Biol 1999; 43(6):567–570.

29. Luo J, Pasceri P, Conlon RA, et al. Mice lacking all isoforms of retinoic acid receptor beta develop normally and are susceptible to the teratogenic effects of retinoic acid. Mech Dev 1995; 53(1):61–71.

30. Matt N, Ghyselinck NB, Wendling O, et al. Retinoic acid-induced developmental defects are mediated by RARbeta/RXR heterodimers in the pharyngeal endoderm. Development 2003; 130(10):2083–2093.

31. Lohnes D, Kastner P, Dierich A, et al. Function of retinoic acid receptor gamma in the mouse. Cell 1993; 73(4):643–658.

32. Iulianella A, Lohnes D. Contribution of retinoic acid receptor gamma to retinoid-induced craniofacial and axial defects. Dev Dyn 1997; 209(1):92–104.

33. Wubah JA, Ibrahim MM, Gao X, et al. Teratogen-induced eye defects mediated by p53-dependent apoptosis. Curr Biol 1996; 6(1):60–69.

34. Moallem SA, Hales BF. The role of p53 and cell death by apoptosis and necrosis in 4-hydroperoxycyclophosphamide-induced limb malformations. Development 1998; 125(16):3225–3234.

35. Pekar O, Molotski N, Savion S, et al. p53 regulates cyclophosphamide teratogenesis by controlling caspases 3, 8, 9 activation and NF-kappaB DNA binding. Reproduction 2007; 134(2):379–388.

36. Boreham DR, Dolling JA, Misonoh J, et al. Teratogenic effects of mild heat stress during mouse embryogenesis: Effect of Trp53. Radiat Res 2002; 158(4):443–448.

37. Gottlieb E, Haffner R, King A, et al. Transgenic mouse model for studying the transcriptional activity of the p53 protein: Age- and tissue-dependent changes in radiation-induced activation during embryogenesis. Embo J 1997; 16(6):1381–1390.

38. Norimura T, Nomoto S, Katsuki M, et al. p53-dependent apoptosis suppresses radiation-induced teratogenesis. Nat Med 1996; 2(5):577–580.

39. Nomoto S, Ootsuyama A, Shioyama Y, et al. The high susceptibility of heterozygous p53(+/−) mice to malformation after foetal irradiation is related to sub-competent apoptosis. Int J Radiat Biol 1998; 74(4):419–429.

40. Kato F, Ootsuyama A, Nomoto S, et al. Threshold effect for teratogenic risk of radiation depends on dose-rate and p53-dependent apoptosis. Int J Radiat Biol 2001; 77(1):13–19.

41. Laposa RR, Henderson JT, Xu E, et al. Atm-null mice exhibit enhanced radiation-induced birth defects and a hybrid form of embryonic programmed cell death indicating a teratological suppressor function for ATM. FASEB J 2004; 18(7):896–898.

42. Wang B, Ohyama H, Haginoya K, et al. Prenatal radiation-induced limb defects mediated by Trp53-dependent apoptosis in mice. Radiat Res 2000; 154(6):673–679.

43. Wang B. Involvement of p53-dependent apoptosis in radiation teratogenesis and in the radioadaptive response in the late organogenesis of mice. J Radiat Res (Tokyo) 2001; 42(1):1–10.

44. Mitchel RE, Dolling JA, Misonoh J, et al. Influence of prior exposure to low-dose adapting radiation on radiation-induced teratogenic effects in fetal mice with varying Trp53 function. Radiat Res 2002; 158(4):458–463.

45. Heyer BS, MacAuley A, Behrendtsen O, et al. Hypersensitivity to DNA damage leads to increased apoptosis during early mouse development. Genes Dev 2000; 14(16):2072–2084.

46. Martinez JD, Pennington ME, Craven MT, et al. Free radicals generated by ionizing radiation signal nuclear translocation of p53. Cell Growth Differ 1997; 8(9):941–949.

47. Ao A, Erickson RP, Bevilacqua A, et al. Antisense inhibition of beta-glucuronidase expression in preimplantation mouse embryos: A comparison of transgenes and oligodeoxynucleotides. Antisense Res Dev 1991; 1(1):1–10.

48. Bevilacqua A, Erickson RP, Hieber V. Antisense RNA inhibits endogenous gene expression in mouse preimplantation embryos: Lack of double-stranded RNA "melting" activity. Proc Natl Acad Sci U S A 1988; 85(3):831–835.

49. Zehentner BK, Haussmann A, Burtscher H. The bone morphogenetic protein antagonist Noggin is regulated by Sox9 during endochondral differentiation. Dev Growth Differ 2002; 44(1):1–9.

50. Ochiya T, Terada M. Antisense approaches to in vitro organ culture. Methods Enzymol. 314, 401–411, 2000.

51. Ochiya T, Sakamoto H, Tsukamoto M, et al. Hst-1 (FGF-4) antisense oligonucleotides block murine limb development. J Cell Biol 1995; 130(4):997–1003.

52. Jiang H, Soprano DR, Li SW, et al. Modulation of limb bud chondrogenesis by retinoic acid and retinoic acid receptors. Int J Dev Biol 1995; 39(4):617–627.

53. Motoyama J, Eto K. Antisense retinoic acid receptor gamma-1 oligonucleotide enhances chondrogenesis of mouse limb mesenchymal cells in vitro. FEBS Lett 1994; 338(3):319–322.

54. Motoyama J, Eto K. Antisense c-myc oligonucleotide promotes chondrogenesis and enhances RA responsiveness of mouse limb mesenchymal cells in vitro. FEBS Lett 1994; 338(3):323–325.

55. Kanwar YS, Liu ZZ, Kumar A, et al. Cloning of mouse c-ros renal cDNA, its role in development and relationship to extracellular matrix glycoproteins. Kidney Int 1995; 48(5):1646–1659.

56. Sainio K, Saarma M, Nonclercq D, et al. Antisense inhibition of low-affinity nerve growth factor receptor in kidney cultures: Power and pitfalls. Cell Mol Neurobiol 1994; 14(5):439–457.

57. Rothenpieler UW, Dressler GR. Differential distribution of oligodeoxynucleotides in developing organs with epithelial-mesenchymal interactions. Nucleic Acids Res 1993; 21(21):4961–496.

58. Rothenpieler UW, Dressler GR. Pax-2 is required for mesenchyme-to-epithelium conversion during kidney development. Development 1993; 119(3):711–720.

59. Durbeej M, Soderstrom S, Ebendal T, et al. Differential expression of neurotrophin receptors during renal development. Development 1993; 119(4):977–989.

60. Vukicevic S, Kopp JB, Luyten FP, et al. Induction of nephrogenic mesenchyme by osteogenic protein 1 (bone morphogenetic protein 7). Proc Natl Acad Sci U S A 1996; 93(17):9021–9026.

61. Liu ZZ, Wada J, Kumar A, et al. Comparative role of phosphotyrosine kinase domains of c-ros and c-ret protooncogenes in metanephric development with respect to growth factors and matrix morphogens. Dev Biol 1996; 178(1):133–148.

62. Arend LJ, Smart AM, Briggs JP. Mouse beta(6) integrin sequence, pattern of expression, and role in kidney development. J Am Soc Nephrol 2000; 11(12):2297–2305.

63. Quaggin SE, Vanden Heuvel GB, Igarashi P. Pod-1, a mesoderm-specific basic-helix-loop-helix protein expressed in mesenchymal and glomerular epithelial cells in the developing kidney. Mech Dev 1998; 71(1–2):37–48.

64. Quaggin SE, Yeger H, Igarashi P. Antisense oligonucleotides to Cux-1, a Cut-related homeobox gene, cause increased apoptosis in mouse embryonic kidney cultures. J Clin Invest 1997; 99(4):718–724.

65. Kanwar YS, Kumar A, Ota K, et al. Identification of developmentally regulated mesodermal-specific transcript in mouse embryonic metanephros. Am J Physiol Renal Physiol 2002; 282(5):F953–F965.

66. Enomoto H, Yoshida K, Kishima Y, et al. Hepatoma-derived growth factor is highly expressed in developing liver and promotes fetal hepatocyte proliferation. Hepatology 2002; 36(6):1519–1527.

67. Yang Q, Tian Y, Wada J, et al. Expression characteristics and relevance of sodium glucose cotransporter-1 in mammalian renal tubulogenesis. Am J Physiol Renal Physiol 2000; 279(4):F765–F777.

68. Li Z, Stuart RO, Qiao J, et al. A role for Timeless in epithelial morphogenesis during kidney development. Proc Natl Acad Sci U S A 2000; 97(18):10038–10043.

69. Shimada M, Yamamoto M, Wakayama T, et al. Different expression of 25-kDa heat-shock protein (Hsp25) in Meckel's cartilage compared with other cartilages in the mouse. Anat Embryol (Berl) 2003; 206(3):163–173.

70. Amano O, Yamane A, Shimada M, et al. Hepatocyte growth factor is essential for migration of myogenic cells and promotes their proliferation during the early periods of tongue morphogenesis in mouse embryos. Dev Dyn 2002; 223(2):169–179.

71. Abe M, Tamamura Y, Yamagishi H, et al. Tooth-type specific expression of dHAND/Hand2: Possible involvement in murine lower incisor morphogenesis. Cell Tissue Res 2002; 310(2):201–212.

72. Chai Y, Zhao J, Mogharei A, et al. Inhibition of transforming growth factor-beta type II receptor signaling accelerates tooth formation in mouse first branchial arch explants. Mech Dev 1999; 86(1–2):63–74.

73. Amano O, Koshimizu U, Nakamura T, et al. Enhancement by hepatocyte growth factor of bone and cartilage formation during embryonic mouse mandibular development in vitro. Arch Oral Biol 1999; 44(11):935–946.

74. Shum L, Sakakura Y, Bringas P Jr, et al. EGF abrogation-induced fusilli-form dysmorphogenesis of Meckel's cartilage during embryonic mouse mandibular morphogenesis in vitro. Development 1993; 118(3):903–917.

75. Slavkin HC, Sasano Y, Kikunaga S, et al. Cartilage, bone and tooth induction during early embryonic mouse mandibular morphogenesis using serumless, chemically-defined medium. Connect Tissue Res 1990; 24(1):41–51.

76. Foerst-Potts L, Sadler TW. Disruption of Msx-1 and Msx-2 reveals roles for these genes in craniofacial, eye, and axial development. Dev Dyn 1997; 209(1):70–84.

77. Chen B, Hales BF. Antisense oligonucleotide down-regulation of E-cadherin in the yolk sac and cranial neural tube malformations. Biol Reprod 1995; 53(5):1229–1238.

78. Augustine KA, Liu ET, Sadler TW. Interactions of Wnt-1 and Wnt-3 a are essential for neural tube patterning. Teratology 1995; 51(2):107–119.

79. Augustine KA, Liu ET, Sadler TW. Antisense inhibition of engrailed genes in mouse embryos reveals roles for these genes in craniofacial and neural tube development. Teratology 1995; 51(5):300–310.

80. Augustine K, Liu ET, Sadler TW. Antisense attenuation of Wnt-1 and Wnt-3a expression in whole embryo culture reveals roles for these genes in craniofacial, spinal cord, and cardiac morphogenesis. Dev Genet 1993; 14(6):500–520.

81. Corey DR, Abrams JM. Morpholino antisense oligonucleotides: Tools for investigating vertebrate development. Genome Biol 2001; 2(5):1015.1–1015.3.

82. Heasman J. Morpholino oligos: Making sense of antisense? Dev Biol 2002; 243(2):209–214.

83. Hyrup B, Nielsen PE. Peptide nucleic acids (PNA): Synthesis, properties and potential applications. Bioorg Med Chem 1996; 4(1):5–23.

84. Summerton J. Morpholino antisense oligomers: The case for an RNase H-independent structural type. Biochim Biophys Acta 1999; 1489(1):141–158.

85. Summerton J, Weller D. Morpholino antisense oligomers: Design, preparation, and properties. Antisense Nucleic Acid Drug Dev 1997; 7(3):187–195.

86. Stein D, Foster E, Huang SB, et al. A specificity comparison of four antisense types: Morpholino, 2'-O-methyl RNA, DNA, and phosphorothioate DNA. Antisense Nucleic Acid Drug Dev 1997; 7(3):151–157.

87. Summerton J, Stein D, Huang SB, et al. Morpholino and phosphorothioate antisense oligomers compared in cell-free and in-cell systems. Antisense Nucleic Acid Drug Dev 1997; 7(2):63–70.

88. Wickstrom E, Choob M, Urtishak KA, et al. Sequence specificity of alternating hydroyprolyl/phosphono peptide nucleic acids against zebrafish embryo mRNAs. J Drug Target 2004; 12(6):363–372.
89. Dmochowski IJ, Tang X. Taking control of gene expression with light-activated oligonucleotides. Biotechniques 2007; 43(2):161, 163, 165 passim.
90. Tang X, Dmochowski IJ. Regulating gene expression with light-activated oligonucleotides. Mol Biosyst 2007; 3(2):100–110.
91. Abes R, Arzumanov AA, Moulton HM, et al. Cell-penetrating-peptide-based delivery of oligonucleotides: An overview. Biochem Soc Trans 2007; 35(Pt 4):775–779.
92. Hunter ES III, Dix DJ. Heat shock proteins Hsp 70–71 and Hsp 70–73 are necessary and sufficient to prevent arsenite-induced dysmorphology in mouse embryos. Mol Reprod Dev 2001; 59(3):285–293.
93. Augustine-Rauch KA, Zhang QJ, Leonard JL, et al. Evidence for a molecular mechanism of teratogenicity of SB-236057, a 5-HT1B receptor inverse agonist that alters axial formation. Birth Defects Res A Clin Mol Teratol 2004; 70(10):789–807.
94. Hartig SM, Greene RR, Dikov MM, et al. Multifunctional nanoparticulate polyelectrolyte complexes. Pharm Res 2007; 24(12):2353–2369.
95. Pouton CW, Seymour LW. Key issues in non-viral gene delivery. Adv Drug Deliv Rev 2001; 46(1–3):187–203.
96. Tachibana R, Harashima H, Shinohara Y, et al. Quantitative studies on the nuclear transport of plasmid DNA and gene expression employing nonviral vectors. Adv Drug Deliv Rev 2001; 52(3):219–226.
97. De Laporte L, Cruz Rea J, Shea LD. Design of modular non-viral gene therapy vectors. Biomaterials 2006; 27(7):947–954.
98. Miyazaki M, Obata Y, Abe K, et al. Gene transfer using nonviral delivery systems. Perit Dial Int 2006; 26(6):633–640.
99. Zuhorn IS, Engberts JB, Hoekstra D. Gene delivery by cationic lipid vectors: Overcoming cellular barriers. Eur Biophys J 2007; 36(4–5):349–362.
100. Lavigne MD, Gorecki DC. Emerging vectors and targeting methods for nonviral gene therapy. Expert Opin Emerg Drugs 2006; 11(3):541–557.
101. Louise C. Nonviral vectors. Methods Mol Biol 2006; 333:201–226.
102. Dobson J. Gene therapy progress and prospects: Magnetic nanoparticle-based gene delivery. Gene Ther 2006; 13(4):283–287.
103. Mancheno-Corvo P, and Martin-Duque P. Viral gene therapy. Clin Transl Oncol 2006; 8(12):858–867.
104. Schepelmann S, Springer CJ. Viral vectors for gene-directed enzyme prodrug therapy. Curr Gene Ther 2006; 6(6):647–670.
105. Ivics Z, Izsvak Z. Transposons for gene therapy! Curr Gene Ther 2006; 6(5):593–607.
106. Foster K, Foster H, Dickson JG. Gene therapy progress and prospects: Duchenne muscular dystrophy. Gene Ther 2006; 13(24):1677–1685.
107. Wu Z, Asokan A, Samulski RJ. Adeno-associated virus serotypes: Vector toolkit for human gene therapy. Mol Ther 2006; 14(3):316–327.
108. Zhang X, Godbey WT. Viral vectors for gene delivery in tissue engineering. Adv Drug Deliv Rev 2006; 58(4):515–534.

109. Wang Y, Yuan F. Delivery of viral vectors to tumor cells: Extracellular transport, systemic distribution, and strategies for improvement. Ann Biomed Eng 2006; 34(1):114–127.

110. Griffiths RA, Boyne JR, Whitehouse A. Herpesvirus saimiri-based gene delivery vectors. Curr Gene Ther 2006; 6(1):1–15.

111. Wong LF, Goodhead L, Prat C, et al. Lentivirus-mediated gene transfer to the central nervous system: Therapeutic and research applications. Hum Gene Ther 2006; 17(1):1–9.

112. Young LS, Searle PF, Onion D, et al. Viral gene therapy strategies: From basic science to clinical application. J Pathol 2006; 208(2):299–318.

113. Turner JJ, Fabani M, Arzumanov AA, et al. Targeting the HIV-1 RNA leader sequence with synthetic oligonucleotides and siRNA: Chemistry and cell delivery. Biochim Biophys Acta 2006; 1758(3):290–300.

114. Harrop R, Carroll MW. Viral vectors for cancer immunotherapy. Front Biosci 2006; 11:804–817.

115. Fischer AJ, Stanke JJ, Omar G, et al. Ultrasound-mediated gene transfer into neuronal cells. J Biotechnol 2006; 122(4):393–411.

116. Mohr JC, de Pablo JJ, Palecek SP. Electroporation of human embryonic stem cells: Small and macromolecule loading and DNA transfection. Biotechnol Prog 2006; 22(3):825–834.

117. Sato Y, Kasai T, Nakagawa S, et al. Stable integration and conditional expression of electroporated transgenes in chicken embryos. Dev Biol 2007; 305(2):616–624.

118. Hartig PC, Hunter ES III. Gene delivery to the neurulating embryo during culture. Teratology 1998; 58(3–4):103–112.

119. Baldwin HS, Mickanin C, Buck C. Adenovirus-mediated gene transfer during initial organogenesis in the mammalian embryo is promoter-dependent and tissue-specific. Gene Ther 1997; 4(11):1142–1149.

120. Hunter ES III, Hartig P. Transient modulation of gene expression in the neurulation staged mouse embryo. Ann N Y Acad Sci 2000; 919:278–283.

121. Leconte I, Fox JC, Baldwin HS, et al. Adenoviral-mediated expression of antisense RNA to fibroblast growth factors disrupts murine vascular development. Dev Dyn 1998; 213(4):421–430.

122. Schachtner S, Buck C, Bergelson J, et al. Temporally regulated expression patterns following in utero adenovirus-mediated gene transfer. Gene Ther 1999; 6(7):1249–1257.

123. Kotch LE, Chen SY, Sulik KK. Ethanol-induced teratogenesis: Free radical damage as a possible mechanism. Teratology 1995; 52(3):128–136.

124. Tolia NH, Joshua-Tor L. Slicer and the argonautes. Nat Chem Biol 2007; 3(1):36–43.

125. Zamore PD, Haley B. Ribo-gnome: The big world of small RNAs. Science 2005; 309(5740):1519–1524.

126. Mattick JS, Makunin IV. Small regulatory RNAs in mammals. Hum Mol Genet 2005; 14(Spec No 1):R121–R132.

127. Prasanth KV, Spector DL. Eukaryotic regulatory RNAs: An answer to the "genome complexity" conundrum. Genes Dev 2007; 21(1):11–42.

128. Carthew RW, Molecular biology. A new RNA dimension to genome control. Science 2006; 313(5785):305–306.

129. Lewis BP, Burge CB, Bartel DP. Conserved seed pairing, often flanked by adenosines, indicates that thousands of human genes are micro RNA targets. Cell 2005; 120(1):15–20.
130. Huppi K, Martin SE, Caplen NJ. Defining and assaying RNAi in mammalian cells. Mol Cell 2005; 17(1):1–10.
131. Mittal V. Improving the efficiency of RNA interference in mammals. Nat Rev Genet 2004; 5(5):355–365.
132. Kunath T, Gish G, Lickert H, et al. Transgenic RNA interference in ES cell-derived embryos recapitulates a genetic null phenotype. Nat Biotechnol 2003; 21(5):559–561.
133. Nagy A, Rossant J, Nagy R, et al. Derivation of completely cell culture-derived mice from early-passage embryonic stem cells. Proc Natl Acad Sci U S A 1993; 90(18):8424–8428.

7

Use of Mammalian In Vitro Systems, Including Embryonic Stem Cells, in Developmental Toxicity Testing

Terence R. S. Ozolinš

Developmental and Reproductive Toxicology Center of Emphasis, Pfizer Drug Safety Research and Development, Groton, Connecticut, U.S.A.

In the past three decades, numerous in vitro developmental toxicity models have been described, but this chapter focuses on those assays currently in use for industrial screening, or that have been proposed for regulatory acceptance with respect to human risk assessment. The following models will be discussed: both rodent and avian micromass cultures, embryonic stem cells (ESCs), and rodent whole embryo culture. The zebra fish model is described in another chapter. A brief overview of the history and rationale for in vitro screens is provided, but rather than restating themes of earlier excellent reviews (1–14), the critical events of the last decade will be emphasized. The most significant occurrence during this time has been the large international effort funded by European Committee for the Validation of Alternative Methods (ECVAM) to validate three embryotoxicity tests. The biology, strengths, and weaknesses of each test are compared and the important contributions made by ECVAM with respect to the generation of prediction models, the validation process, and possible regulatory acceptance are highlighted. Examples are also given pertaining to the application of in vitro developmental toxicity tests in both industrial screening and the risk assessment process. ECVAM has recommended the regulatory acceptance of these in vitro tests, and this is discussed in the context of European legislation that strives to prohibit animal testing for certain chemical product classes. Finally, several workshops have been held to consider the limitations and possible improvements of the embryotoxicity models. Their recommendations serve as the basis for what in vitro embryotoxicity testing might look like in the future.

INTRODUCTION

The Need for In Vitro Tests

In general, simplicity and low cost make most in vitro systems useful for the study of biological systems. Developmental toxicity is particularly amenable to such modeling because in vitro studies are free from confounding maternal influences, and they permit the examination of direct treatment effects on embryo/fetal tissue. These attributes are useful to investigate two areas of interest: (1) the elucidation of mechanisms of normal and abnormal embryogenesis and (2) the assessment of developmental toxicity hazard. The former point will not be covered in this chapter, but is topic of an excellent review (15). Instead, this chapter focuses on the application of in vitro systems to screen for developmental toxicity.

Before proceeding, it is appropriate to briefly clarify some terms. In vitro will be used to encompass all culture models, including that of whole embryos, even though it is, strictly speaking, an ex utero model in which the entire intact embryo is used rather than just some portion of it. Although micromass and whole embryo culture require donor animals, the numbers used are dramatically reduced when compared to comparable in vivo studies. By using nonsentient organisms (nervous development begins after the termination of embryo culture) and by not directly treating sentient animals with test chemical, these in vitro and ex utero models are considered "alternative methods." Thus, for the purposes of this chapter "in vitro," "ex utero" and "alternative method" will be used interchangeably.

A compelling case for the use of in vitro developmental toxicity testing has been well argued elsewhere (16), but several key points warrant reiteration. It has been estimated that only 3000 of the approximately 60,000 to 90,000 commercial chemicals have been tested for their potential to adversely affect development (16). Unfortunately, it has also been recognized that, although current in vivo regulatory testing methods have been largely successful at protecting fetal health, it would require unjustifiably vast numbers of experimental animals, and too large an expenditure of time and resources to be used on every new chemical entity. Moreover, there are a number of legislative mandates to "categorize" tens of thousands of agents with subsequent screening and full assessment where warranted (17–21). This presents the uncomfortable predicament of trying to meet the obligation to protect public health in a cost-effective manner. Over the years, a number of strategies have been proposed to meet this challenge (22–26,159). The current thinking is to integrate in silico and alternative tests to prioritize chemicals for more comprehensive embryo/fetal toxicity evaluation. Although in silico developmental toxicity structure–activity relationship models are cost-effective, significant challenges remain, in part, due to the diverse array of mechanisms of dysmorphogenesis (17). In vitro tests are reasonably accurate and cheap when compared to in vivo toxicity studies, and therefore, it has been suggested that they may be useful as "prescreens," to prioritize chemicals for more comprehensive regulatory testing (7,16,25).

Each year, the pharmaceutical industry synthesizes thousands of novel chemical entities with an array of biological activities, of which just a fraction will ultimately make it to market (26). It has been estimated that about 7% of pharmaceutical product failures are due to reproductive/developmental toxicity concerns (26). Initially, this does not appear significant, but embryo/fetal toxicity testing is generally conducted relatively late in product development, after significant investments of time and resources have been made. Therefore, late-stage drug candidate attrition is catastrophic from an economic perspective, indicating an undeniable need to identify developmental risks much earlier. The implementation of in vitro developmental screens may facilitate the nomination of less embryotoxic candidates into the product pipeline (16,21,27,28).

Taken together, the advancement of efficient and predictive in vitro tests may serve both public health and industrial interests. First, they may help to prioritize the testing of chemical contaminants for more comprehensive regulatory testing. Second, they may ease drug development costs by reducing late-stage pharmaceutical failures caused by embryo/fetal toxicity.

History

There has been a formal interest in the use of in vitro systems for teratogenicity testing since 1975, when the "International Conference on Tests of Teratogenicity In Vitro" brought together investigators with experience in this field (29). Although the applicability of in vitro tests for screening purposes did not receive significant attention at that time, this aspect has become increasingly important to a number of stakeholders, including people concerned about animal use in toxicity testing and the legislators who translate these concerns into law, as well as those directly involved in safety assessments, namely, industrial users and their regulators. As the field of in vitro developmental toxicity testing matured, one of the problems that arose was the number of different platforms that had been proposed and "validated" with some kind of chemical test set (Table 1). These models have varied complexity and originate from a wide range of organisms including poxvirus, hydra, frog, fish, chick, rodent, and human. The proposed applicability of these diverse organisms to human/mammalian risk assessment is not surprising in view of the fact that most animals (metazoa) share 17 intercellular signaling pathways that are critical to embryogenesis (30). Disturbing any one of these signaling pathways is likely to be universally disruptive to development, even though the resultant phenotypes may be species specific. The proposed tests have a broad range of complexity including simple poxvirus replication in cells, differentiation of micromass cultures into neuronal, retinal or chondrogenic lineages, the differentiation of pluripotent ESCs into beating cardiomyocytes, and the most complex, the development of whole

Table 1 Alternative Embryotoxicity Models for Which Validation Sets Have Been Analyzed

Model	No. of agents validated	Reference
Nonmammalian whole organism		
Chick embryotoxicity screening test (CHEST)	130	(31)
Frog embryo teratogenesis assay: *Xenopus* (FETAX)	5	(32)
Hydra regeneration	24	(33)
Drosophila embryo	100	(34)
Zebra fish	12	(35)
Mammalian whole organism		
Rat whole embryo culture (WEC)	**25**	(36)
Mouse whole embryo culture (WEC)	10	(37)
Primary cell cultures		
Micromass Mouse embryo limb mesenchyme	27; 23	(38,39)
Rat embryo limb mesenchyme	**51; 25 retinoids**	(40,41)
Chick embryo limb mesenchyme	14	(42)
Rat embryo midbrain/limb	46	(43)
Rat embryo, in vivo dosed, CNS (central nervous system)/limb	31	(44)
Chick embryo neural crest	15	(45)
Chick embryo neural retina cell culture	45	(79)
Established cell lines		
Mouse ovarian tumor—attachment	102	(47)
Human embryonal palatal mesenchyme—proliferation	55	(48)
Neuroblastoma—differentiation	57	(49)
Embryonal carcinoma cells—differentiating	5	(50)
Poxvirus infected cells—poxvirus replication	49	(51)
Murine embryonic stem cells (ESC)	**16; 25**	(52,53)

Notes: A number of platforms for alternative tests have been proposed as developmental toxicity screens, and those which had been subjected to some kind of validation test set are listed. With the exception of the zebra fish, this represents the landscape in the mid 1990s as ECVAM began its validation efforts. The three tests that were selected for further improvement and validation are bolded.

intact embryos ex utero. This was the status of field until about the 1990s when a large international effort was initiated in by the ECVAM to develop and validate a series of embryotoxicity tests, which today represent the current state of the art. The purpose of this review will be to focus on these newer developments.

The political climate that led to the aforementioned validation efforts is interesting and deserves some elaboration. For some time, there has been a growing antivivisectionist sentiment which became institutionalized and gained legitimacy via the formation of such organizations as Fund for the Replacement of Animals in Medical Research (FRAME) in the United Kingdom, Interagency

Coordinating Committee of the Validation of Alternative Methods (ICCVAM) in the United States, and the Zentralstelle zur Erfassung und Bewertung von Ersatz-und Ergänzungsmethoden zum Tierversuch (ZEBET; translation: Centre for Documentation and Evaluation of Alternative Methods to Animal Experiments) in Germany. The creation of the European Union (EU) subsequently led to the creation of the pan-European ECVAM. The purpose of ECVAM, as defined in 1993 by its Scientific Advisory Committee, is to reduce, refine, or replace the use of laboratory animals in the biological sciences, by promoting the scientific and regulatory acceptance of alternative in vitro methods (11). These goals were not novel, and were in fact very similar to those of FRAME, ZEBET, and ICCVAM, but their incorporation into the mandate of a large international organization had the advantage of an abundance of economic and administrative resources. Driving ECVAM's aggressive validation efforts were several circumstances unique to Europe. First was the protection, to varying degrees, of nonhuman animal rights in the constitutions of E.U. member nations (54). In addition, E.U. Cosmetics Directive (55) will prohibit animal testing for cosmetics, and the REACH initiative mandates the generation of toxicity data for thousands of chemicals for which there is currently no such information (56). This presented an interesting dilemma: In the face of an increased need for developmental toxicity testing, how does one simultaneously protect animals from toxicity testing and consumers, on the other hand, from developmental toxicity hazard? The strategy was to invest heavily in the development and validation of alternative methods.

Overview of the ECVAM Validation Efforts

In spite of the divergent opinions about the merits of in vitro developmental toxicity tests (16,57,58), the ECVAM efforts were able to achieve a number of important milestones that had eluded the field of in vitro teratogenicity screening for 25 years. In about 7 years, at a cost of € 1.6 million (59), consensus had been achieved in several key areas including the selection of the in vitro test models, harmonized protocols employing the principles of good laboratory practice (GLP), a validation chemical test set, and blinded validation studies with appropriate statistical analysis. These are discussed briefly later.

In Vitro Test Models

The approach of ECVAM was to implement a process with which to become better informed about the state-of-the-art nonanimal test development and validation. The format used was to organize ECVAM-sponsored workshops on specific topics, where invited experts from industry, academia, and regulatory agencies would review the current status of in vitro tests and make recommendations concerning a strategy forward. The twelfth such workshop, held in 1994, was jointly organized with the European Teratology Society and its report was entitled "Screening Chemicals for Reproductive Toxicity: The Current Alternatives (11)." At that time, almost 20 different models had been "validated" with some kind of chemical test set (see Table 1), and it was concluded that four in vitro systems were capable of detecting substances likely to exert potent effects on the physical development of the embryo, and that ECVAM should be responsible for their comparison and validation. The four tests were frog embryo teratogenesis assay-*Xenopus* (FETAX), chick embryotoxicity screening test (CHEST), micromass cultures, and whole embryo culture. It was also recommended that methods using mammalian ESCs (52,53) demonstrated sufficient promise to warrant further development and validation (11). Subsequent meetings and workshops ultimately culminated in the final definitive validation study of three models: rat limb bud micromass, the murine embryonic stem cell test, and rat whole embryo culture (60), which are discussed in more detail later. With this apparent success, ECVAM continues to fund research into novel cellular platforms and improved end points, and these are discussed in the section entitled "Future Directions."

Harmonized Protocols

Another notable achievement by ECVAM was the harmonization of very divergent protocols for each of the three recommended in vitro assays. Whole embryo culture will be used for illustrative purposes, but bear in mind similar considerations apply to the other models. The first issue is species selection. Rat and mouse whole embryo culture is most common (2,3,61,65), but rabbit whole embryo culture, an important species from a regulatory perspective (62), has also been used (63,64,100). The duration of incubations is also highly variable with 26, 30, 44, and 48 hours being most commonly reported (14,61,65,66). The composition of the growth medium, in particular the serum content and source, may also vary widely between labs and has been reviewed elsewhere (14). One study suggests that such changes are not significant over 26 hours of culture (61), although over 44 hours we have found significant differences (67). Lastly, there are a number of end points to be assessed that measure embryonic growth and development (3,68–70). Following input from a number of workshops and incorporating the validation principles compliant with GLP (71,72), standardized protocols for all three tests were agreed upon and made publicly available (73–75).

Chemical Validation Set

Numerous test chemicals have been proposed and used with varying degrees of overlap for the validation of in vitro developmental toxicity screens (Table 1), and not surprisingly their merits and shortcomings have been debated. This topic has been reviewed in considerable detail by others (5,16,36,76–79), but it is a critical aspect of the validation process (72) justifying a brief overview of several fundamental points. The first is a philosophical question: What is a teratogen? One definition, used by a panel of teratologists (76) when generating a 47 chemical test set, was that a teratogen is a compound, which in the absence of maternal toxicity induces embryolethality, growth retardation, structural abnormality, or prenatal or postnatal functional deficit. Others contend that this is actually the definition of a developmental toxicant, not specifically a teratogen (5). It has also been suggested that since these in vitro models are ultimately designed for human hazard identification, only human teratogens should be used, but such an approach is hampered because there are no properly controlled human teratogenicity trials, and as a result the clinical data set is primarily retrospective studies or case reports that do not lend themselves to rigorous assessment of "true" teratogenicity as in the case of animal experiments (5).

There are also different strategies to bin validation test agents. One is dichotomous or binary, in which chemicals are segregated into either teratogens or nonteratogens (76), irrespective of the relative degree of hazard. Another is to stratify chemicals using the A/D ratio, in which the relative toxicity in adult versus developmental tissue is compared. Either the lowest observed adverse effect level (LOAEL) or the no observed adverse effect level (NOAEL) may be used for the A/D ratio, although LOAEL is less susceptible to error (78). One drawback to A/D ratios is that they vary across species (80). It has been suggested that the best strategy is to make qualitative assessments about the developmental hazard; terms like "weakly positive" or "strongly positive" are used (5). Unfortunately, there is no consensus on the estimates of mammalian hazard in vivo (5), but, ultimately, it was this approach that was endorsed by the ECVAM management team (78).

Several other points must also be considered. There are divergent views for the inclusion/exclusion criteria for the admissibility of in vivo teratogenicity data sets, particularly with respect to clinical reports (5,78). In addition, the appropriateness of certain chemicals has also been debated. For example, in early chemical validation sets, many "nonteratogens" were endogenous or xenobiotic agents that are generally nontoxic such as glutamate, lysine, ascorbic acid, and saccharin, whereas, in contrast, "teratogens" were potent antimetabolites, alkylating agents, and hormones. Thus, any biological system, not just embryonic tissue, would respond very differently to each chemical group (5,78). Another consideration is that some agents are proteratogens and must be bioactivated, generally in the maternal compartment, to become teratogenic. In such cases, the metabolite, not the parent compound, must be used; for example, salicylic acid in lieu of acetylsalicylic acid and dimethadione in lieu of trimethadione (78). Another consideration is whether to include agents such as hormones which mediate their effects through well-characterized receptor-mediated pathways that may be detected with other nonembryonic in vitro assays.

After considering the factors discussed earlier, other theoretical aspects of validation test chemical selection (72,80,81), and a survey of earlier validation lists (76,82–86), a test set with 309 potential chemicals was proposed by Nigel Brown (78). A draft candidate list from this diverse array of chemicals was submitted to the ECVAM study management team for consensus of the final 20 agents, with the final endorsement made by ZEBET, the contractors to the European Commission. Briefly, the following three criteria were used for chemical selection (78): (1) No distinction was made between different manifestations of developmental toxicity. That is, structural malformation, functional deficits, intrauterine growth retardation, and in utero death were considered equal as long as they were observed in

Table 2 Twenty Test Chemicals used in the ECVAM Validation Study

Nonembryotoxic	Weakly embryotoxic	Strongly embryotoxic
Acrylamide	Boric acid	5-bromo-2'deoxyuridine
Isobutyl-ethyl-valproic acid	Pentyl-4yn-valproic acid	Methyl mercury chloride
D(+)-Camphor	Valproic acid	Hydroxyurea
Dimethyl phthalate	Lithium chloride	Methotrexate
Diphenylhydramine hydrochloride	Dimethadione	All *trans*-retinoic acid
Penicillin G sodium salt	Methoxyacetic acid	6-Aminonicotinamide
Saccharin sodium hydrate	Salicylic acid sodium salt	

Notes: The twenty chemical validation test set as determined by Nigel Brown (78). Embryotoxicity was defined to include any of the four signs of developmental toxicity: intrauterine growth retardation, decreased fetal viability, functional, or structural deficits. All manifestations were considered to have equal weighting as long as they were noted in the absence of maternal toxicity. The three classes of hazard were described as unequivocal nonembryotoxicants; unequivocal embryotoxicants, referred to as strongly embryotoxic; and a class in between, termed weakly embryotoxic, in which the developmental effects may have been species specific.

the absence of maternal toxicity. (2) The validation test set was divided into three classes of developmental toxicity: unequivocal nonembryotoxicants; unequivocal embryotoxicants, referred to as strongly embryotoxic; and a class in between, termed weakly embryotoxic (Table 2). Thus, the toxicity models would be tested in their ability to discriminate between these three classes. (3) Importantly, the test articles represent both pharmaceutical agents and industrial contaminants, theoretically making any validated test broadly applicable to a range of chemicals.

Validation Trial

Details about the ECVAM validation processes are reviewed elsewhere (60,73), but some points warrant mention. First, ECVAM, ICCVAM, and the Organisation for Economic Co-operation and Development (OECD) collaborated to develop specific criteria for the validation of all alternative toxicity tests (72). These principles were applied to the validation of the three embryotoxicity tests and include some of the points discussed later. First, each of the three in vitro tests was conducted in four different laboratories across Europe, adhering to the principles of Good Laboratory Practice (GLP) (87). In addition, test chemicals were coded prior to distribution, ensuring that each laboratory was applying the test articles blindly, and the results were returned to ECVAM to be analyzed "at arm's length" by a biostatistician (81). There was a preliminary prevalidation phase (88) with 6 of the 20 recommended test articles, permitting the derivation of biostatistical prediction models that could be validated in the definitive phase with the remaining 14 agents. This use of mathematical prediction models is one of the fundamental tenets of the validation process because in this way in vivo outcomes may be predicted with in vitro results (72).

BIOLOGY OF ASSAYS

Prior to describing the outcomes from the ECVAM validation trial, an overview of the technical and biological aspects of the alternative tests will be provided. The in vitro tests validated by ECVAM were based upon three distinct platforms: micromass, ESCs, and whole embryo culture. The central thesis is that at least one basic developmental process is represented in each model, and that its disruption in vitro represents an in vivo developmental toxicity risk.

Micromass

The appeal of micromass cultures lays in their technical simplicity and low cost. In this approach, an intact embryonic organ is harvested and dissociated into a single cell suspension via mechanical forces and enzymatic digestion. These cells are placed into culture, where they are allowed to replicate, migrate, re-aggregate, and differentiate into a specific cell types. The capacity of a test article to interfere with these processes reflects, at least in theory, its in vivo teratogenic potential.

The micromass protocol used by ECVAM was the in vitro toxicity (INVIT-TOX) protocol 114 (74), modified from earlier investigators (40,43,89). It is based upon cultures of dissociated cells from gestation day 14 rat embryo limb buds, which are seeded into 96-well plates as high-density spots. Within 5 days, the dis-aggregated cells differentiate into chondrocytes, which are detected using Alcian Blue, a specific stain for cartilage. The inhibition of differentiation potential is determined by the relative decrease in Alcian Blue staining. Despite the aggressive cell preparation techniques, it is encouraging to note that micromass limb bud cultures of rat and rabbit retain their respective species-specific resistance and sensitivity to the effects of thalidomide (90).

In addition to the mammalian-based micromass assays discussed earlier, a chick embryo neural retina cell culture model has also been proposed (79,91). Although not validated by ECVAM, it is an alternative test that is currently in use for chemical screening in at least one industrial setting (46). Here, chick embryonic retina is harvested and dissociated. Using a rotating suspension culture, the single cells form multicellular aggregates that eventually express a histologic and biochemical phenotype similar to the in situ retina (92,93). In the proposed developmental toxicity screen, three end points are measured: the number of aggregates formed, their protein content, and glutamine synthetase activity, which reflect the capacity for cell–cell interactions, growth, and differentiation, respectively (91). A given test article may affect each of the three end points to different extents, but in this assay only one end point needs to be affected for it to be considered at risk for developmental toxicity (79). A schematic representation of the micromass method is depicted in Figure 1.

Embryonic Stem Cells

Recent advances in embryonic stem cell technology have made these cells available for a variety of toxicity models (94–96). The use of murine stem cells for

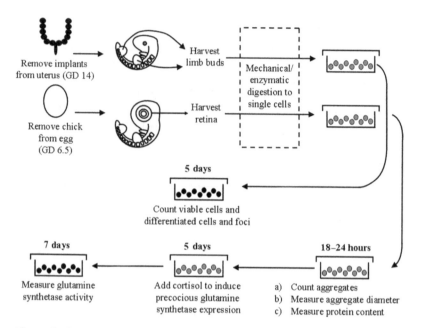

Figure 1 The highlights of the rat limb bud and chick embryonic neural retina micromass cultures are depicted. The relevant tissues are harvested on the appropriate gestation day (GD) and digested to a single-cell suspension. For rat limb cultures, high-density spots are plated and allowed to grow for approximately 5 days. Cytotoxicity is calculated by assessing the cell number with Neutral Red stain. The degree of differentiation is determined by measuring the intensity of Alcian Blue dye, a specific stain for mesenchymal cartilage production. Four factors are considered in the chick retina test. The first three are measured after 24 hours of culture and include (a) the number of aggregates that are produced, (b) the size of the aggregates, and (c) their protein content. After 5 days, cortisol is added to precociously induce glutamine synthetase activity. This is measured 2 days later, after a total of 7 days of culture. Each parameter is uniquely sensitive to different agents, but a decrease in any one parameter is considered to be toxicologically relevant.

developmental toxicity testing is based upon the observation that in culture these pluripotent cells, derived from the inner cell mass of a blastocyst, may be induced to differentiate into cell types from the three primary germ layers. Their gene expression patterns reflect a rough concordance with the gene expression patterns observed during the differentiation of early, preimplantation embryos in vivo (94,97). These observations suggest that in vitro embryonic stem cell differentiation may replicate many processes that occur during in vivo embryogenesis, and therefore may be an appropriate surrogate for the embryo with respect to developmental toxicity testing. Moreover, from an ethical perspective, unlike other in vitro protocols described in this chapter, which require the sacrifice of donor animals for test tissue, embryonic stem cell lines are immortal necessitating no further animal sacrifice. That said, this protocol is not entirely "animal friendly" due to the

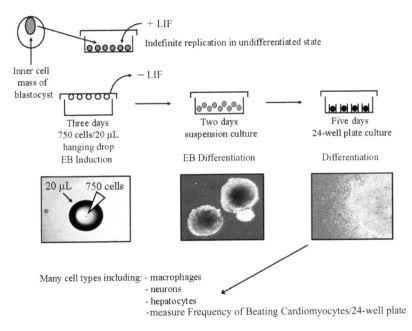

Figure 2 The notable points of stem culture are depicted. ESCs are isolated from the inner cell mass of a blastocyst. When placed into culture in the presence of LIF, they will both maintain their pleuripotency and replicate indefinitely. The ECVAM stem cell model begins with the removal of LIF from a permanent stem cell line, which induces stem cells to differentiate. Three days of hanging drop cultures relies on gravity to force stem cells into close proximity resulting in the induction of spheroidal aggregates, termed EB. Two days of suspension culture permit EB to further differentiate. A single EB is then placed into each well of a 24-well plate to differentiate a further 5 days. Although a variety of factors may be added to the media to drive stem cell differentiation down specific pathways, the ECVAM protocol allows stem cells to spontaneously produce a variety of cell types including beating cardiomyocytes. A decrease in the number of wells which contain beating cardiomyocytes is a measure of developmental toxicity. *Source*: Photographs graciously provided by Donald B. Stedman.

fetal bovine serum needed to support ESC growth and differentiation. As a result, efforts are underway to identify the critical components of fetal bovine serum to facilitate development of an artificial serum using recombinant technologies (98).

In the ECVAM INVITTOX protocol, 113 stem cells are grown under conditions that produce, as the end point of differentiation, beating cardiomyocytes (73) (Fig. 2). In brief, the procedure first requires the removal of leukemia inhibitory factor (LIF) from the culture media; this allows murine ESCs to start differentiating. Cells are grown for 3 days in "hanging drop" culture, where small drops are placed onto Petri dish covers which are inverted, allowing gravity to aggregate the cells at the nadir of the drop of medium. These roughly spheroidal aggregates, termed embryoid bodies (EB), are then brought into a standard suspension culture for two more days, promoting further differentiation. This is followed by the seeding of a single embryoid body per well in a 24-well plate for a further 5 days (10 days in total). At the termination of culture, these wells contain a variety of cell types from all three germ layers including chondrocytes; neuronal and macrophagic precursors; and short primitive vascular beds, red blood cells, and beating cardiomyocytes. The formation of contracting cardiomyocytes is relatively complex, dependant upon a variety of fundamental processes such as cellular differentiation, migration, cell–cell recognition, and ultimately the communication of synchronized electric impulses across a large surface area (as these cells tend to beat in unison). In principle, developmental toxicants reduce the frequency of wells (within a 24-well plate) that contain beating cells, whereas nontoxicants do not.

Whole Embryo Culture

Although rodent whole embryo culture has existed for at least half a century, it only became popularized after Dennis New published a simplified protocol in 1978 (99). The reader is referred to several excellent reviews on rodent whole embryo culture and its use as a tool for in vitro toxicity screening (4,8,13).

The simplified procedure was based upon the explantation of presomitic or early somitic rodent embryos with intact visceral yolk sacs, their growth for approximately 48 hours in rotating bottles containing heat inactivated rat serum, and exposure to increasing concentrations of oxygen (99). If explanted on day 10 and grown for about 48 hours, embryos have a beating heart and circulation, closed neural tubes, limb buds, and express many of the major *anlagen* for major organs such as eyes, ears, maxilla, and mandible. Moreover, in cultures that are terminated at the 30 to 32 somite stage, the in vitro growth and development of embryos are visually indistinguishable from in vivo, save a small decrease total embryonic protein (3,99), and a transient induction of immediate/early response genes (66). Prolonged cultures up to 72 hours with more frequent media changes have been reported, and although this extended duration has clear advantages for addressing queries related to developmental biology or mechanisms of teratologic action, its utility for developmental toxicity screening is yet unproven (14). Taking these factors into consideration, the ECVAM protocol starts with gestation day 10 (0–6 somites) rat embryos grown for approximately 48 hours (75), according to INVITTOX protocol 68, depicted schematically in Figure 3.

The embryonic stem cell and micromass tests (limb bud and chick retina) are each based upon a single process or organ; beating cardiomyocytes, chondrogenesis, or retinal cell generation, respectively. In contrast, the entire embryo and all of its *anlagen* are assessed in whole embryo culture. The parameters measured may be divided into four categories: functional assessment, growth, development, and the presence or absence of deformities. The functional evaluation considers embryo viability as determined by heartbeat, an operative blood circulation in

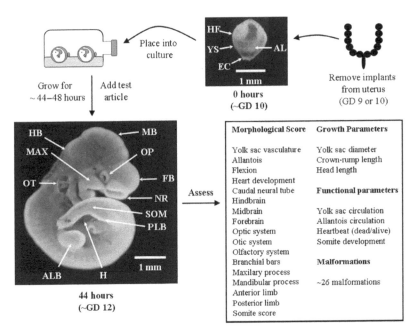

Figure 3 The highlights of rat whole embryo culture. Implants are removed from the gestation day (GD) 10 uterus and microdissected to yield an embryo of about 0 to 6 somites with intact visceral yolk sac. These are grown in a rat serum–based medium with intermittent exposure to increasing oxygen levels. After approximately 44 to 48 hours, the embryo is virtually indistinguishable from an in vivo grown GD 12 embryo. A number of *anlagen* have developed at this time including hindbrain (HB), midbrain (MB), optic region (OP), forebrain (FB), nasal ridge (NR), somites (SOM), posterior limb buds (PLB), anterior limb buds (ALB), otic region (OT), and maxillary process (MAX). Four end points are evaluated: functional parameters, growth (size), abnormal development (approximately 26 malformations), and speed of development (total morphological score; sum total development of the 17 *anlagen*).

the yolk sac, and allantois and somite integrity. Growth is gauged by the physical size of the embryo consisting of measurements of the yolk sac diameter, crown-rump length, and head length. The developmental progress is quantified in a fairly consistent and unbiased manner by using a morphological scoring system that ascribes numerical scores for various developmental landmarks in at least 17 distinct organ/*anlagen* systems. The sum yields a total morphological score (TMS) which increases linearly with advancing age during the middle widow of organogenesis, thereby providing a reasonable measure of the speed of development (68). The first and most widely used scoring system was developed for rat by Brown and Fabro (68). The utility of a morphological scoring system was recognized and modifications were made to accommodate mouse (69) and rabbit embryo (64,100). Lastly, similar to an in vivo embryo/fetal study, the presence and frequency of deformities can also be determined, and in many cases, the phenotype of chemically induced malformations in vitro correlate with those found following in vivo administration (3). Taken together, these end points make whole embryo culture very data rich and the most in vivo like. Interlaboratory validation studies with eight blinded chemicals confirmed the relative consistency of whole embryo culture (65), but nevertheless, this model exhibited the greatest degree of interlaboratory variability in the ECVAM study (60,101).

Although the most lifelike of the assays described, whole embryo culture is also the shortest, which raises an important question: What is the practical utility of a 2-day assay spanning but 10% of rodent gestation? As illustrated in Figure 4, the sensitivity to teratogenic insult varies during gestation, with organogenesis being the most sensitive. Thus, the critical determinant of teratogenicity is the window, not the duration, of exposure. Serendipitously, the time period during which whole embryo culture is conducted (gestation day 9–11 or 10–12 in the rat) is the most sensitive window of organogenesis. In fact, the developmental events that occur during whole embryo culture and the embryonic stem cell test correspond at, or near, the peak of sensitivity, an ideal scenario for a short-term in vitro toxicity test. In contrast, the rat limb bud micromass test covers the tail of the sensitivity curve (Fig. 4). This may explain, in part, why during the ECVAM validation trial the micromass test did not perform, as well as the stem cell test or whole embryo culture. The aforementioned test characteristics are summarized in Table 3.

INTERPRETATION OF IN VITRO EMBRYOTOXICITY DATA

Classical Approach

These alternative models all represent a surrogate embryo against which to test chemically mediated developmental toxicity directly on embryonic tissue without confounding maternal influences. However, in their simplicity, these models also present a toxicological conundrum, as underscored by a translation from Philipus Aureolus Paracelsus. "All things are poison and nothing is without poison; only the dosage makes it a poison." The "dose" that may be administered to these in vitro systems is limited only by the solubility of the test article, and, therefore, virtually any agent will induce embryotoxicity if the concentration is high enough. Thus, these in vitro/ex utero assays differ significantly from in vivo fetal toxicity testing, where maternal toxicity will be dose limiting. In fact, three parameters are used to gauge teratogenic risk in vivo: maternal toxicity, embryo/fetal toxicity, and malformations (Table 4). For example, there is relatively little concern of developmental toxicity if an agent only produces structural malformation at doses that cause maternal toxicity and overt embryotoxicity such as intrauterine growth retardation or in utero death. In contrast, there is unease if malformations occur in the absence of maternal toxicity, and added concern if they occur in the absence of embryotoxicity. Thus, in vivo, the response of the conceptus is normalized against

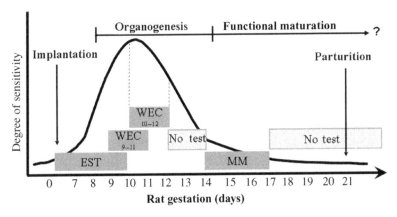

Figure 4 The sensitivity to a theoretical teratogen during rat gestation [modified from (102)]. The most susceptible window is organogenesis with low levels of vulnerability at the time of implantation and the period of functional maturation. Superimposed are the approximations of when the biological events that are represented in the three in vitro tests occur. The developmental events occurring during the mouse embryonic stem cell test (EST) correspond roughly to the period of gestation day (GD) 6–10 in the rat, near the peak of sensitivity. Whole embryo culture (WEC) recapitulates the window at the peak of sensitivity, between GD 9–11, GD 10–12, and GD 9–12 if 72-hour culture is conducted. The rat limb bud micromass (MM) cultures represent chondrogenic development of approximately GD 14–17, at the tail of the sensitivity curve. No validated in vitro tests reflect processes that occur between GD 12–14 and GD 17 to parturition.

another toxicity benchmark. As described later, a number of solutions were offered to overcome this issue in vitro.

Using whole embryo culture as an example, a number of retrospective studies were conducted to show that either the concentration needed to induce in vitro malformations was similar to the in vivo circulating maternal plasma levels (3) or that specific relationships existed between in vitro concentrations and in vivo dose (36). These approaches provided comfort in some aspects of the validity of the whole embryo culture, but were not particularly useful when testing unknown chemicals with little to no in vivo exposure data, as would be the case with industrial early screening programs. Alternatively, one may normalize the occurrence of in vitro malformations against toxicity measures of the yolk sac or embryonic development (morphological score). That is, a chemical would be classified as a potential teratogen if malformations occurred without yolk sac effects, thereby ensuring that effects were not secondary to yolk sac cytotoxicity. The TMS may be considered the in vitro surrogate to fetal body weight (Table 4), and by extension to the in vivo scenario, there would be teratogenic concern if malformations occurred at concentrations where the TMS was unchanged. This approach was helpful, but not without its shortcomings, as illustrated in the example below.

Table 3   Summary of the Characteristics of the ECVAM—Validated Tests

Characteristic	Micromass	Embryonic stem cells	Whole embryo culture
Duration of the in vitro test	5 days	10 days	2 days
Gestational days (GD) represented	~14–17	~ 5–10	10–12
Approximate gestational processes represented	Cartilage formation	Blastocyst formation to beating heart tube	Early/mid organogenesis of entire embryo
Approximate location on the teratogen sensitivity curve (Fig. 4)	Tail of curve (low sensitivity)	Near peak (higher sensitivity)	Peak (highest sensitivity)
End point measured	Intensity of Alcian Blue staining of cartilage	Frequency of wells containing beating cardiomyocytes	(1) Functional (2) Growth (3) Speed of development (TMS of 17 organ *anlagen*) (4) ~26 discrete malformations
Technical difficulty (overall)	+	+++	+++++
Experimental conduct	+	+++++	++++
End point assessment	+	+	+++++
Cost	+	+++++	+++++
Animal requirements:			
Tissue donors	Yes	No	Yes
Serum in Media	Yes	Yes	Yes
Current species used	Rat	Mouse	Rat
Other possible species	Rabbit	Rabbit Human	Mouse Rabbit Hamster

Notes: Various characteristics of the ECVAM validated tests are summarized. The technical difficulty and costs are rated as high (+++++) or low (+).

Briefly, using mouse whole embryo culture, it was determined that the selective serotonin receptor uptake inhibitors, fluoxetine and sertraline (Prozac™ and Zoloft™, respectively), were at risk for producing in vivo craniofacial malformations (103). This was a logical conclusion because craniofacial malformations were observed at in vitro concentrations that did not appreciably reduce the overall rate of embryo development as determined by TMS, and, furthermore, the yolk sacs appeared healthy and unaffected. However, the in vivo regulatory studies conducted in rat and rabbit for both fluoxetine and sertraline found no evidence of malformations (104), and a recent comprehensive National Toxicity Program Center for the Evaluation of Risks to Human Reproduction (NTP CERHR) retrospective review of all clinical and experimental data relating to the developmental toxicity for fluoxetine (105) concluded that there might be a slight increase in the risk of spontaneous abortion, but no added risk for malformations. A similar review has not been conducted with sertraline, but the current literature also suggests no added risk of malformations (106,107). Why the discordance between the in vitro prediction and the in vivo outcome? Quite simply, the maternal component of developmental toxicity data interpretation was not accounted for. It will be recalled that the presence of malformations is only a concern in the absence of maternal toxicity. In the sertraline example, the in vitro drug concentration needed to produce malformations was considerably higher than the in vivo maternal plasma levels at a lethal dose (103), indicating the in vitro drug levels had no in vivo relevance. The biostatistical prediction model developed by ECVAM attempts to address this concern.

Novel ECVAM Approach to Data Interpretation

The outcome of the example above underscores the dilemma faced when screening for in vitro developmental toxicity, namely, how best to predict or model maternal toxicity? One strategy may be to use in silico absorption, distribution, metabolism, excretion (ADME) predictions of therapeutic concentrations, but the accuracy of such approaches has not been demonstrated to date (108). In an entirely novel approach, ECVAM employed a general cytotoxicity measure as determined in a terminally differentiated 3T3 fibroblast cell line. From Table 4, it will be noted that this may be viewed as an in vitro surrogate to the in vivo maternal toxicity measure. Its importance is underscored by several observations. First, in the original ECVAM whole embryo culture prediction model, referred to as PM1, only embryonic malformations and TMS were considered. This model reflected the earlier philosophy of considering malformations toxicologically relevant only if the TMS was relatively unaffected, and its accuracy was 68%, but when the 3T3 cytotoxicity measure was included in the prediction model (PM2), the accuracy increased to 82% (60,101). In addition, the acceptance of this general cytotoxicity measure in nondifferentiating tissue has been recognized by others as they develop new embryotoxicity models independent of ECVAM (67,109).

Another novel approach underpinning the ECVAM initiative was the strong influence of biostatistical methods. It was determined that one of the criteria needed for "validation" and regulatory acceptance was the derivation of statistical prediction models from which in vivo toxicity outcomes could be predicted from in vitro data sets (71,72,88,110,111,160). Briefly, for the three embryotoxicity tests, in vitro surrogates were identified for each of maternal toxicity, embryotoxicity, and frank malformations (Table 4). The three respective toxicity curves were generated and using an empirical approach, the goal was to identify points (IC values) that could be used to distinguish between non embryotoxicants, weak embryotoxicants,

Table 4 Determinants of Developmental Toxicity Risk

In Vivo	Maternal Toxicity	Embryo Toxicity	Structural/Functional Deficits	Determination of developmental hazard
	- death - ↓ absolute body weight - >10% ↓ in body weight gain - exaggerated pharmacology of test article (e.g., convulsion)	- IUGR (intrauterine growth retardation) - ↑ postimplantation loss - ↓ litter size - ↓ fetal viability	- visceral malformation - skeletal malformations - ↑ skeletal variation - behavioral effects	- Occurrence of structural/functional deficits in the absence of maternal or embryo toxicity - Determined in "Consensus" meetings in which consensus is seldom achieved.
In vitro WEC Pre-ECVAM	Surrogate Not assessed	Surrogate - ↓ total morphological score - ↓ somite number - ↓ crown rump or head length - yolk sac compromise	Surrogate - ↑ dysmorphogenesis - petechia - surface blistering/swelling - tissue destruction - absence of circulation/heart beat	Surrogate Functional/ structural deficits in the absence of embryo-toxicity (No change in TMS or yolk sac integrity)

ECVAM MM	Not assessed	Not assessed	- ↓ intensity of Alcian Blue staining of chondrocytes	Biostatistical Prediction Model
ECVAM EST	3T3 cytotoxicity	Embryonic stem cell cytotoxicity	- ↓ frequency of wells with beating cardiomyocytes	Biostatistical Prediction Model
ECVAM WEC	3T3 cytotoxicity	↓ TMS	- ↑ dysmorphogenesis	Biostatistical Prediction Model

Notes: The relationship between the various determinants of developmental toxicity risk are compared between in vivo and in vitro tests. The risk of in vivo developmental toxicity is low if neither embryotoxicity nor structural deficits occur in the face of maternal toxicity. Conversely, the risk is high if embryotoxicity or structural/functional deficits are observed in the absence of maternal toxicity. In real life, the distinction is never this clear requiring "consensus" meetings with multiple "experts" to make the determination, although consensus is seldom achieved. The in vitro surrogates for maternal toxicity, embryotoxicity, and structural/functional deficits are described. In Pre-ECVAM whole embryo culture (WEC), the developmental hazard was high if structural defects were noted without a reduction in total morphological score (TMS) or yolk sac toxicity. The ECVAM micromass (MM), embryonic stem cell test (EST), and WEC use biostatistical prediction models to determine whether the relationship between the in vitro surrogates warrants concern.

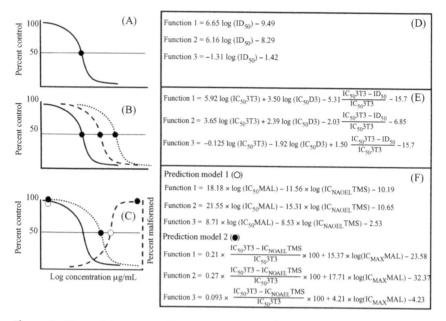

Figure 5 The toxicity curves, IC values, and discriminate function equations are used in the three validated ECVAM embryotoxicity prediction models. The single curve in the micromass test (**A**) reflects the inhibition of chondrocyte differentiation as determined by a decrease in Alcian Blue staining. The concentration that inhibited differentiation to 50% control levels is identified (ID_{50}; ●) and substituted into the corresponding linear discriminate functions (**D**). In the stem cell test (**B**), three parameters are measured: the inhibition of cardiomyocyte beating (*solid line*), stem cell viability (cytotoxicity) (*dashed line*), and viability (cytotoxicity) 3T3 fibroblast cells (*dotted line*). The concentrations that inhibited each parameter to 50% control levels are determined. Respectively, they are ID_{50} (*solid line*), $IC_{50}D3$ (*dashed line*), and $IC_{50}3T3$ (*dotted line*). These values are applied to the function equations in panel **E**. Two prediction models were developed for the whole embryo culture test (**C**). In Prediction Model 1 (○), the no observed adverse effect level for the total morphological score was determined (*solid line*; ○; $IC_{NOAEL}TMS$), as well as the concentration that produced a 50% malformation rate (*dashed line*; ○; $IC_{50}MAL$). These two variables values were plugged into the equations describing Prediction Model 1 (○; panel **F**). Prediction Model 2 (●) also used the $IC_{NOAEL}TMS$ (●; *solid line*). The lowest concentration that produced 100% malformations (*dashed line*) is identified (●; $IC_{MAX}MAL$). Like the stem cell test, the $IC_{50}3T3$ is used (●; *dotted line*). All three values are inserted into Prediction Model 2 (●) in panel **F**.

and strong embryotoxicants (Fig. 5). Without delving into the mathematical intricacies, it is known that the in vivo relationship between maternal toxicity, embryotoxicity, and frank malformations was different for each of the three classes of embryotoxicity. The objective of the biostatistician was to "describe" these relationships with linear discriminate function analysis for the in vitro data set (Fig. 5).

In this way, unknowns could be applied to the test system, and the embryotoxicity class to which they most closely respond could be identified (60).

The prediction model works as follows: In all three in vitro models, Function 1, 2, and 3 describe, respectively, the non embryotoxicants, weak embryotoxicants and strong embryotoxicants. After conducting the toxicity tests for an unknown, the corresponding inhibitory concentration (IC) values are substituted into the three functions and the numeric value calculated. The largest value identifies the embryotoxicity class. For example, if Function 1 is > both Function 2 and Function 3, the unknown test article is predicted to be nonembryotoxic.

In summary, ECVAM's practical contributions to the field of in vitro embryotoxicity testing were broad and influential and include the consensus on the validation test set, the evolution of the embryonic stem cell test, the incorporation of Standard Operating Procedures (SOP), and the incorporation of GLPs (87). These steps pale in comparison to the giant leap forward in the philosophical approach to in vitro data interpretation, namely, the inclusion of a surrogate for maternal toxicity and the evolution of the concept of biostatistical prediction models.

APPLICATION OF ECVAM PREDICTION MODELS

Having outlined the theoretical and practical aspects of the prediction models, we now proceed to their performance in the validation trial and their practical application. This section is divided into three parts. The first summarizes the validation trial outcomes, followed by the performance of the in vitro tests as screening tools in the pharmaceutical and cosmetics industry. The ultimate goal of ECVAM was to validate these tests for regulatory acceptance, and the status of this effort is examined.

ECVAM Validation Results

In keeping with the formal validation process, ECVAM determined, a priori, the criteria necessary to evaluate the performance of the tests. Because there were three classes of embryotoxicity, 33% correct classification was expected by chance alone. Based upon this, the ECVAM management team decided the criteria were (60):

33% by chance
<65% insufficient
≥65% sufficient
≥75% good
≥85% excellent

This evaluation is purported to take into account the inherent variability of the in vivo data (78). Therefore, excellent performance was defined as ≥85% for each of the performance criteria: predictivity, performance, and accuracy (defined in Table 5).

Table 5 Features of the 3 × 3 Contingency Table

"True" in vivo toxicity	In vitro predicted embryotoxicity		
	Nonembryotoxic	Weakly embryotoxic	Strongly embryotoxic
Nonembryotoxic	*a*	b	c
Weakly embryotoxic	d	*e*	f
Strongly embryotoxic	g	h	*i*
Predictivity for nonembryotoxic chemicals	a / (a + d + g) × 100		
Predictivity for weakly embryotoxic chemicals	e / (b + e + h) × 100		
Predictivity for strongly embryotoxic chemicals	i / (c + f + i) × 100		
Precision for nonembryotoxic chemicals	a / (a + b+ c) × 100		
Precision for weakly embryotoxic chemicals	e / (d + e + f) × 100		
Precision for strongly embryotoxic chemicals	i / (g + h + i) × 100		
Accuracy	(a + e + i) / n × 100		

Notes: Contingency tables (3 × 3) permit analysis of in vitro predicted embryotoxicity class relative to the "true" in vivo embryotoxicity class. In this format, precision is defined as the proportion of correctly classified strong (weakly or non-) embryotoxic compounds from the in vitro test that are truly strongly (weakly or non-) embryotoxic in vivo. Predictivity for strongly (weakly or non-) embryotoxicants is the likelihood that a positive prediction in the test correctly identifies the strongly (weakly or non-) embryotoxicant. Accuracy is the mean overall predictivity and precision.

With respect to correctly categorizing non-, weak, and strongly teratogenic agents, the overall accuracies of the embryonic stem cell test and the whole embryo culture test were each approximately 80%: between good and excellent (Table 6). The results from the micromass test were approximately 70%: sufficient. Therefore, although validated, it was concluded that the micromass test, in its current form, was not sufficiently robust to serve as a useful screening tool (60).

Application in Industry

The ECVAM embryonic stem cell test and whole embryo culture test have found favor within the pharmaceutical and cosmetics industries. From the limited data set that has been publicly disclosed, it can be concluded that in the industrial setting both the embryonic stem cell test (112–115) and the whole embryo culture (28) have an accuracy of about 80 ± 5% (Table 6). This approximates the performance initially disclosed by ECVAM in their validation study, and is not too surprising since some of the test chemicals were similar. Based upon the literature reports and personal communications, there is no evidence to suggest that the ECVAM micromass test has been similarly adopted, likely reflecting its poorer performance relative to the other tests. However, the chick embryonic retina micromass test,

Table 6 Summary of In Vitro Embryotoxicity Test Performance

Parameter[a]	MM[b] ECVAM validation	EST[c] ECVAM validation	EST[d] Pfizer	EST[e] Pfizer	EST[f] Hofmann La Roche	EST[g] Pfizer	WEC[h] ECVAM (PM-1)	WEC[i] ECVAM (PM-2)	WEC[j] Pfizer (PM-2)	CERC[k] Proctor & Gamble
Predictivity for nonembryotoxic%	57	72	50	100	86	88	56	70	100	19 False +ve
Predictivity for weakly embryotoxic%	71	70	63	63	67	59	75	76	67	
Predictivity for strongly embryotoxic%	100	100	86	100	83	71	79	100	53	
Precision for nonembryotoxic%	80	70	50	57	100	70	70	80	64	14 False −ve
Precision for weakly embryotoxic%	60	83	79	100	67	77	45	65	75	
Precision for strongly embryotoxic%	69	81	100	100	71	83	94	100	100	
Accuracy%	70	78	73	87	81	75	68	80	77	82
Number of chemicals	20	20	18	18	16	53	20	20	40	45

[a] Contingency tables (3 × 3) permit analysis of in vitro predicted embryotoxicity class relative to the "true" in vivo embryotoxicity class. The definitions of predictivity, precision, and accuracy are defined in Table 5.

[b] The micromass test (MM) as conducted in the ECVAM validation (60) with the Brown 20 chemical test set (78).

[c] Embryonic stem cell test (EST) as conducted in the ECVAM validation (60).

[d] EST conducted at Pfizer in compliance with the ECVAM protocol on chemicals that were mostly part of the Brown chemical set (112).

[e] Same experiment as in [d], except instead of using beating cardiomyocytes, the changes in gene expression using a Mahalanobis distance model was used (112).

[f] EST conducted at Hoffmann La Roche in compliance with the ECVAM protocol using a mix of Brown chemicals and proprietary agents (113).

[g] EST conducted at Pfizer in compliance with the ECVAM protocol using a mix of Brown chemicals and proprietary agents (114).

[h] Whole embryo culture (WEC) as conducted in the ECVAM validation test (60). Results are shown for prediction model one (PM-1), in which only two embryonic parameters, total morphological score and malformations, were assessed.

[i] WEC as conducted in the ECVAM validation test (60), but using prediction model two (PM-2). Here total morphological score, malformations, and a 3T3 cytotoxicity component are assessed.

[j] WEC conducted at Pfizer in accordance with the ECVAM WEC PM-2 protocol. A mix of Brown compounds and proprietary agents were used (28).

[k] The chick embryo neuronal retina cell (CERC) model as validated at its site of origin, Proctor & Gamble (79). A dichotomous classification was used, therefore performance is expressed as percent false positive (+) or negative (−).

developed independently from ECVAM, is currently used to screen cosmetics candidates (46).

Taken together, the performance of the current tests (28,112–115) favors their further optimization (109,112,116,117), and suggests that they may be useful as prescreens for drug and cosmetics development. Due to proprietary concerns, precise details about how these tests have been employed are unavailable, so the author can only provide limited insights about the 3 years experience at Pfizer. These tests tend to be part of a larger process, in which a variety of factors are examined to identify a lead target. Moreover, due to the large number of potential drug targets and programs searching for leads, these in vitro tests are only used in two circumstances: (1) when there is a theoretical concern about the target based on known biology or a literature precedent or (2) if the first lead encountered unanticipated developmental toxicity. These assays are relatively new, and it is unclear what the inherent advantages of each test are. Consequently, most programs are tested in both the stem cell test and whole embryo culture, so we hope to gather sufficient data to make this determination in the future.

Two specific drug development programs will be used to illustrate how the whole embryo culture test has been employed at Pfizer. In the first, unexpected specific craniofacial malformations were encountered with the first generation lead compound, PF-1. To verify that this target was applicable for whole embryo culture, the initial lead was tested, and classified correctly as "strongly embryotoxic," and fortuitously the specific craniofacial anomaly was replicated in vitro. Upon testing, the remaining four backups were all "weakly embryotoxic," presenting an interesting problem of how to discriminate between them. This was done by considering three factors. First, based upon the concentrations for the $IC_{NOAEL}TMS$ and $IC_{MAX}MAL$, the chemical we will identify as PF-4 was within a few $\mu g/mL$ of being classified as "nonembryotoxic," whereas the others were close to being "strongly embryotoxic." Second, all the chemicals, except PF-4, had the in vitro manifestations of the malformation initially observed in the in vivo study. Lastly, PF-4 was of a different chemical backbone that did not yield a metabolite common to the other four and, importantly, this metabolite was classified as strongly embryotoxic in vitro. Putting these pieces together, the backup PF-4 was nominated. In this instance, we did not use these tests to blindly bin chemicals into embryotoxicity categories, but rather we used a variety of end points to create a rank ordering. We are not alone in this approach, as the chick embryo retinal culture is used similarly (46).

Keeping to the example above, in addition to selecting chemical leads, these alternative tests are a useful tool for making business decisions. Due to the history of this program, there was still the risk of a late-stage product failure due to developmental toxicity. Therefore, in spite of the clean in vitro signal, the in vivo embryo/fetal toxicity study was conducted earlier than usual to mitigate the risk of wasting 2 years of drug development cost on a late-stage failure.

Although there are other success stories for these tests at Pfizer, it must be acknowledged that there have been instances in which neither test was helpful, as in the second example we will describe. Here, six compounds were directed at a common, but unprecedented pharmacologic target. Four of the agents were used for "benchmarking," but the in vivo data were of varied quality. Several compounds had regulatory compliant embryo/fetal toxicity studies, others had single species "range-finder" data with low sample size at maternally toxic doses, and there was the perception that at least one was not developmentally toxic. The mRNA of the targeted receptor was expressed both in stem cells and in whole embryos. Unfortunately, in the stem cells all agents, including those used for benchmarking, were classified as weakly embryotoxic whereas in embryo culture all were classified as "strong" (67). Did stem cells underestimate the risk or did embryo culture overestimate the risk? In this instance, the question may never be resolved, but this example underscores several key points.

First, each pharmacologic target will probably predispose these tests uniquely to over- or underestimate developmental risk; it has been our experience that embryo culture tends to overestimate risk compared to the stem cell test (28). Second, the quality of the in vivo data used for "calibration" is critical. In this case, product development of the nonembryotoxic "negative control" was later halted due to potent hepatotoxicity. We hypothesize that it was nonembryotoxic in vivo due to dose-limiting maternal toxicity, resulting in lower in vivo embryonic exposures than in vitro, but pharmacokinetic analysis has not been done to confirm this.

Complicating this situation, a nonbenchmarking agent in this same program was classified as strongly embryotoxic, yet caused no malformations in regulatory compliant embryo/fetal toxicity studies in two species. Drug development teams perceived this outcome as a false positive. We would argue that this was not the case because the chemical did produce significant declines in fetal birth weight and increased postimplantation loss in the absence of maternal toxicity. It will be recalled that in defining embryotoxicity, the ECVAM model considers all manifestations of developmental toxicity equally. In this instance, the development team was more concerned about frank malformations than low fetal birth weight. This reflects the fact that based upon the therapeutic indication; the risk–benefit ratio for pharmaceuticals often differs significantly from agricultural chemicals or cosmetics.

The Use of In Vitro Models in Risk Assessment

One long-term goal of ECVAM, ICCVAM, and FRAME is the replacement of in vivo regulatory toxicity tests with appropriate in vitro alternatives, and to use these data for risk assessment. To facilitate this vision ECVAM and ICCVAM have developed specific Performance Standards. These include reference chemicals, essential test method components, and statistical performance results, all conducted in compliance with the principles of the OECD's GLP to ensure that the in vitro tests are reproducible, credible, and acceptable (72,87,88,110,111). The ECVAM Scientific Advisory Committee has concluded that the three embryotoxicity tests were successfully validated (118), in accordance with the ECVAM/ICCVAM/OECD recommendations for the validation of in vitro toxicity tests (110). In addition, the ECVAM Scientific Committee has indicated these three tests are ready for regulatory consideration (118). In practical terms what does this mean? Not surprisingly, the answer is far from simple, as discussed later.

The framework in which alternatives tests are standardized or formally validated is determined globally as discussed earlier, but public attention and awareness can shape legislative outcomes. People affected by health problems are more accepting of animal experiments to develop potential therapies than those who are healthy (96). In addition, laboratory animal testing becomes less acceptable for products, such as cosmetics, not perceived to be vital to human health. These factors together influence the debate on legislation in which consumer health and safety is pitted against animal rights, and the use of animals in testing. Two current examples of such legislation are The Seventh Amendment of the Cosmetics Directive 76/768/EEC (European Economic Community) (55) and the E.U. Chemicals Policy (the REACH system) (56). Underlying these relatively recent European directives is a common theme within all the agencies in North America and Europe to include provisions for the use of improved methodologies that reduce and refine animal use, although these provisions are vague (19,20,96).

The Seventh Amendment of the Cosmetics Directive 76/768/EEC projects the phasing out of in vivo testing for developmental and reproductive toxicity by 2013 (55,119). Although ECVAM has promoted the use of its embryotoxicity tests for regulatory testing (59,60,119), the Scientific Committee on Cosmetics and Non-Food Products (SCCNFP), the group that decides the appropriateness of toxicity tests, does not agree (20). Nevertheless, it has been suggested that the embryotoxicity tests may be used in the near future in the regulatory testing of cosmetics (20).

In Europe, a new Chemical Policy (REACH) was launched in June 2007. Briefly, it requires a stepwise Registration, Evaluation, Authorisation, and Restriction of CHemicals, based upon perceived/actual toxicity and the quantity manufactured; cosmetics, detergents, and pharmaceuticals are governed by separate legislation. A major implication is the need to generate embryotoxicity data for a large number of previously untested agents, leading to the belief that this will significantly increase the number of animals used in toxicity testing (120). ECVAM has suggested its validated in vitro embryotoxicity tests be incorporated as part of a larger strategy to reduce animal use in this initiative (25,118,121,122), but there has been no guidance for how registrants are to incorporate in vitro testing in this process (120). Although these three tests purportedly meet the regulatory criteria set forth within the OECD test guideline 414, and Annex V, Part B of the EU–Dangerous Substance Directive, for reducing and/or refining the use of animal procedures, they are not listed as acceptable toxicity detection methods in Annex V of the Dangerous Substance Directive (119). Moreover, when the European Commission established Annex IX of this same Directive, to list validated alternative methods, the Commission published an empty table (119). Thus, in

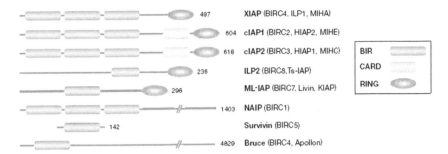

XIAP (BIRC4, ILP1, MIHA) — 497

cIAP1 (BIRC2, HIAP2, MIHE) — 604

cIAP2 (BIRC3, HIAP1, MIHC) — 616

ILP2 (BIRC8, Ts-IAP) — 236

ML-IAP (BIRC7, Livin, KIAP) — 298

NAIP (BIRC1) — 1403

Survivin (BIRC5) — 142

Bruce (BIRC4, Apollon) — 4829

BIR
CARD
RING

Figure 1.7 Comparison of conserved motifs (BIR, CARD, and RING) among members of the IAP family.

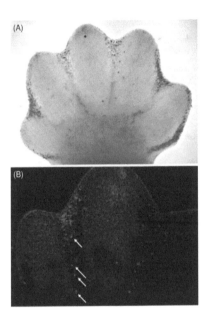

(A)

(B)

Figure 1.8 (**A**) E13 mouse embryo limb bud stained with Neutral Red showing interdigital PCD. (**B**) E13 mouse limb bud immunohistochemically stained showing caspase-3 is activated in apoptotic cells within the interdigital mesenchyme.

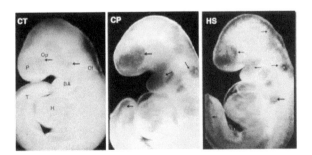

CT   CP   HS

Figure 1.9 Neutral red stained E 9 mouse embryos showing cell death (*arrows*) in untreated (CT), CP-treated or HS-treated embryos.

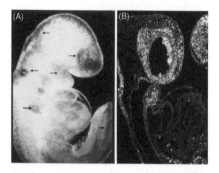

Figure 1.12 Whole mount (**A**) and parasagittal section (**B**) of a day 9 mouse embryo 5 hours after exposure to HS and then stained for activated caspase-3 (*red fluoresence*) and DNA fragmentation (*green fluorescence*). (See page 28 for complete legend.)

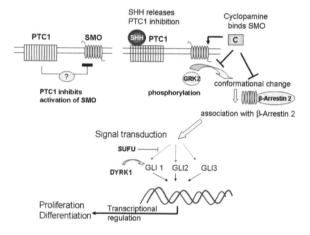

Figure 2.3 In the Shh signaling pathway, Ptc1 inhibits Smo. (See page 53 for complete legend.)

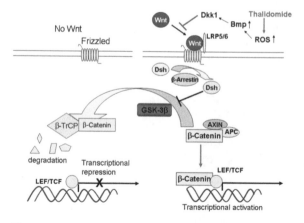

Figure 2.4 In the absence of Wnt, ß-catenin is phosphorylated by GSK-3ß, associates with ß-TrCP, and is targeted for degradation by ubiquitination. (See page 56 for complete legend.)

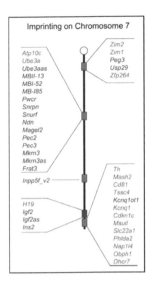

```
Imprinting on Chromosome 7

Atp10c          Zim2
Ube3a           Zim1
Ube3aas         Peg3
MBII-13         Usp29
MBI-52          Zfp264
MB-I85
Pwcr
Snrpn
Snurf
Ndn
Magel2
Pec2
Pec3
Mkrn3
Mkrn3as
Frat3           Th
                Mash2
Inpp5f_v2       Cd81
                Tssc4
H19             Kcnq1ot1
Igf2            Kcnq1
Igf2as          Cdkn1c
Ins2            Msuit
                Slc22a1
                Phlda2
                Nap1l4
                Obph1
                Dhcr7
```

Figure 4.4 The five imprinted domains on chromosome 7 in the mouse. This figure demonstrates the lack of random distribution in imprinted genes, which occur in clusters on a chromosome. *Source*: From Ref. 81.

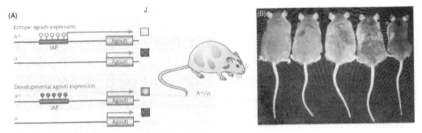

Figure 4.5 (**A**) The murine A^{vy} metastable epiallele in which a cryptic promoter in the IAP promotes constitutive *Agouti* expression. (**B**) The degree of CpG methylation in the A^{vy} IAP correlates with ectopic *Agouti* expression. *Source*: From Ref. 95.

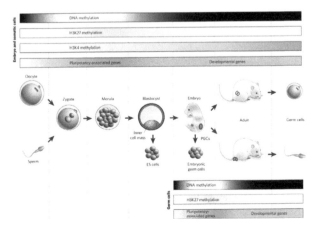

Figure 4.6 Critical periods of epigenetic regulation during development. *Source*: From Ref. 82.

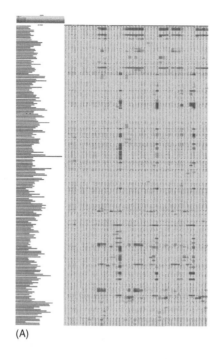

(A)

. . . GO:0006955 immune response	79	92	116	137	65	86	113	138
. . . . GO:0001816 cytokine production	9	10	11	14	7	9	14	16
. . . . GO:0006953 acute-phase response	5	6	8	8	5	5	5	11
. . . . GO:0006954 inflammatory response	26	32	38	47	27	34	43	61
. . . . GO:0006959 humoral immune response	27	32	38	40	23	28	35	42
. . . . . GO:0006956 complement activation	15	15	16	16	11	14	15	16
. . . . . . GO:0006957 complement activation, alternative pathway	4	4	4	4	3	4	4	4
. . . . . . GO:0006958 complement activation, classical pathway	9	9	9	9	7	9	9	9
. . . . . GO:0016064 humoral defense mechanism (sensu Vertebrata)	17	20	24	26	16	19	24	27
. . . . . . GO:0019735 antimicrobial humoral response (sensu Vertebrata)	7	8	11	13	6	7	12	14
. . . . . GO:0019730 antimicrobial humoral response	8	9	12	14	7	8	13	16
. . . . GO:0006968 cellular defense response	13	15	19	23	14	17	21	23
. . . . . GO:0016066 cellular defense response (sensu Vertebrata)	7	8	11	14	7	9	13	15
. . . . . . GO:0042087 cell-mediated immune response	6	7	8	10	5	7	10	11
. . . GO:0019882 antigen presentation	6	9	10	12	8	8	9	10
. . . . GO:0019883 antigen presentation, endogenous antigen	3	5	5	7	4	4	5	5
. . . . GO:0019884 antigen presentation, exogenous antigen	3	4	4	4	4	4	4	4
. . . GO:0030333 antigen processing	9	10	10	13	7	8	9	10
. . . . . GO:0019885 antigen processing, endogenous antigen via MHC class I	4	5	5	8	4	4	5	5
. . . . . GO:0019886 antigen processing, exogenous antigen via MHC class II	4	4	4	4	3	4	4	4

(B)

Figure 11.1   Gene expression segregated by GO categories. GO terms are arranged hierarchically into groups according to molecular function, cellular component, or biological process. In this figure, the intensity of color in each grid indicates the number of genes affected in that GO category, and the columns represent a unique experimental condition, such as increasing dose levels, time points, or statistical condition, particularly decreasing stringency for statistical significance from left to right. Figure 11.1 **A** is a heat map of the entire GO classification for molecular function for an experiment evaluating the effects of an estrogen on fetal uterus. Figure 11.1 **B** is an enlargement of one part of the heat map in 1 **A**.

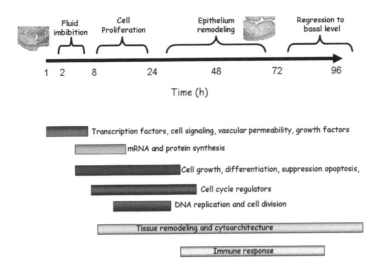

Figure 11.2 Temporal sequence of changes in gene expression as part of the uterotrophic response in rats. (See page 306 for complete legend.)

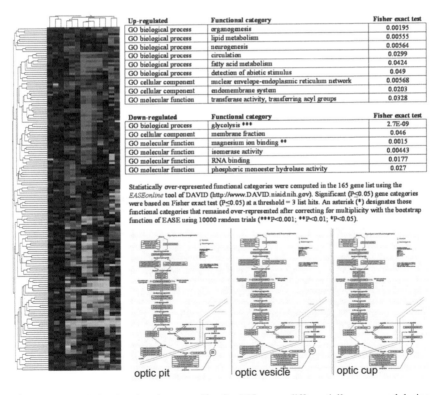

Up-regulated	Functional category	Fisher exact test
GO biological process	organogenesis	0.00195
GO biological process	lipid metabolism	0.00555
GO biological process	neurogenesis	0.00564
GO biological process	circulation	0.0299
GO biological process	fatty acid metabolism	0.0424
GO biological process	detection of abiotic stimulus	0.049
GO cellular component	nuclear envelope-endoplasmic reticulum network	0.00568
GO cellular component	endomembrane system	0.0203
GO molecular function	transferase activity, transferring acyl groups	0.0328

Down-regulated	Functional category	Fisher exact test
GO biological process	glycolysis ***	2.7E-09
GO cellular component	membrane fraction	0.046
GO molecular function	magnesium ion binding **	0.0015
GO molecular function	isomerase activity	0.00443
GO molecular function	RNA binding	0.0177
GO molecular function	phosphoric monoester hydrolase activity	0.027

Statistically over-represented functional categories were computed in the 165 gene list using the *EASEonline* tool of DAVID (http://www.DAVID.niaid.nih.gov). Significant (P≤0.05) gene categories were based on Fisher exact test (P≤0.05) at a threshold = 3 list hits. An asterisk (*) designates those functional categories that remained over-represented after correcting for multiplicity with the bootstrap function of EASE using 10000 random trials (***P<0.001; **P<0.01; *P<0.05).

optic pit | optic vesicle | optic cup

Figure 12.1 Molecular abundance profiles for 165 genes differentially expressed during optic cup morphogenesis in the mouse and rat embryo. (See page 317 for complete legend.)

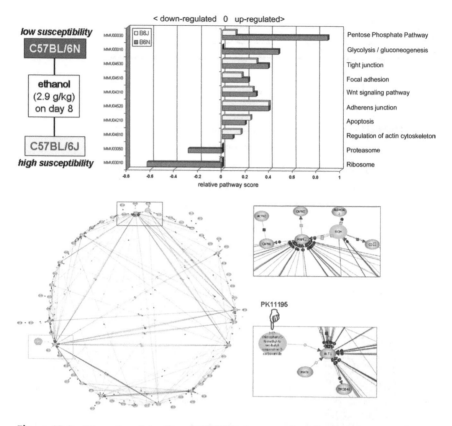

Figure 12.2 Hierarchy of significant KEGG Pathways and predicted computational gene network defining FAS induction. (See page 321 for complete legend.)

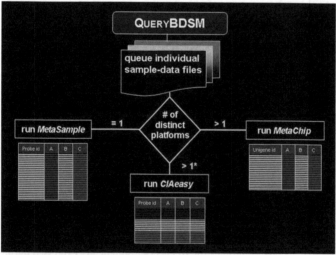

Figure 12.3 Workflow schema for the QueryBDSM module of BDSM. (See page 324 for complete legend.)

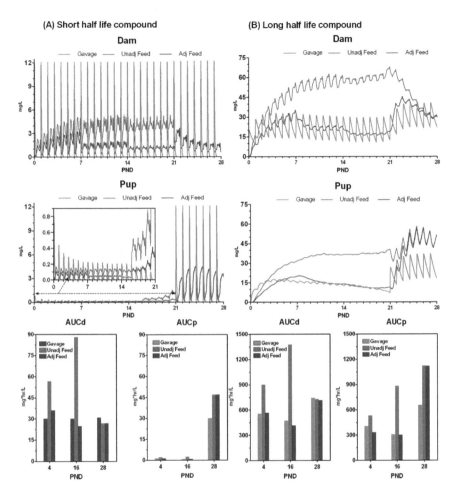

Figure 12.8 Predicted concentrations and area under the curve (AUC) values for base case compounds in the pup versus dam as computed with a PBPK model. Pup levels tended to be lower than in the dam for short half-life compounds whereas the reverse was true for long half-life chemicals. *Source*: From Ref. 114.

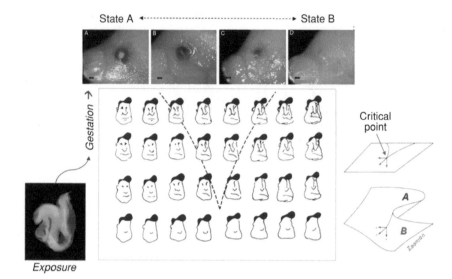

Figure 12.12 Catastrophe theory: projecting output of computational models to phenotype-space. Top panel: A range of malformations can often be detected in a prenatal developmental toxicity study. For example, consider the range of ERDs in mouse fetuses following maternal exposure to ethanol (12). ERD phenotypes range from normophthalmia to different degrees of anophthalmia and complete apparent anophthalmia, representing a transition of the embryo from normal (State A) to abnormal (State B) as development advances from exposure on GD 8 to the fetus at GD 17. The phenotype transition can be modeled through a range of features. Consider, for example, the diagram showing features shifting columnwise from a man's face to a woman (135) [reprinted from (132)]. Each row of images reflects a progression of features that would be drawn and, likewise, would develop over time following maternal alcohol exposure on GD 8, leading to an anatomically complete (**A**) or incomplete (**B**) eye on GD 17. Features in the formal system are added general to specific, much like during natural morphogenesis. Each column represents a possible solution from CC3D; each row represents morphing of features across states A and B. Sweeping different parameters in the model would determine the range of probabilities (%) in each outcome.

spite of suggestions that the tests are validated and acceptable tests, they are not used in such a capacity at this time.

It should be emphasized that the ECVAM management team did not conclude that these tests represent a complete replacement of preclinical whole animal developmental toxicity testing (Segment II test) for pharmaceutical products. Nevertheless, an ECVAM/ZEBET workshop, which included experts from the pharmaceutical and chemical industries as well as scientists from academia and regulatory agencies, suggested that these validated tests could be used as screens in the pharmaceutical industry (21). Indeed, as discussed earlier, the tests are already being used in such a way. As reviewed elsewhere (8,13–15), in vitro methodologies are an excellent tool to study developmental biology or mechanisms of teratogenesis, and it is in this capacity that whole embryo culture has been used as a tool to gain insights that were used in to help human risk assessment. Two such examples are described below.

In the first, rat whole embryo culture was used to investigate the teratogenicity of three newly registered drug combinations. The drugs had already been tested and marketed separately, but were now being considered as combination therapies. By relating in vitro culture drug concentrations to clinical plasma levels in vivo, the key findings were the identification of synergistic and additive interactions that were submitted as part of the toxicology registration package (123). In another example, a common industrial chemical, ethylene glycol, was developmentally toxic in rat, but not rabbits. The issue was which animal species more accurately reflected the human risk? Briefly, a series of pharmacokinetic experiments were conducted in vivo as well as in vitro using whole embryo culture of rat and rabbit (124–126). These data, when combined with the known differences in embryonic pH, yolk sac structure and function, allowed the investigators to conclude that the rabbit more accurately reflected the human risk of ethylene glycol whereas the rat overestimated that risk (124,126). From these examples, it should be noted that neither ESCs nor micromass have been used in a similar capacity and that the ECVAM prediction model was not used.

In summary, the implications from the REACH initiative and the Cosmetics Directive are that in vitro developmental toxicity testing will need to be acceptable in specific regulatory environments as part of a larger strategy for hazard identification. A major obstacle in achieving this end is that legislative policy has underestimated the time required to develop such tests, and it has failed to provide sufficient guidance to participants about how to incorporate these in vitro alternatives. With respect to the regulatory requirements of the pharmaceutical industry, in vitro embryotoxicity tests will not be replacing standard in vivo toxicity testing for some time. However, in vitro prediction models are currently used to guide drug-development strategy and as a screening and lead optimization tool. Whole embryo culture of rodents and rabbit, in the absence of ECVAM prediction models, are valuable adjuncts to facilitate data interpretation and extrapolation of human risk.

FUTURE DIRECTIONS

Following the formal report of the successful ECVAM validation trials (60), work-shops have taken place in Europe (21) and North America (98) to bring together assay users and other interested parties with the expressed purpose of defining the limits of these assays and proposing recommendations for further optimization. What follows is an overview of the meetings and the reader is referred to the original reports for more details. *Limitations* are considered first, setting the stage for *Paths Forward*.

Limitations

As described in Figure 4, all three assays represent limited windows of develop-ment, and it is therefore unclear whether they can detect teratogens which mediate their effects beyond the specified time periods or differentiation processes they represent. Studies are underway to understand this potential limitation because it may be possible to extend the window of detection in whole embryo culture with the use of biomarkers. For example, although the morphology of thalidomide-mediated limb defects is only discernible several days after the termination of whole embryo culture, the gene expression changes that precede the expression of the phenotype are readily detectible within the culture window (90).

A further layer of complexity relates to the issue of species specificity. Developmental toxicity test species have traditionally been the rat and rabbit, but with the increasing importance of biologics in the pharmaceutical industry, the mouse and nonhuman primate have become increasingly important. We have found that, controlling for gestational age, the predicted severity of embryotoxicity of some compounds is markedly different in rat versus mouse whole embryo culture (28). Does the replacement of the rat by the mouse as the rodent tests species require derivation of a murine-specific model?

Embryo/fetal development is a series of morphogenic events, yet only the whole embryo culture model, and to a limited extent the micromass test, capture such information. Despite this data richness, in its current format much of the whole embryo culture data are untapped. For example, all 17 *anlagen* comprising the TMS and all 26 malformations have been reduced to a single point on a toxicity curve. It has been suggested that more detailed examination of morphology may help to discriminate non- from weak embryotoxicants (21). Indeed, this assertion was corroborated in the previously described case study in which all agents were classified as "weakly" embryotoxic, but the biological severity of the malformations seen in embryo culture was used for lead selection.

For mechanistic studies, the isolation of in vitro test systems from mater-nal influences is a clear advantage, but for broad screening the inability to cap-ture maternal influences that may contribute to embryotoxicity may be a serious shortcoming. For example, the xenobiotic biotransformation activities of all these assays are undefined, but presumed to be limited when compared to the maternal compartment in vivo. As a result, these models currently only assess the intrinsic embryotoxic potential of the parent compound. That said, this may be an advantage in the pharmaceutical industry because medicinal chemists are able to synthesize in vivo–generated metabolites enabling the discrimination of proximate from proteratogens, further enhancing the ability to steer teams toward favorable chemistry. Therefore, the desirability of a drug metabolizing system mimicking in vivo biotransformation depends upon the specific purpose of the test; mechanistic studies and high-throughput screens may have differing and sometimes conflicting requirements.

The use of in vitro models as a tool to conduct human risk assessment is contingent upon the ability to correlate relevant in vitro concentrations with anticipated and actual in vivo exposures. There are instances in which there has been concordance between, for example, the in vitro concentrations that induce malformations in whole embryo culture and the in vivo circulating levels or dose (27,127,128). However, more comprehensive studies using benchmark dose levels of the ECVAM validation test set have failed to demonstrate such a relationship with whole embryo culture (129). Returning to the sertraline example, malformations were only induced at in vitro concentrations that were higher than plasma levels achieved at a maternally lethal dose (103). This suggests in vitro and in silico ADME approaches will need to be integrated with in vitro toxicity models to optimize estimates of in vivo exposures, leading to better risk estimates, particularly for unknown agents (21,115,130).

All three in vitro prediction models struggled in discriminating between non- and weak embryotoxicants (21,98). The solution to this problem is unclear, but it has been suggested that the ability to separate between these two classes of embryotoxicity may depend upon biological assessments rather than statistical analyses (21). This is supported by the fact that a combined approach using gene expression markers and alternative statistical methods (Mahalanobis Distance model in lieu of linear discriminate analysis) failed to significantly improve the separation of non- and weakly embryotoxic chemicals in the embryonic stem cell test, although overall predictivity was somewhat enhanced (112).

The three prediction models were designed specifically to detect strong embryotoxicants, but it is unclear whether they are capable of detecting strong teratogens outside the validation test set. For example, most of the strongs used in the validation test set were anticancer medicines that inhibited proliferation. Newer antineoplastic therapies have very different mechanisms of action. Thus, the prediction models may need to be adapted to specific chemical classes or mechanisms of action (21,28).

Another concern relates to some unusual features of the whole embryo culture prediction model wherein it is impossible for a test compound to be classified as nonembryotoxic if the $IC_{MAX}MAL$ is less than 5 μg/mL (28). Such exposure levels are extremely high for biologically active materials, resulting in a previously discussed tendency to overpredict embryotoxicity. To a large degree, this is the result of the chemical test set used to develop the prediction model. It will be recalled that the ECVAM test set used approximately equal numbers of industrial chemicals and pharmaceutical agents, an excellent strategy because the intended purpose was to apply these models to a broad chemical base. However, the extent to which these tests were embraced by the pharmaceutical sector was impossible to foresee. This is an important consideration as pharmaceuticals are, by their very nature, extremely biologically active whereas, in contrast, many of the agents in the chemical industry are relatively inert. Furthermore, many of the representative pharmaceutical agents were antibiotics, which are designed to interact with bacterial macromolecules, rather than with mammalian tissue, or they were cytotoxic chemotherapeutic agents, designed to interact irreversibly with cellular macromolecules (alkylating agents), in contrast to current trends of focusing on reversible and specific receptor-mediated therapies. Therefore, the pharmaceutical agents used in the ECVAM validation set may contrast current strategies used in pharmaceutical drug design, suggesting that changes be made to the test set, with corresponding changes to the prediction models (21,28,110,112). Although the validation test set used for the derivation of the prediction models was less than ideal for pharmaceutical agents, the fact that acceptable accuracy rates are still obtainable speaks both to the careful thought of the chemicals proposed by Brown and to the robustness of the prediction models.

In the ECVAM tests, 3T3 cytotoxicity parameters were included as a surrogate of maternal toxicity in an attempt to normalize for this important determinant of in vivo developmental toxicity. However, it must be recognized that these cells are only a limited surrogate (21). In addition, developmental toxicity may be mediated by cytotoxic events, and therefore it is difficult to separate these parameters.

Paths Forward

Based upon the challenges faced by the end users of these assays, a number of recommendations have been made to optimize these tests. It should be recognized that the workshops attracted varied interest groups, and therefore some suggested courses of action may appear to be in conflict, or have differing rationales. Certain recommendations are generally applicable to both whole embryo culture and ESCs, while others are assay specific.

General Recommendations

Both the embryo and stem cell culture should be made more "animal friendly." Embryo culture requires both serum and embryo donors and ESCs require fetal bovine serum. The strategy is to use recombinant technology to manufacture the critical components thus allowing for the manufacture of "synthetic" serum (98). Earlier attempts to do this for rodent whole embryo culture failed, but newer technologies and aggressive financial support should increase the chances of success. With the eventual use of human ESCs, it is important to note that some human stem cell lines need additional animals to supply the required murine embryonic fibroblast feeder layer. Initiatives are underway to develop xeno-free human embryonic stem cell cultures to end the need for additional animals (98). In addition, a purely synthetic medium would be a further step toward the standardization of cultures between laboratories.

Both stem cells and embryo culture may benefit from enhanced embryotoxicity end points. For embryo culture, these may include markers of cell fate, cellular identity, and signaling patterns (98). In addition, using heart development as example, atrial isomerism is undetected with standard gross morphological assessment, but readily identified using more advanced imaging techniques (131). Imaging has also been proposed for use in the embryonic stem cell test (132). Two-dimensional electrophoresis of protein from cultured rat embryos has revealed toxicologically relevant protein changes (133). Studies are ongoing to understand whether addition of analysis of gene expression patterns may enhance predictivity (98,109,112,117,134–139). Together, such end points may increase the sensitivity of the test by revealing events not normally detected with gross morphological examination, or they may enable the early detection of events that have become visible only after the termination of culture.

Many agents are proteratogens that must be bioactivated by P-450 class enzymes to be rendered teratogenic. Because rodent embryonic tissue, including stem cells, have low P-450 activities, teratogens requiring metabolic transformation are incorrectly classified as nonembryotoxic (140). Thus, for broad scale chemical screening, the incorporation of adult-like biotransformation activity into each of the embryonic stem cell test, micromass, and whole embryo culture model would be advantageous. Exogenous bioactivating sources have been added to the micromass test (141) as well whole embryo culture in the forms of S-9 drug metabolizing fractions (142) or hepatocytes (143–146). These have the disadvantage of requiring increased whole animal use but an engineered cell line has been created expressing CYP2B1. When added to the embryonic stem cell test, supernatants from this cell line result in the correct embryotoxic prediction of the proteratogen, cyclophosphamide (140). As a result of this proof of concept, it has been recommended that adult-like biotransformation activity be added to the stem cell test (21,138).

It is widely acknowledged that exposure characteristics of in vivo and in vitro toxicity models may be quite different, and it has been suggested that the inability to adequately model metabolism is the bottleneck in in vitro toxicological test development (147). Whereas in vitro test article exposures tend to be constant, in vivo levels fluctuate. This is an important consideration because, for example, it has been demonstrated that some teratogens mediate their deleterious effects based upon total embryonic exposure, whereas others are dependent upon peak concentrations (148). Therefore, to improve the accuracy of these in vitro tests, mathematical modeling and pharmacokinetics parameters must be applied to define better correlations between effective in vitro concentrations and in vivo maternal concentrations in test animals and humans (21,129,130). Indeed, a meeting has been held with the expressed purpose of applying physiologically based kinetic modeling to alternative toxicity models (149).

Another factor impacting both embryo/fetal exposure and metabolite generation is the placenta. More effective models of placental transfer will be required as well as the ability to integrate the placenta-specific biotransformation capacity, which differs from the more frequently modeled liver (150).

Another recommendation common to both assays is to update the prediction models to include a number of additional validation chemicals (21,28,98). More compounds with known in vivo embryotoxic activity must also be added to make prediction models more robust. In addition, it has been suggested that different prediction models may need to be developed based upon the compound classes (21) or the intended purpose of the test (21,28). For example, the risk–benefit ratio of a chemical exposure is a critical consideration in regulatory risk assessment. Consequently, due to the potential lifesaving benefits, the tolerance for some kinds of developmental toxicity risk (such as lower birth weight) may be higher for pharmaceutical compounds than agents generated by the chemical or cosmetic industry. It will be recalled that in developing the ECVAM embryotoxicity classes, low birth weight carries the same weight as a structural malformation (78), and for this reason we have trying to develop ways of discriminating between these manifestations of developmental toxicity (28). One recommendation from the 2007 HESI conference on In Vitro Developmental Toxicity Testing was that the pharmaceutical industry make available its large, high-quality developmental toxicity data base of "failed" chemicals (112), thus increasing the data set for model generation. This may also facilitate the derivation of a robust pharmaceutical-specific prediction model.

One disadvantage of the ECVAM-validated tests is that, in comparison to many other in vitro toxicity tests, they are relatively low throughput. To mitigate this weakness, there have been attempts to develop different models of shorter duration (67,116,151). The application of automated technologies is another strategy being investigated (152). A combination of both approaches may provide the desired breakthrough.

Stem Cell Test–Specific Recommendations

Currently, the end point of the embryonic stem cell test is the beating cardiomyocyte, but there is need to differentiate stem cells into other lineages to make it more representative of an embryo (137). Theoretically, a battery of stem cell–based assays each representing different lineages would achieve this goal and, moreover, may recapitulate some processes which are not represented by the alternative tests (gestation day 12–14; Fig. 4). This would broaden its applicability and create a battery of embryonic target tissue types against which to test chemicals, theoretically improving test accuracy. With different cell types comes the need to use tissue-specific tissue markers to establish more quantitative end points. A number of techniques including the use of microarrays, reverse transcriptase polymerase chain reaction (RT-PCR), reporter constructs, and immunohistochemistry have all been proposed (98,109,112,117,132,134–139,153). The use of fluorescent activated cell sorting (FACS) to quantify changes in reporter construct expression may make such tests suitable for true high-throughput screening (154). Whether these changes improve the accuracy of the tests remains to be seen. For example, in one effort the use of gene expression markers and alternative statistical approaches were unable to improve the discrimination between non- and weak embryotoxicants, although overall accuracy increased by 5% (112).

These tests were designed for human risk assessment, and with the advent of human stem cell technology, there is hope that the use of such cells may increase the human relevance. Efforts in to create human stem cell–based embryotoxicity tests have begun (155).

Whole Embryo Culture–Specific Recommendations

As part of larger proposed initiatives of standardization is the creation of an atlas of malformations to ensure uniform terminology across laboratories (98). This is a necessary prelude to the creation of a centralized database for those agents having both in vivo and in vitro developmental toxicity data. It is intended that this database be accessible to, and updated by, scientists in academia, industry, and government. In this way, a more diverse array of chemicals and statistical approaches may lead to enhanced prediction models (98). One obstacle to this initiative is the protection of industrial proprietary interests, although they would have the most to gain by such an effort.

Despite the large number of morphological parameters that are assessed in whole embryo culture, they do not appear to be used to their maximum potential, and it is not clear if newer imaging techniques which reveal internal, previously unused, parameters would improve the test (131). Furthermore, a comprehensive analysis is required to determine if a smaller cohort of end points would improve the assay throughput and/or accuracy. Such investigations are ongoing in the author's laboratory using a 40 chemical test set (67). In addition, other efforts are underway to shorten the whole embryo culture assay by limiting the testing to only three concentrations. In one approach, specific IC values from the 3T3 chemical cytotoxicity curve are used to identify the three concentrations to be used for whole embryo culture (67). The rationale here is that when normalized for "maternal toxicity," non- and strong embryotoxicants should affect embryos differently. In an alternative approach, the three test concentrations are predetermined to be 0.1, 1.0, and 10 μM (98,116). The embryos are examined with an in-house derived scoring system, which is heavily biased toward neural tube/craniofacial effects. Using a limited test set, effects at 0.1 μM appear to be 80% predictive.

ReProTect

The implementation of the aforementioned recommendations is anticipated to significantly improve the risk assessment potential of these in vitro tests. Indeed, as outlined, much of this work has already begun, and the importance of these recommendations for further improvements is underscored by the fact that many of these initiatives have been incorporated into the mandate of the ECVAM-led integrative project, ReProTect. Its goal is to facilitate the further development and optimization of a myriad of in vitro tests into a series of testing batteries and strategies that, when fully integrated, will provide detailed risk assessment of chemical exposures to the entire mammalian developmental and reproductive cycle (156). To facilitate the ReProTect objectives, a comprehensive list of toxicity tests that are applicable to a variety of processes has been published (157). In all likelihood, the advances made in the ReProTect project may make the second-generation embryotoxicity tests more robust. Applicable to developmental toxicity are in vitro assays for implantation, physiologically based pharmacokinetic modeling, prediction of placental transfer and metabolism, neurodevelopmental toxicity, and adverse effects to additional embryonic target organs and tissues. If successful, this initiative may allow ECVAM to achieve its goal of regulatory acceptance of in vitro embryotoxicity testing.

CONCLUSIONS

Models for in vitro developmental toxicity testing have existed for several decades, but their broader implementation and acceptance were hampered most notably by fragmented interests and a lack of funding. The creation of ECVAM in the early 1990s changed these circumstances. Under its leadership, the field of in vitro developmental toxicity testing made significant leaps forward by bringing together a variety of stakeholders to reach consensus on several key areas including the selection of an appropriate chemical test set, identifying in vitro test models, compliance to GLP standards, and a rigorous validation process. This culminated in an ambitious international effort to validate the micromass, embryonic stem cell, and whole embryo culture toxicity models. Scientifically, this process made three revolutionary contributions to in vitro embryotoxicity modeling. First was the use of mouse ESCs as an in vitro test. Second was the incorporation of an in vitro surrogate to maternal toxicity which is used as a "normalizer" of general toxicity, thus more closely reflecting how in vivo developmental toxicity is assessed. Last was their use of biostatistical prediction models, which permitted the extrapolation of an in vivo outcome from an in vitro data set. Together, these enhancements produced a prediction accuracy of approximately 80% in two of the three tests, and ECVAM has recommended these tests for regulatory acceptance. They have not been implemented in such a capacity thus far, but changes in political climate, coupled with legislative pressures, suggest that their regulatory implementation with respect to the E.U. Cosmetics Directive is imminent. Nevertheless, such implementation may require a number of enhancements that have been discussed at various workshops. One is the possible use of human ESCs, which would enable the risk assessment to be conducted in embryonic tissues of the toxicologically most relevant species. The second is the ReProTect initiative in which multiple in vitro models, sensor technologies, and in silico approaches are being integrated to assess the entire mammalian reproductive cycle including further developmental end points. The success or failure of ReProTect will be a bell weather of the broader acceptance of in vitro regulatory developmental toxicity testing. Although the replacement of in vivo embryo/fetal toxicity tests with in vitro surrogates is not likely in the near term within the pharmaceutical industry, in vitro tests (the ECVAM tests as well as those developed independently) are valuable tools in animal-friendly lead optimization, and in strategic decision making in drug development. It has been suggested that a radical change in thinking will be required to ensure that future regulatory toxicity testing has a greater relevance than current animal-based procedures. This will entail a greater reliance on rigorously tested and scientifically relevant in vitro models (158), and on the recent availability human ESCs. An underlying variable is the differing perception of the importance of alternative in vitro testing in Europe versus North America (161). How all this will impact in vitro embryotoxicity testing is unclear, but what is certain is that this field will remain exciting, both scientifically and politically.

REFERENCES

1. Kochhar DM. Embryo explants and organ cultures in screening of chemicals for teratogenic effects. In Developmental Toxicology, Kimmel CA, Buelke-Sam J, eds. Raven Press, New York, NY, 1981:303–319.
2. Faustman EM. Short-term tests for teratogens. Mut Res 1988; 205:355–384.
3. Sadler TW, Horton WE, Warner CW. Whole embryo culture: A screening technique for teratogens. Teratog Carcinog Mutagen 1982; 2:243–253.
4. Flynn TJ. Teratological research using *in vitro* systems. I Mammalian whole embryo culture. Environ Health Perspect 1987; 72:203–210.
5. Brown NA. Teratogenicity testing *in vitro*: Status of validation studies. Arch Toxicol 1987; 1987(Suppl):11105–11114.
6. Hood RD. Tests for developmental toxicity. In Developmental Toxicology Risk Assessment and the Future, Hood RD, ed. Van Nostrand Reinhold, New York, NY, 1989:155–175.
7. Kimmel GL, Kochhar DM. *In vitro* methods in developmental toxicology: Use in defining mechanisms and risk parameters. CRC Press, Boca Raton, FL, 1990.
8. New DAT. Whole embryo culture, teratogenesis, and the estimation of teratological risk. Teratology 1990; 42:635–642.
9. Schwetz BA, Morrisey RE, Welsch F, et al. *In vitro* teratology. Environ Health Perspect 1991; 94:265–268.
10. Steele CE. Whole embryo culture and teratogenesis. Human Reprod 1991; 6:144–147.
11. Brown NA, Spielmann H, Bechter R, et al. Screening chemicals for reproductive toxicity: The current alternatives. The report and recommendations of an ECVAM/ETS workshop (ECVAM Workshop 12). ATLA 1995; 23:868–882.
12. Welsch F. *In vitro* approaches to the elucidation of mechanisms of chemical teratogenesis. Teratology 1992; 46:3–14.
13. Webster WS, Brown-Woodman PDC, Ritchie HE. A review of the contribution of whole embryo culture to the determination of hazard and risk in teratogenicity testing. Int J Dev Biol 1997; 41:329–335.
14. Flick B, Klug S. Whole embryo culture; An important tool in developmental toxicology today. Curr Pharmaceut Design 2006; 12:1467–1488.
15. Harris C, Hansen JM. *In Vitro* methods for the study of mechanisms of developmental toxicity. In Developmental and Reproductive Toxicology, a Practical Approach, 2nd ed. Hood RD, ed. CRC Press, Boca Raton, FL, 2006:647–695.

16. Daston GP. The theoretical and empirical case for *in vitro* developmental toxicity screens, and potential application. Teratology 1996; 53:339–344.

17. Julien E, Willhite CC, Richard AM, et al. Challenges in constructing statistically based structure activity relationship models for developmental toxicity. Birth Defects Res A 2004; 70:902–911.

18. Indans I. The use and interpretation of *in vitro* data in regulatory toxicology: Cosmetics, toiletries and household products. Toxicol Lett 2002; 127:176–182.

19. Garthoff B. Alternatives to animal experimentation: The regulatory background. Toxicol Appl Pharmacol 2005; 207:S388–S392.

20. Pauwels M, Rogiers V. Safety evaluation of cosmetics in the EU reality and challenges for the toxicologist. Toxicol Lett 2004; 151:7–17.

21. Spielmann H, Seiler A, Bremer S, et al. The practical application of three validated *in vitro* embryotoxicity tests. ATLA 2006; 34:527–538.

22. Kimmel GL, Smith K, Kochhar DM, et al. Overview of *in vitro* teraotgenicity testing: Aspects of validation and application to screening. Teratog Carcinog Mutagen 1982; 2:221–229.

23. Donald JM, Monserrat LE, Hooper K, et al. Prioritizing candidate reproductive/developmental toxicants for evaluation. Reprod Toxicol 1992; 2(6):99–108.

24. Green S, Goldberg A, Zurlof J. TestSmart-high production volume chemicals: An approach to implementing alternatives into regulatory toxicology. Toxicol Sci 2001; 63:6–14.

25. Bremer S, Hartung T. The use of embryonic stem cells for regulatory developmental toxicity testing in vitro-the current status of test development. Curr Pharm Des 2004; 10:2733–2747.

26. Kola I, Landis J. Can the pharmaceutical industry reduce attrition rates? Nat Rev Drug Discov 2004; 3:711–715.

27. Cicurel L, Schmid BP. Postimplantation embryo culture for the assessment of the teratogenic potential and potency of compounds. Experientia 1988; 44:833–840.

28. Ozoliṅš TRS. Use of WEC in pharmaceutical screening/testing, HESI (Health and Environmental Sciences) Workshop on *In Vitro* testing. Cary, NC, 2007, http://www.hesiglobal.org/Committees/Committees/technicalCommittees/DART/DART+Alt.+Assays+Workshop.html (Accessed 2007–09–22).

29. Proceedings on the international conference on tests of teratogenicity *in vitro*. Ebert JD, Marios M., Wolf E. North Holland Publishing Company, New York, NY, 1976.

30. National Research Council. Scientific frontiers in developmental toxicology and risk assessment. National Academy Press, Washington, DC, 2000.

31. Jelinek R, Peterka M, Rychter Z. Chick embryotoxicity screening test—130 substances tested. Ind J Exp Biol 1985; 23:588–595.

32. Dawson DA, Fort DJ, Newell DL, et al. Developmental toxicity testing with FETAX: Evaluation of five compounds. Drug Chem Toxicol 1989; 12:67–75.

33. Johnson EM, Bagel BE. An artificial "embryo" for detection of abnormal developmental biology. Fundam Appl Toxicol 1983; 3:243–249.

34. Bournias-Vardiabasis N, Teplitz RL, Chernoff GF, et al. Detection of teratogens in the drosophila embryonic cell culture test: Assay of 100 chemicals. Teratology 1983; 28:109–122.

35. Seng WL, Augustine KA. Zebrafish: A predictive model for assessing developmental toxicity. HESI (Health and Environmental Sciences) Workshop on *In*

Vitro testing. Cary, NC, 2007, www.hesiglobal.org/NR/rdonlyres/264DCF81-OE61–4A37-A72D-E45341BC59FB/0/REVSengHESIWORKSHOP.pdf (Accessed 2008–09-22).

36. Cicurel L, Schmid BP. Post-implantation embryo culture: Validation with selected compounds for teratogenicity testing. Xenobiotica 1988; 18:617–624.

37. Van Maele-Fabry G, Picard JJ. Evaluation of the embryotoxic potential of ten chemicals in the whole mouse embryo culture. Teratology 1987; 36:95–106.

38. Guntakatta M, Matthews EJ, Rundell JO. Development of a mouse embryo limb bud cell culture system for the estimation of chemical teratogenic potential. Teratog Carcinog Mutagen 1984; 4:349–364.

39. Wise LD, Clark RL, Rundell JO, et al. Examination of a rodent limb bud micromass assay as a pre-screen for developmental toxicity. Teratology 1990; 41:341–351.

40. Kistler A. Limb bud cell cultures for estimating the teratogenic potential of compounds. Arch Toxicol 1987; 60:403–414.

41. Renault J-Y, Melcion C, Cordier A. Limb bud cell culture for *in vitro* teratogen screening: Validation of an improved assessment using 51 compounds. Teratog Carcinog Mutagen 1989; 9:83–96.

42. Wilk AL, Greenberg JH, Horigan EA, et al. Detection of teratogenic compounds using differentiating embryonic cells in culture. *In Vitro* 1980; 16:269–276.

43. Flint OP, Orton TC. An *in vitro* assay for teratogens with cultures of rat embryo midbrain and limb cells. Toxicol Appl Pharmacol 1984; 76:383–395.

44. Flint OP, Orton TC, Ferguson RA. Differentiation of rat embryo cells in culture: Response following acute maternal exposure to teratogens and non-teratogens. J Appl Toxicol 1984; 4:109–116.

45. Greenberg JH. Detection of teratogens by differentiating embryonic neural crest cells in culture: Evaluation as a screening system. Teratog Carcinog Mutagen 1982; 2:319–323.

46. George Daston, personal communication.

47. Braun AG, Buckner CA, Emerson DJ, Nichinson BB. Quantitative correspondence between the *in vivo* and *in vitro* activity of teratogenic agents. Proc Natl Acad Sci U S A 1982; 79:2056–2060.

48. Pratt RM, Willis WD. *In vitro* screening assay for teratogens using growth inhibition of human embryonic cells. Proc Natl Acad Sci U S A 1985; 82:5791–5794.

49. Mummery CL, van den Brick CE, van der Saag PT, et al. A short-term screening test for teratogens using differentiating neuroblastoma cells *in vitro*. Teratology 1984; 29:271–279.

50. Piersma AH, Haakmat AS, Hagenaars AM. *In vitro* assays for the developmental toxicity of xenobiotic compounds using differentiating embryonal carcinoma cells in culture. Toxicol *In Vitro* 1993; 7:615–621.

51. Keller SJ, Smith MK. Animal virus screens for potential teratogens. I. Poxvirus morphogenesis. Teratog Carcinog Mutagen 1984; 2:361–374.

52. Laschinski G, Vogel R, Spielmann H. Cytotoxicity test using blastocyst-derived eupoid embryonal stem cells: A new approach to *in vitro* teratogenesis screening. Reprod Toxicol 1991; 5:57–64.

53. Newall DR, Beedles KE. The stem-cell test: An *in vitro* assay for teratogenic potential. Results of a blind trial with 25 compounds. Toxicol Vitro 1996; 10:229–240.

54. Editorial. Constitutional protection for animals. Nat Neurosci 2002; 5:611.

55. Anon. Directive 2003/15.EC of the European Parliament and the Council directive of 27 February 2003 amending Council directive 76/768/EEC on the approximation of the laws of the Member States relating to cosmetics products. Official J Eur Union 2003; L66:26–35.

56. Anon. White paper on a strategy for a future chemical policy (COM(2001)88 final) The commission of European communities; 2001, pp. 32. http://europa. eu.int/comm./environment/chemicals/whitepaper.htm. (Accessed 2008-09-25)

57. Mirkes PE. Prospects for the development of validated screening tests that measure developmental toxicity: View of one skeptic. Teratology 1996; 53:334–338.

58. Guest I, Buttar HAS, Smith S, et al. Evaluation of the rat embryo culture system as a predictive test for human teratogens. Can J Physiol Phamacol 1994; 72:57–62.

59. Spielmann H, Liebsch M. Lessons learned from validation of in vitro toxicity test: From failure to acceptance into regulatory practice. Toxicol Vitro 2001; 15:585–590.

60. Genschow E, Spielmann H, Scholz G, et al. The ECVAM international validation study on *in vitro* embryotoxicity tests: Results of the definitive phase and evaluation of prediction models. ATLA 2002; 30:151–176.

61. Van Maele-Fabry G, Picard JJ, Attenon P, et al. Interlaboratory evaluation of three culture media for postimplantation rodent embryos. Reprod Toxicol 1991; 5:417–426.

62. Foote RH, Carney EW. The rabbit as a model for reproductive and developmental toxicity studies. Reprod Toxicol 2000; 14:476–493.

63. Naya M, Kito Y, Eto K, et al. Development of rabbit whole embryo culture during development. Cong Anom 1991; 31:153–156.

64. Carney EW, Tornesi B, Keller C, et al. Refinement of a morphological scoring system for postimplantation rabbit conceptuses. Birth Defects Res B 2007; 80:213–222.

65. Piersma AH, Attenon P, Bechter R, et al. Interlaboratory evaluation of embryotoxicity in the postimplantation rat embryo culture. Reprod Toxicol 1995; 9:275–280.

66. Ozolinš TRS, Hales BF. Posttranslational modification of the AP-1 response in the rat conceptus during oxidative stress. Mol Pharmacol 1999; 56:537–544.

67. Ozolinš TRS. Unpublished results.

68. Brown NA, Fabro S. Quantitation of rat embryonic development *in vitro*: A morphological scoring system. Teratology 1981; 24:65–78.

69. Van Maele-Fabry G, Delhaise F, Picard JJ. Morphogenesis and quantification of the development of postimplantation mouse embryos. Toxicol Vitro 1990; 4:149–156.

70. Klug S, Lewandowski C, Neubert D. Modification and standardization of the culture of early postimplantation embryos for toxicological studies. Arch Toxicol 1985; 58:84–88.

71. Balls M, Blaauboer B, Brusik D, et al. Report and recommendations of the CAAT/ERGATT workshop on the validation of toxicity test procedures. ATLA 1990; 18:313–337.

72. Balls M, Blaauboer BJ, Fentem J, et al. Practical aspects of the validation of toxicity test procedures. The report and recommendations of ECVAM Workshop 5. ATLA 1995; 23:129–147.

73. Anon. INVITTOX protocol no. 113. Embryonic Stem Cell Test (EST), List of INVITTOX protocols, 2002, Web site: http://ecvam–dbalm.jrc.ec.eu (Accessed 2007-09-02).

74. Anon. INVITTOX protocol no. 114. *In vitro* Micromass Teratogen Assay. List of INVITTOX protocols, 2002, Web site: http://ecvam–dbalm.jrc.ec.eu (Accessed 2007–09-02).

75. Anon. INVITTOX protocol no.68. Embryotoxicity testing using a whole embryo culture WEC procedure, List of INVITTOX protocols, 2002, Web site: http://ecvam–dbalm.jrc.ec.eu (Accessed 2007–09–02).

76. Smith MK, Kimmel GL, Kochhar DM, et al. A selection of candidate compounds for *in vitro* teratogenesis test validation. Teratog Carcinog Mutagen 1983; 3:461–480.

77. Schwetz BA. Criteria for judging the relative toxicity of chemicals from developmental toxicity data: A workshop summary. Teratology 1992; 45:337–339.

78. Brown NA. Selection of test chemicals for the ECVAM international validation study on *in vitro* embryotoxicity tests. ATLA 2002; 30:176–198.

79. Daston GP, Baines D, Elmore E, et al. Evaluation of chick embryo neural retina cell culture as a screen for developmental toxicity. Fund Appl Toxicol 1995; 26:203–210.

80. Daston GP, Rogers JM, Versteeg DJ, et al. Interspecies comparisons of A/D ratios: A/D ratios are not constant across species. Fund Appl Toxicol 1991; 17:696–722.

81. Holzhügtter HG, Archer G, Dami N, et al. Recommendations for the application of biostatistical methods during the development and validation of alternative toxicological methods. ECVAM biostatistics task force report. ATLA 1996; 24:511–530.

82. Schwetz BA, Harris MW. Developmental toxicology: Status of the field and contribution of the National Toxicology Program. Environ Health Perspect 1993; 100:269–282.

83. Frankos VH. FDA perspectives on the use of teratology data for human risk assessment. Fund Appl Toxicol 1985; 5:615–625.

84. Schardein JL. Chemically induced birth defects, 3rd ed. Marcel Dekker Inc, New York, NY, 2000.

85. Shepard T. Catalogue of teratogenic agents, 7th ed. Johns Hopkins University Press, Baltimore, MD, 1992.

86. Spielmann H, Steinhoff R, Schaefer C, et al. Arzneiverordnung in schwangerschaft und stillzei, 5th ed. Gustav Fischer, Stuttgart, Germany, 1998.

87. Cooper-Hannan C, Harbell JW, Coecke S, et al. The principles of good laboratory practice: Application to *in vitro* toxicology studies. The report and recommendations of ECVAM workshop 37. ATLA 1999; 27:539–576.

88. Curren RD, Southee JA, Spielmann H, et al. The role of pre-validation in the development, validation and acceptance of alternative methods. ECVAM prevalidation task force report 1. ATLA 1995; 23:211–217.

89. Flint OP. *In vitro* tests for teratogens: Desirable endpoints, test batteries and current status of the micromass teratogen test. Reprod Toxicol 1993; 7:103–111.

90. Hansen JM, Gong SG, Philbert M, et al. Misexpression of gene expression in the redox-sensitive NF-{kappa} b-dependent limb outgrowth pathway by thalidomide. Devel Dynam 2002; 225:186–194.

91. Daston GP, Baines D, Yonkers JE. Chick embryo neural retina cell culture as a screen for developmental toxicity. Toxicol Appl Pharmacol 1991; 109:352–366.

92. Saad AD, Moscona AA. Cortisol receptors and inducibility of glutamine synthetase in embryonic retina. Cell Differ 1985; 16:241–250.

93. Moscona A. Rotation-mediated histogenic aggregation of dissociated cells. Exp Cell Res 1961; 22:455–475.

94. Rohwedel J, Guan K, Hegert C, et al. Embryonic stem cells as *in vitro* model for mutagenicity, cytotoxicity and embryotoxicity studies: Present state and future prospects. Toxicol *In Vitro* 2001; 15:741–753.
95. Davila JC, Cezar GG, Thiede M, et al. Use and application of stem cells in toxicology. Toxicol Sci 2004; 79:214–223.
96. Goodman S. Race is on to find alternative tests. Nature 2002; 418:116.
97. Guan K, Rohwedel J, Wobus AM. Embryonic stem cell differentiation models: Cardiogenesis, myogenesis, neurogenesis, epithelial and vascular smooth muscle cell differentiation *in vitro*. Cytotechnology 1999; 30:211–226.
98. Chapin RE, Augustine-Rauch K, Beyer B, et al. The state of the art in developmental toxicity screening methods, and the way forward. Birth Defects Res B 2008 (in press).
99. New DAT. Whole-embryo culture and the study of mammalian embryos during organogenesis. Biol Rev 1978; 53:81–122.
100. Pitt JA, Carney EW. Development of a morphologicaly-based scoring system for postimplantation New Zealand White rabbit embryos. Teratology 1999; 59:88–101.
101. Piersma AH, Genshow E, Verhoef A, et al. Validation of the postimplantation rat whole-embryo culture test in the international ECVAM validation study on three *in vitro* embryotoxicity tests. ATLA 2004; 32:275–302.
102. Wilson JG. Environment and birth defects. Academic Press, New York, NY, 1973.
103. Shuey DL, Sadler TW, Lauder JM. Serotonin as a regulator of craniofacial morphogenesis: Site specific malformations following exposure to serotonin uptake inhibitors. Teratology 1992; 46:367–378.
104. Physician's desk reference, 55th ed. Medical Economics, Montville, NJ, 2001.
105. Hines RN, Adams J, Buck GM, et al. NTP-CERHR Expert panel report on the reproductive and developmental toxicity of fluoxetine. Birth Defects Res B 2004; 71:193–280.
106. Einarson TR, Einarson A. Newer antidepressants in pregnancy and rates of major malformations: A meta-analysis of prospective comparative studies. Pharmacoepidemiol Drug Saf 2005; 14:823–827.
107. Way CM. Safety of newer antidepressants in pregnancy. Pharmacotherapy 2007; 27:546–552.
108. Stouch TR, Kenyon JR, Johnson SR, et al. *In silico* ADME/Tox: Why models fail. J Comput Aided Mol Des 2003;17:83–92.
109. Festag M, Viertel B, Steinberg P, et al. An *in vitro* embryotoxicity assay based on the disturbance of the differentiation of murine embryonic stem cells into endothelial cells. I: Testing of compounds. Toxicol *In Vitro* 2007; 8:1631–1640.
110. OECD (Organisation for Economic Co-operation and Development). Final report of the OECD workshop on harmonisation of validation and acceptance criteria for alternative toxicological test methods. OECD Publications Office, Paris, France, 1996.
111. NIEHS (National Institute of Environmental Health Sciences). Validation and regulatory acceptance of toxicological test methods: A report of the ad hoc interagency co-ordinating committee o the validation of alternative methods. NIEHS NIH Publication No. 97–3981, Research Triangle Park, NC, 1997.
112. Chapin RE, Stedman D, Paquette J, et al. Struggles for equivalence: *In vitro* developmental toxicity model evolution in pharmaceuticals in 2006. Toxicol *In Vitro* 2007; 21:1545–1551.

113. Whitlow S, Bürgen H, Clemann N. The embryonic stem cell test for the early selection of pharmaceutical compounds. ALTEX 2007; 24:3–7.
114. Paquette JA, Kumpf SW, Streck RD, et al. Assessment of the embryonic stem cell test and application and use in the pharmaceutical industry. Birth Def Res B 2008; 83:104–111.
115. Vanparys P. ECVAM and pharmaceuticals. ATLA 2002; 30(Suppl 2):221–223.
116. Sutherland V. Novel applications of rodent WEC for teratogenic assessment of test compounds, HESI (Health and Environmental Sciences) Workshop on *In Vitro* testing, Cary, NC, 2007, http://www.hesiglobal.org/Committees/TechnicalCommittees/DART/DART+Alt.+Assays+Workshop.html (Accessed 2007–09-25).
117. Festag M, Sehner C, Steinberg P, et al. An *in vitro* embryotoxicity assay based on the disturbance of the differntiation of murine embryonic stem cells into endothelial cells. I: Establishment of the differentiation protocol. Toxicol *In Vitro* 2007; 8:1619–1630.
118. ESAC. The use of scientifically validated *in vitro* tests for embryotoxicity. ATLA 2002; 30:265–273.
119. Ruhdel IW. EU-Cosmetics: Timetables for the replacement of animal experiments. ALTEX 2005; 22:117–119.
120. Combes R, Dandrea J, Balls M. A critical assessment of the European Commission's proposals for the risk assessment and registration of chemical substances in the European Union. Altern Lab Anim 2003; 31:354–364.
121. Worth AP, Balls M. The principles of validation and the ECVAM validation process. ATLA 2002; 30(Suppl 2):15–21.
122. Liebsch M, Spielmann H. Currently available *in vitro* methods used in the regulatory toxicology. Toxicol Lett 2002; 127:127–134.
123. Bechter R, Schön H. Use of the whole embryo culture system in drug safety assessment. Toxicol *In Vitro* 1988; 2:195–203.
124. Carney EW, Pottenger LH, Johnson KA, et al. Significance of 2-methoxypropionic acid formed from beta-propylene glycol monomethyl ether: Integration of pharmacokinetic and developmental toxicity assessment in rabbits. Toxicol Sci 2003; 71:217–228.
125. Corley RA, Bartels MJ, Carney EW, et al. Development of a physiologically-based pharmacokinetic model for ethylene glycol and its metabolite, glycolic acid, in rats and humans. Toxicol Sci 2005; 85:479–490.
126. Carney EW, Tornesi B. Alternative species for whole embryo culture, Health and Environmental Sciences (HESI) *In Vitro* Assays Workshop, Cary, NC, 2007. http://www.hesiglobal.org/%20Committees/%20TechnicalCommittees/DART/%20DART+Alt.+Assays+Workshop.html (Accessed 2007–09-22).
127. Warner CW, Sadler TW, Shockey J, et al. A comparison of the *in vivo* and *in vitro* response of mammalian embryos to a teratogenic insult. Toxicology 1983; 28:271–282.
128. Schmid BP, Cicurel L. Application of the post-implantation rat embryo culture system to *in vitro* teratogenicity testing. Food Chem Toxicol 1986; 24:623–626.
129. Piersma AH, Janer G, Wolterink G, et al. Quantitative extrapolation of *in vitro* whole embryo culture embryotoxicity data to developmental toxicity *in vivo* using the benchmark dose approach. Toxicol Sci 2007; 101:91–100.

130. Verwei M, van Burgsteden JA, Krul CAM, et al. Prediction of *in vivo* embryotoxic effect levels with a combination of *in vitro* studies and PBPK modeling. Toxicol Lett 2006; 165:79–87.

131. Brown NA. Overview of whole embryo culture assay, HESI (Health and Environmental Sciences) Workshop on *In Vitro* testing. Cary, NC, 2007, http://www.hesiglobal.org/Committees/TechnicalCommittees/DART/DART + Alt. + Assays + Workshop.html (Accessed 2007–09–25).

132. Paparella M, Kolossov E, Fleischmann BK, et al. The use of quantitative image analaysis in the assessment of *in vitro* embryotoxicity endpoints based on a novel embryonic stem cell clone with endoderm-related GFP expression. Toxicol *In Vitro* 2002; 16:589–597.

133. Usami M, Mitsunaga K, Nakazawa K. Two-dimensional electrophoresis of protein from cultured postimplantation rat embryos for developmental toxicity studies. Toxicol *In Vitro* 2007; 21:521–526.

134. Seiler A, Buesen R, Hayess K, et al. Current status of the embryonic stem cell test: The use of recent advances in the field of stem cell technology and gene expression analysis. ALTEX 2006; 23:393–399.

135. Seiler A, Visan A, Buesen R, et al. Improvement of the embryonic stem cell test (EST) for developmental toxicity by establishing molecular endpoints of tissue-specific development. Reprod Toxicol 2004; 18:231–240.

136. zur Nieden NI, Ruf LJ, Kempka G, et al. Molecular markers in embryonic stem cells. Toxicol *In Vitro* 2001; 15:455–461.

137. zur Nieden NI, Kepka G, Ahr HJ. Molecular multiple endpoints embryonic stem cell test—a possible approach to test for the teratogenic potential of compounds. Toxicol Appl Pharmacol 2004; 194:257–269.

138. Pellizzer C, Bremer S, Hartung T. Developmental toxicity testing from animal towards embryonic stem cells. ALTEX 2005; 22:47.

139. Pellizzer C, Bello E, Adler S, et al. Detection of tissue-specific effects by methotrexate on differentiating mouse embryonic stem cells. Birth Defects Res B 2004; 71:331–341.

140. Bremer S, Pellizzer C, Coecke S, et al. Detection of the embryotoxic potential of CPA by using a combined system of metabolic competent cells and embryonic stem cells. Altern Lab Anim 2002; 30:76–85.

141. Parsons JF, Rockley J, Richold M. *In vitro* micromass teratogen test: Interpretation of results from a blind trial of 25 compounds using three separate criteria. Toxicol *in Vitro* 1990; 4:609–611.

142. Fantel AG, Greenaway JC, Juchau MR, et al. Teratogenic bioactivation of cyclophosphamide *in vitro*. Life Sci 1979; 25:67–72.

143. Ogelsby LA, Ebron MT, Beyer PE, et al. Co-culture of rat embryos and hepatocytes: *In vitro* detection of a proteratogen. Teratog Carcinog Mutagen 1986; 6:120–138.

144. Ozolins TRS, Oglesby LA, Wiley MJ, et al. *In vitro* murine embryotoxicity of cyclophosphamide in embryos co-cultured with maternal hepatocutes: Development and application of a murine embryo-hepatocyte co-culture model. Toxicology 1995; 102:259–274.

145. Ozolins TRS, Wiley MJ, Wells PG. Phenytoin covalent binding and embryopathy in mouse embryos co-cultured with maternal hepatocytes from mouse rat and rabbit. Biochem Pharmacol 1995; 50:1831–1840.

146. Piersma AH, van Aerts LA, Verhoef A, et al. Biotransformation of cyclophosphamide in post-implantation rat embryo culture using maternal hepatocytes in co-culture. Pharmacol Toxicol 1991; 69:47–51.

147. Coecke S, Ahr H, Blaauboer BJ, et al. Metabolism: A bottleneck in *in vitro* toxicological test development. The report and recommendations of ECVAM workshop 54. ATLA 2006; 34:49–84.

148. Nau H. Pharmacokinetic aspects of *in vitro* teratogenicity studies: Comparison to *in vivo*. In *In Vitro* Methods in Developmental Toxicology: Use in Defining Mechanisms and Risk Parameters, Kimmel GL, Kochhar DM, eds. CRC Press, Boca Raton, FL, 1990:29–43.

149. Bouvier d'Yvoire M, Prieto P, Blaauboer BB, et al. Physiologicaly-bsaed kinetic modeling (PBK modelling): Meeting the 3Rs agenda. ATLA 2007; 35:661–671.

150. Hakkola J, Pelkonen O, Pasenen M, et al. Xenobiotic-metabolizing cytochrome P450 enzymes in the human feto-placental unit: Role in intrauterine toxicity. Crit Rev Toxicol 1998; 28:35–72.

151. Adler S, Paparella M, Pellizzer C, et al. The detection of differentiation-inducing chemicals by using green fluorescent protein expression in genetically engineered teratocarcinoma cells. ATLA 2005; 33:91–103.

152. Walmod PS, Gravemann U, Nau H, et al. Discriminative power of an assay for automated *in vitro* screening of teratogens. Toxicol *In Vitro* 2004; 18:511–525.

153. Buesen R, Visan A, Genschow E, et al. Trends in improving the embryonic stem cell test (EST): An overview. ALTEX 2004; 21:15–22.

154. Bremer S, Worth AP, Paparela M, et al. Establishment of the an *in vitro* reporter gene assay for developmental cardiac toxicity. Toxicol *In Vitro* 2001; 15:215–223.

155. Adler S, Pellizzer C, Hareng L, et al. First steps in establishing a developmental toxicity test method based on human embryonic stem cells. Toxicol *In Vitro* 2008; 22:200–211.

156. Hareng L, Pellizzer C, Bremer S, et al. The integrated project ReProTect: A novel approach in reproductive toxicity hazard assessment. Reprod Toxicol 2005; 20:441–452.

157. Bremer S, Cortvrindt R, Daston G, et al. Reproductive and Developmental Toxicity. Altern Lab Anim 2005; 33(Suppl 1):183–209.

158. Balls M, Combes R. The need for a formal invalidation process for animal and non-animal tests. ATLA 2005; 33:299–308.

159. Grindon C, Combes R, Cronin MT, et al. Integrated decision-tree testing strategies for environmental toxicity with respect to the requirements of the EU REACH legislation. Altern Lab Anim 2006; 34:651–664.

160. Genschow E, Scholz G, Brown N, et al. Development of prediction models for three *in vitro* embryotoxicity tests. *In Vitro* Mol Toxicol 2000; 13:51–65.

161. Becker RA, Borgert CJ, Webb S, et al. Report of an ISRTP workshop: Progress and barriers to incorporating alternative toxicological methods in the U.S. Reg Toxicol Pharmacol 2006; 46:18–22.

8

Zebrafish: A Nonmammalian Model of Developmental Toxicology

Kimberly C. Brannen and Julieta M. Panzica-Kelly
Discovery Toxicology, Pharmaceutical Candidate Optimization,
Bristol-Myers Squibb Company, Pennington, New Jersey, U.S.A.

Jeffrey H. Charlap
Preclinical Services, Charles River Laboratories, Horsham, Pennsylvania, U.S.A.

Karen A. Augustine-Rauch
Discovery Toxicology, Pharmaceutical Candidate Optimization,
Bristol-Myers Squibb Company, Pennington, New Jersey, U.S.A.

INTRODUCTION

The zebrafish (*Danio rerio*), a small freshwater fish originating from Asia, is a common pet shop favorite. The species is generally not aggressive, can thrive in simple home aquarium conditions, and can be bred relatively easily. Zebrafish are oviparous (fertilization and development occur externally) and can be prolific egg producers, with between 50 and 300 fertilized eggs produced by a breeding pair at least once per week (1). Zebrafish embryos lend themselves particularly well to the in-life exploration of vertebrate development since the embryos are transparent and their development can be followed up to adulthood. Adult zebrafish husbandry and zebrafish embryo culture are not resource intensive, especially when compared to other animals commonly used in studies of developmental toxicology.

These are among the reasons that the zebrafish has also become a model system of choice for developmental biologists and has achieved popularity as a useful model for conducting functional genomics research. George Streisinger, considered by many the founding father of zebrafish genetics research, chose to work with zebrafish primarily because they are an organism that is relatively easy to manipulate genetically (2). For many years, this model has been used for environmental testing, but more recently, it has also become a promising, emerging model in the field of developmental toxicology and teratology.

Zebrafish offer many practical and scientific advantages as a developmental toxicology model species. Development has been well characterized, especially during the embryonic and early larval periods (3–5). In addition, embryonic development is very rapid, with organogenesis reaching completion by approximately 3 days postfertilization (dpf). Given the small size, ease of maintenance, and fecundity of adult zebrafish, a rotation of breeding pairs or groups can readily be established to provide researchers with fertilized eggs daily if desired. The embryos are quite amenable to natural growth and development in tubes or culture plates. Many of them can be grown at one time in multi-well culture plates, even in wells as small as those of a standard 96 or 384-well plate. Furthermore, the developing fish can be maintained in culture from the onset of fertilization, through organogenesis, and on into larval stages, given the appropriate conditions. For at least the first 5 days of development, they require no external nourishment as they draw nutrients from their yolk sac, and they grow and develop quite well in simple 28.5°C incubation ovens. Importantly, while zebrafish embryos or larvae are usually maintained "in culture" for experimental manipulation or toxicity testing, the conditions under which the zebrafish develop are essentially similar to those in an aquarium or in the wild, making this in reality an in vivo model system. The developing zebrafish can be readily viewed and manipulated. For much of early development, zebrafish are transparent, enabling easy assessment of developing organ systems with base-illumination stereomicroscopy. Diverse anatomical structures and organ systems can be evaluated in the developing zebrafish including morphology of the central nervous system, cardiovascular system, craniofacial morphology, skeletal system, gastrointestinal system, axial development, and fin morphology. The larvae (and, to some extent, later-stage embryos) can also be assessed for a spectrum of motor activities and reflexes.

The zebrafish genome has been sequenced by the Sanger Institute, and genome annotation is nearly finished (6–8). Release of the complete, manually annotated genome is slated for late 2008 (7). Forward genetics approaches, including 2 large-scale mutagenesis screens (9,10), have produced thousands of zebrafish mutants, many of which have been described in the literature. Such mutations have affected organogenesis, physiology, and/or behavior. Successfully linking genotypes to many zebrafish mutant phenotypes has contributed significantly to a better understanding of gene function. Conservation of molecular and cellular physiology of many embryological processes between fish and mammals and the ease of the analysis of phenotype and genotype have been key factors in the zebrafish coming to be viewed as a valuable developmental model system.

In general, the advantages of using the zebrafish as a developmental model system outweigh the disadvantages. However, for researchers interested in modeling mammalian developmental toxicology, zebrafish present the potentially important and obvious disadvantage of being a nonmammalian species. Compared to the more traditional mammalian models of developmental toxicology, zebrafish will present some differences in genetics, development, and physiology. Such diversity should be considered when zebrafish are studied for these purposes, particularly with respect to hazard identification and risk assessment. In addition, the zebrafish embryo-larvae are cultured in an aqueous environment where challenges could arise in evaluating lipophilic or aqueous insoluble compounds. However, microinjection of such compounds into the yolk ball may be able to circumvent this problem. Zebrafish embryo-larval development is sensitive to pH, and deviations from the optimal pH caused by test compound treatments could result in associated toxicity. Thus, establishing pH measurements as one of the controls in testing strategies is recommended.

The focus of this chapter is to introduce the developmental toxicologist to the zebrafish as a valuable model for identification of compounds with mammalian teratogenic potential and for characterization of teratogenic mechanisms. Zebrafish development, use of zebrafish for developmental toxicity testing, methods for evaluating molecular mechanisms underlying teratogenicity and developmental toxicity, and selected development toxicity findings in zebrafish are reviewed.

OVERVIEW OF ZEBRAFISH DEVELOPMENT

Zebrafish embryonic and larval developments have been thoroughly described elsewhere (3–5), and a brief overview summarized from those sources is provided here. Embryonic development of zebrafish occurs within the first 3 days following fertilization, with larval development beginning at 72 hours postfertilization (hpf). This section reviews the seven periods of embryonic zebrafish development (zygote, cleavage, blastula, gastrula, segmentation, pharyngula, and hatching periods) and early larval development, with focus on aspects that can be microscopically observed (Fig. 1).

Day 1

A phenomenal amount of development occurs within the first day of zebrafish development, with fertilized eggs forming an embryo with a recognizable vertebrate body plan. Development begins with the zygote period (0–$\frac{3}{4}$ hpf), which lasts until the first cleavage occurs. The main event that occurs during this period is the segregation of the oocyte cytoplasm from the yolk through cytoplasmic streaming, in which the non-yolk cytoplasm moves toward the animal pole and the yolk-rich cytoplasm toward the vegetal pole of the cell. This segregation of the animal and vegetal poles continues through several cleavages.

The cleavage period ($\frac{3}{4}$–$2\frac{1}{4}$ hpf) consists of the first cleavage through the 64-cell embryo stage. The first cleavage is oriented vertically, passing through only the blastodisc and leaving the yolk undivided. Subsequent blastomere divisions occur synchronously approximately every 15 minutes with each perpendicular to

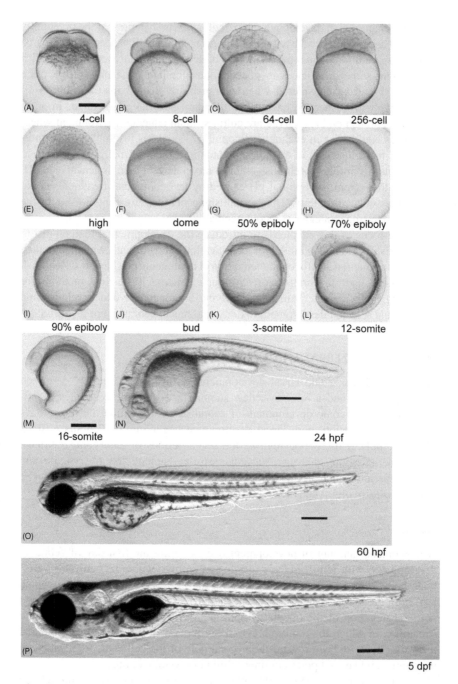

Figure 1 Overview of zebrafish development from early embryonic through early larval periods. Scale bar = 250 μm.

the last. These early cleavages are not complete; the cells are still connected with each other and the yolk cell. After the fifth cleavage, cellular divisions start to run horizontally. This makes it difficult to distinguish between a 32-cell and a 64-cell embryo, but when observed laterally, one can see that the cells of the 64-cell embryo are smaller and mounded higher than those of the 32-cell embryo.

The blastula period ($2^{1}/_{4}$–$5^{1}/_{4}$ hpf) begins with the 128-cell embryo stage and ends at about the 14th round of cell division (the onset of gastrulation). Although the term "blastula" is used to describe the embryo during this period, the zebrafish embryo does not develop a blastocoele. The main events of the blastula period include the mid-blastula transition, the development of the yolk syncytial layer, and the commencement of epiboly. The mid-blastula transition occurs at about the 512-cell stage ($2^{3}/_{4}$ hpf) and consists of the compression of the ball of cells, an increase in cell-cycle duration, asynchrony of cell divisions, and finally the start of zygotic transcription. At about the same stage, the marginal blastomeres collapse and release their contents into the cytoplasm of the adjacent yolk cell. This event forms the yolk syncytial layer, which contains a ring of nuclei at the margin of the blastodisc. Consequently, by the 1000-cell stage (3 hpf), the blastomeres no longer have any cytoplasmic bridges to other cells. At this point, the yolk syncytial layer will divide 3 times and stop abruptly (dome stage; $4^{1}/_{3}$ hpf) prior to the beginning of epiboly, in which the nuclei start to expand in size and the blastodisc layer thins and spreads out, with the yolk syncytial layer and blastodisc eventually covering the entire yolk surface by the end of epiboly. Starting around the 1000-cell stage, cell divisions gradually become asynchronous.

The gastrula period ($5^{1}/_{4}$–10 hpf) begins at about 50% epiboly, includes gastrulation, and ends after completion of epiboly and the formation of the tail bud. Gastrulation constitutes the processes of involution, convergence, and extension. Involution begins at the 50%-epiboly stage, and within minutes the germ ring is formed as a thickening at the margin due to involution of the marginal deep cell layer of the blastoderm. Convergence of germ ring cells forms the embryonic shield which consists of two cell layers, the outer epiblast and the inner hypoblast, and marks the dorsal side of the embryo. Near the end of epiboly (90%), the neural plate is formed at the dorsal region of the embryo by anterior thickening of the epiblast. The neural plate extends along the whole axis of the embryo eventually giving rise to the spinal cord (posterior portion of plate) and to the brain (more prominent anterior portion). At this stage, postmitotic cells are present that will form the notochord, somite-derived muscles, and specific neurons of the hindbrain. Soon after epiboly is complete, the tail bud is formed by thickening of the caudal region of the embryo that will contribute primarily to the development of the tail.

The main events of the next period of development, the segmentation period (10–24 hpf), include somite development; embryo elongation along the anterior–posterior axis; primary organ rudiment development; and the first body movements.

Somite development begins around $10^{1}/_{3}$ hpf with formation of the posterior boundary of the first somite, and the remaining somites develop at a rate of approximately 2 to 3 per hour. The majority of the interior cells of each somite give rise to a muscle segment, the myotome (or myomere). Elongation of the muscle fibers in the myotome is later responsible for the prominent V (or chevron) shape of the somites. Another major somite derivative, the sclerotome, gives rise to vertebral cartilage. Sclerotome cells delaminate, adopt a mesenchymal appearance, and migrate along the path between the myotome and the notochord. Finally, at the end stages of the segmentation period, the muscle cells in the somites begin to contract, and the first body movements can be detected. As the embryo develops, these movements become more intense and synchronized.

Notochord development is also prominent during this period. Some of the presumptive notochord cells vacuolate and swell differentiating into the notochord in an anterior to posterior direction, while other cells differentiate into a notochord sheath composed of an epithelial monolayer. By the middle of the segmentation period, the notochord acquires a "stack of pennies" appearance.

During the segmentation period, major organ rudiments begin to develop, with dramatic progress in brain morphogenesis. Morphogenesis of the brain begins with an invagination of epithelium at the midline of the brain rudiment into a neural keel, which will subsequently round to form a neural rod. A cavity called the neurocoel will eventually develop, forming the neural tube lumen, but even before the neurocoel is formed, segmentation of the brain begins. Ten neuromeres, segmental swellings of the presumptive brain, develop along the anterior–posterior axis. The three most anterior segments give rise to the three ventricles of the brain: the forebrain (telencephalon and diencephalon) and the midbrain (mesencephalon). The remaining seven segments, the rhombomeres, make up the hindbrain. By the end of the segmentation period, all 10 neuromeres can be seen. In addition, the neural tube has developed into the trunk.

Several other major organ rudiments develop during the segmentation period. The pronephric kidneys, which develop subjacent to the third somite, are formed during this time, and the pronephric ducts grow toward the anus where they fuse. At about $16\frac{1}{2}$ hpf, the optic vesicle starts to develop laterally from the diencephalon, and the otic placode develop between the optic vesicle and the first somite. About 3 hours later, the lens placode becomes evident, and the otic vesicle, including its otoliths, develops from the otic placode. The tail detaches from the yolk cell and continues to elongate throughout the segmentation period. The caudal region of the yolk cell is responsible for the development of the yolk extension and elongates along with the tail while the yolk ball becomes smaller throughout embryo development.

Day 2

By the beginning of the pharyngula period (24–48 hpf), the embryo is well organized with an evident body plan, a complete set of somites, a well-developed notochord, and five defined brain lobes. Development during this period is characterized by the formation of the pharyngeal arches which give rise to the jaws and gills. Other structures that develop during this stage are the hatching gland, pigment cells, the fins, and the cardiovascular system.

Seven pharyngeal arches develop from a primordial region that extends between the otic vesicle and yolk sac. A visible boundary between the second and third arch develops. The jaws and operculum will develop anterior, and the gills will develop posterior to this boundary. The two anterior arches are referred to as the mandibular and hyoid arches, and the five posterior arches are referred to as the branchial (or gill) arches.

Another structure that becomes visible in this period is the hatching gland, which is found on the pericardium over the anterior yolk sac. The cells that make up the gland refract light due to their granular composition and contain important hatching enzymes. It is presumed that the cells lose their granular appearance when they release the enzymes during the hatching period.

In addition, three types of pigment cells differentiate during this period: melanophores (black), xanthophores (yellow), and iridophores (iridescent silver). The first area where pigment develops is in the retinal epithelium; later melanophores extend longitudinally from the diencephalon to the end of the tail to form the dorsal stripe. Furthermore, some extend laterally to the trunk and tail forming the lateral stripe, and later some extend ventrally to form the ventral stripe. Melanophores also extend to the dorsal-anterior side of the yolk sac and dorsal midbrain, and sporadically some develop in the forebrain region.

The median and pectoral fins also begin to develop during the pharyngula period. Mesenchymal cells migrate to the bilateral locations where pectoral fins will develop and contribute to formation of the fin buds. As the buds grow outward, they begin to develop an apical tip at their center, and eventually an apical ectodermal ridge forms running obliquely across the prospective fin. Subsequently, additional mesenchymal cells migrate and underlie the ridge to form the fin blades. Over the same period, the median fin fold becomes prominent, and its ventral region begins to underlie the yolk extension and develop into collagenous fin rays. These will develop into segmented bony fin rays later in development.

Another major event in the pharyngula period is the formation of the cardiovascular system, including contraction of the heart. Early in this period, the heart is a cone-shaped tube located below the brain. At first, it beats with no apparent rhythm, but at about 26 hpf, the contractions start synchronizing in a posterior to anterior direction concurrent with elongation of the tube. At about the same time, erythrocytes begin moving toward the region between the notochord and the yolk where the major vessels begin to develop. At about 30 hpf, a single aortic arch can be identified, as well as a full arterial system with flowing blood between the pharynx and the tail. Also, a maze-like system of channels connects the caudal artery and vein toward the mid-end section of the tail, moving toward the tip of the tail by the end of this period. By 36 hpf, the heart loops slightly, and the axial vein, posterior cardinal vein, and common cardinal vein are well established. At 42 hpf, the ventricle and atrium of the heart can be clearly identified with contractions beginning at the atrium followed by the ventricle.

There is also striking neurobehavioral development during the pharyngeal period. The sensory-motor reflexive networks become stronger, and the embryo becomes progressively responsive to touch. In addition, spontaneous trunk-tail contractions occur more rapidly, beginning to appear as swimming movements.

Day 3

The third day of development is referred to as the hatching period (48–72 hpf) due to the fact that embryos hatch from their chorion (the tough membrane that surrounds the embryo and yolk) during this time. Zebrafish embryos from the same clutch may hatch at different times during the day. Therefore, developing zebrafish are referred to as embryos prior to the arbitrary threshold of 72 hpf and as larvae thereafter. Hatching does not affect the progression of embryonic development; consequently, embryos that hatch later than others should not exhibit any developmental delays. At the onset of the hatching period, most of the major organ rudiments have completed morphogenesis, and development consists mainly of growth and maturation. However, significant further development of the jaw, gills, and pectoral fins occurs during this stage.

Significant morphogenesis of the pharyngeal arches occurs throughout the whole period. At the beginning of the hatching period, the mouth is visible ventrally between the eyes, but the mouth shifts location anteriorly throughout the period, protruding beyond the eyes by the end of the third day. Among the pharyngeal arches, cartilage begins to develop in the mandibular (pharyngeal arch 1) and hyoid (pharyngeal arch 2) arches earliest, and by the end of this period distinct dorsal and ventral elements form in these arches. The ventral cartilages in these arches are referred to as Meckel's and ceratohyal cartilages, respectively. These become important support structures for the lower jaw. The dorsal cartilages (the quadrate of the first arch and the hyosymplectic of the second arch) are more delicate and have intricate shapes. The first two pharyngeal arches also bear aortic arches 1 and 2. Pharyngeal arches 3 to 7 are also referred to as branchial arches 1 to 5 since gills are formed in this region. Branchial arches 1 to 4 (pharyngeal arches 3–6) bear gills and aortic arches 3 to 6. Five gill slits develop, one posterior to each pharyngeal arches 2 to 6. Branchial arch 5 (pharyngeal arch 7) forms a supportive cartilage with the pharyngeal teeth, but no gill slit or aortic arch develops in this arch. Each of the branchial arches forms a relatively simple rod of cartilage, ceratobranchial cartilage, during this period.

In this period, the pectoral fin rudiments go from being elongated buds that stick out at right angles from the yolk ball to posteriorly projecting fins. The fin blades develop from the spreading of the distal epithelial fold capping the bud, and collagenous fin rays begin to appear. At about the same time, the fins start developing a circulatory channel, which will later subdivide into the subclavian vein and the subclavian artery.

Pigment cells differentiate and organize themselves throughout the body of the embryo during this period. Melanophores cover much of the embryo early in the hatching period, and by the end of the period, four main pigmentation stripes are formed: the ventral-most yolk sac stripe, the ventral stripe, the dorsal stripe, and the lateral stripe. Iridophores develop initially in the eye, the dorsal tail stripe, and the anterior ventral stripe, with many present in all but the lateral stripe by the end of the third day. Xanthophores develop throughout the body and proliferate, giving the embryo a strong yellowish color by the end of the day.

Early-Mid Larval Period (3–13 dpf)

Since morphogenesis is mostly complete by 3 dpf, the principal developmental event during early-mid larval development is an increase in size. Some morphological changes that do occur during this period are the development and inflation of the swim bladder, additional development of the mouth, increased pigmentation, slight relocation of the gut, and gradual disappearance of the yolk extension. In terms of functional maturation, the larvae begin to swim, to move their jaw and eyes, and eventually to seek and ingest food.

USES OF ZEBRAFISH IN DEVELOPMENTAL TOXICOLOGY

As described earlier, there are many advantages to using the zebrafish as an experimental model in developmental toxicology. Their small size, fecundity, relatively minimal environmental requirements, external development, and low cost make them ideally suited for developmental toxicity screens and studies of molecular mechanisms. In addition, the remarkable optical clarity of zebrafish embryos allows the investigator to examine the entire anatomy readily up to at least the third day of development when pigmentation accelerates. Thus, immunohistochemistry or in situ hybridization in whole mount, rather than sections, is possible with these small, clear embryos. If desired, the period of transparency can be extended further by chemical prevention of pigmentation with phenylthiourea treatment or by use of albino mutant lines. External fertilization prevents the need for euthanasia of adult animals required in mammalian studies, making zebrafish experiments consistent with the three Rs of animal testing, and external development allows repeated observation of individual live embryos or larvae, facilitating time-course experiments and strengthening statistical power. The combined qualities of optical clarity and external development make zebrafish embryos quite easy to manipulate experimentally.

Chemical exposure can be achieved by waterborne delivery, injection, or administration in food. By far, the most common and simplest approach is exposure to a compound diluted in the water or medium in which the developing zebrafish is grown. Zebrafish embryos and larvae can absorb small molecules from the water or embryo medium through their skin and gills, and when they start swallowing 3 or 4 dpf, exposure may also be achieved by the oral route (11–13). For most compounds, waterborne exposure is adequate since even compounds with low water solubility can be tested at sufficiently high concentrations when an organic solvent vehicle such as 0.1% dimethyl sulfoxide (DMSO) is used. An example of this is TCDD, which has been successfully tested in zebrafish embryos with administered concentrations up to 1 ppm despite quite limited aqueous solubility (14–16). Excessively hydrophobic molecules can be injected into the embryonic yolk.

Environmental Toxicity Testing

For many years, developing zebrafish have been used as a model for environmental toxicity testing and research (17–27). Zebrafish are also one of the fish species recommended for the fish early-life-stage toxicity test, an assay designed to determine the lethal and sublethal effects of chemicals for evaluation of the potential environmental effects on other fish species (25,28,29). While there are limitations to the use of assays like these for definitive toxicity testing, they can be quite helpful in setting priorities for which compounds should be assessed in definitive tests.

Such a toxicity test is the zebrafish *Danio rerio* embryo toxicity test (*Dar*T). *Dar*T was developed and proposed for use as an alternative to the acute fish toxicity test for environmental toxicology testing of chemicals and has been accepted in Germany as a method for the testing of wastewater effluent (30). In the *Dar*T, fertilized zebrafish eggs are collected within an hour of fertilization and placed individually into wells of a multi-well plate containing test solution in water. For testing toxicity or teratogenicity of chemicals, multiple concentrations of each chemical are tested along with controls. After 48 hours of treatment, lethal [coagulation of embryos (coagulated eggs are opaque white and may appear dark under a microscope), tail not detached from yolk ball, no somites, and no heartbeat], sublethal (completion of gastrula, formation of somites, development of eyes, spontaneous movement, heartbeat/circulation, pigmentation, and edema), and teratogenic (malformation of the head; malformation of the sacculi/otoliths; malformation of the tail; malformation of the heart; modified structure of the notochord, scoliosis, and rachischisis; deformity of the yolk; and growth retardation) endpoints are assessed.

When LC_{50} values obtained from the *Dar*T assay were compared to LC_{50} data from acute toxicity tests with zebrafish or golden ide (*Leuciscus idus melanotus*) for 58 compounds, a strong correlation was found between the results of the two tests (30). Given that the acute toxicity test is conducted with juvenile or adult fish and has a longer duration than the *Dar*T assay (95 and 48 hours, respectively), there is a good deal of interest in replacing the conventional acute fish toxicity test with the *Dar*T. A similar study was done to test this assay for routine wastewater control with only the lethal endpoints of the assay (described earlier). The test was found to be successful in this application as well (30). These results illustrate the broad utility of the zebrafish embryo for environmental toxicity testing.

DEVELOPMENTAL TOXICITY SCREENS

The zebrafish embryo model seems also to be powerful as a teratogenicity assay to screen new drugs or other products. Although it appears unlikely that zebrafish embryo test methods will be used in pivotal toxicology studies or to replace conventional models in safety testing altogether, the potential for use of the model as a screen for the safety of drug candidates is being discussed by scientists in industry, academia, and government (1,12,13,31,32). A greater number of compounds could be tested in assays using zebrafish earlier in the drug discovery and development process than is possible with conventional testing methods, which are resource intensive and expensive by contrast.

However, no comprehensive analysis has been reported to date to determine what experimental designs or conditions are most appropriate and predictive of mammalian teratogenicity. The developmental stages at which exposure is initiated and effects assessed are undoubtedly critical to the success of the assay. While many researchers have explored the effects of exposures over different periods of development, either the data sets are not large enough to draw conclusions, or the study designs are not similar enough to make comparisons across studies. In addition, there is evidence that the chorion may reduce exposure of zebrafish embryos to some compounds (33), and the environmental conditions in which embryos are grown may affect chorion permeability (34). Therefore, the decision must be made whether to expose intact embryos (chorion in place), dechorionated embryos, or embryos whose chorions have been enzymatically weakened but left in place. However, there is insufficient information to date on which to base such a decision. Other considerations include what zebrafish strain to use, how to interpret the data, whether to use a statistical prediction model, and what endpoints to include (including how finely they should be assessed).

The *Dar*T assay has also been studied for its utility as a screening method for the teratogenic potential of compounds in mammals in addition to its original environmental testing purpose. In this application, lethality and teratogenicity are assessed, and a Teratogenic Index is calculated as the quotient of a test compound's LC_{50} concentration and EC_{50} concentration for induction of teratogenic effects. In a study of 41 chemicals of described teratogenic potential in mammals, including teratogens (some that require metabolic activation), nonteratogens, and compounds with ambiguous results in mammalian tests, there was 88% concordance with outcomes in mammals, with four false positives and one false negative (30). Furthermore, even compounds that are known to require metabolic activation for mammalian teratogenicity were correctly categorized in this study.

A number of contract research labs have begun offering zebrafish developmental toxicity screening as a fee-based service for sponsors, and groups at some pharmaceutical companies, including ours, have begun developing and using their own versions of zebrafish teratogenicity screens for drug discovery lead optimization. Zebrafish embryos can be used in assays with throughput close or equal to that of in vitro assays but have the added benefits of a whole organism model with intact pharmacokinetics and pharmacodynamics (8), making them an attractive model to use in place of or in conjunction with more conventional methods.

Zebrafish Genetics, Functional Genomics, and Mechanistic Studies

This section briefly reviews ways in which genetics and functional genomics have contributed to our understanding of zebrafish development and developmental toxicology. In addition, means of using these approaches to enhance developmental toxicology studies or screens are described. The ability of the zebrafish embryo-larva to be readily assessed for phenotype, in combination with a capability to apply a multitude of gene expression and functional characterization tools, has made this a powerful model for evaluating mechanisms of development, toxicity, and teratogenicity.

Zebrafish orthologs of many mammalian genes have been identified and can be readily accessed through databases of the Zebrafish Information Network (ZFIN) (35,36) and the resources of its sister organization, the Zebrafish International Resource Center (ZIRC) (37). Using such resources enables convenient web-based linkage to the appropriate GenBank accession numbers and cDNA sequences. These websites also provide information critical for generation of reagents such as polymerase chain reaction (PCR) primers or in situ hybridization probes for evaluating gene expression, useful in detecting transcriptional changes following exposure to a test compound. Commercially available zebrafish cDNA microarrays exist, enabling simultaneous evaluation of transcriptional patterns in large numbers of genes. The ease of growing embryos and larvae and manipulating them at various time points provide significant advantages for the use of zebrafish in molecular studies.

Similarly, chemical, radiation, and insertional mutagenesis have produced thousands of mutant zebrafish with intriguing phenotypes. Descriptions of these phenotypes can also be accessed through the ZFIN and ZIRC websites, and many of these mutant strains are available for purchase through ZIRC. Hundreds of these mutants have been characterized to the extent that the affected genes have been identified.

Validation of a suspect target effect is an essential component of a mechanistic teratology study. In the context of evaluating teratogenic mechanisms of pharmaceutical compounds, the underlying question that requires experimental testing is whether the compound's effects are related to its intended pharmacology or to an unintended "off-target" effect. Such questions can be effectively addressed with the zebrafish model. In the case where a zebrafish mutant phenotype with similarities to the effects of a compound exists, knowledge or identification of the mutant genotype may lead to generation of a hypothesis regarding the mechanism of teratogenicity induced by the compound (38–40). Also, gene rescue studies could be conducted in mutant embryos by injecting mRNA or administering small molecules that could either recover suspect target expression or pharmacologically agonize a suspect inactivated gene target or pathway. Loss- or gain-of-function approaches can also be conducted in normal zebrafish embryos in cases where a putative target of teratogenicity has been identified and requires functional targeting for validating the hypothesized mechanism. By producing mutant zebrafish that either overexpress or underexpress a gene, one can observe the effects of perturbing the gene product without having to rely on compound availability. Gene functions have been studied with antisense methods, the generation of transgenic lines, mutation, or overexpression.

Mutagenesis

The zebrafish model allows for large-scale screens for mutations affecting development. These mutagenesis screens are examples of forward genetics approaches, and there are three methods commonly used in the zebrafish: N-ethyl-N-nitrosourea (ENU) treatment, radiation, and insertion of sequences with retroviral vectors.

ENU, which is mutagenic in a number of organisms including zebrafish (38,41,42), results in a modification of a single DNA strand. It can be used in premeiotic male germ cells, in which mutations become fixed in both strands through DNA replication, or in postmeiotic male germ cells where the changes do not become fixed until after fertilization. These mutated genes can then be identified by using positional cloning (43,44). In an interesting recent technical advancement, the technique of targeting induced local lesions in genomes (TILLING) was developed to identify fish-containing mutations in a gene of interest within an ENU-mutagenized population (45–48), offering the first opportunity for gene target-selected mutagenesis studies. In TILLING, a typical ENU approach to mutagenizing males is performed, F1 progeny are produced through breeding of the ENU-treated males with wild-type females, and samples are collected from the progeny and screened for mutations in the target gene.

Radiation is also useful in forward genetics studies of zebrafish. Double-stranded DNA breaks can be induced with ionizing radiation, allowing for chromosomal rearrangements (49). Both mosaic and nonmosaic progeny can be produced with gamma radiation (50), and similar results have been reported with ultraviolet light in presence of psoralen, a compound in the family of furocoumarins which sensitize DNA to UV light-induced strand breaks (51,52).

As with ENU and radiation, insertional mutagenesis has typically been used for forward genetics, but it can also be useful in reverse genetics approaches. Retroviruses have been used as insertional mutagens in large-scale studies of zebrafish (53–55). One such viral mutagen is the murine leukemia virus (56–58). A proviral molecular "tag" at the site of insertion allows for rapid identification of the mutated gene (57,59,60). These mutated genes can then be cloned rapidly in most insertional mutants (49).

An essential element of mutagenesis is the ability to map the genes affected. This is commonly performed on genetic maps. Mutations are placed on maps that are produced by scoring a large number of polymorphic markers on a panel of F2 fish from a reference cross. If two markers are close to one another, there will be a similar pattern of alleles across the panel. Mutations are indirectly placed on the map by scoring previously mapped markers on a panel of mutant embryos and estimating map position of the mutation from observed linkages (49). Another method for rapid mapping of mutations uses small nucleotide polymorphisms

(SNPs) and oligonucleotide microarrays. In 2002, a first generation SNP map of the zebrafish genome was developed to accelerate the identification of mutations (61). This group also developed a method to score hundreds of SNPs in parallel by hybridizing to an oligonucleotide map.

Reverse Genetics Approaches

The most widely used reverse genetics approach in zebrafish research utilizes morpholino antisense oligonucleotides (MOs), which have high specificity for targeting specific transcripts and function by binding to the target RNA and causing steric hindrance of translation. The backbone of a morpholino consists of a six-membered morpholine ring and a nonionic phosphorodiamidate intersubunit linkage [Fig. 2(A)] (62). Unlike other antisense technologies, morpholinos are RNA-induced silencing complex and RNase-H independent, do not degrade RNA, and have been found to be very stable in biological systems (63). MOs are highly aqueous (64), making injections into the yolk of the young zebrafish egg relatively straightforward [Fig. 2(B)]. Morpholinos can also be purchased commercially from a number of suppliers.

Morpholinos have been successfully used to knock down expression of gene targets and evaluate gene function in vivo (65). MOs can also be used in chemical screens by knocking down a target followed by compound treatment. This can offer insight into pathways involved and mechanisms of action of drugs or other toxic chemicals (66). Recently, a large-scale reverse genetic strategy was successfully executed with morpholinos, and this approach yielded a relatively high rate of developmental phenotypes compared to typical results with random gene mutagenesis studies (Fig. 3) (67). However, not all morpholinos necessarily work to knock down the target of interest efficiently, and it is necessary to demonstrate reductions in target protein levels in order to verify gene expression

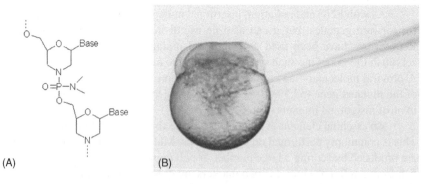

(A)   (B)

Figure 2   Use of morpholinos antisense oligonucleotides in zebrafish functional genomics. (**A**) Morpholino oligonucleotide backbone structure. (**B**) Microinjection into the zebrafish embryo yolk ball is a typical method of MO delivery to early embryos.

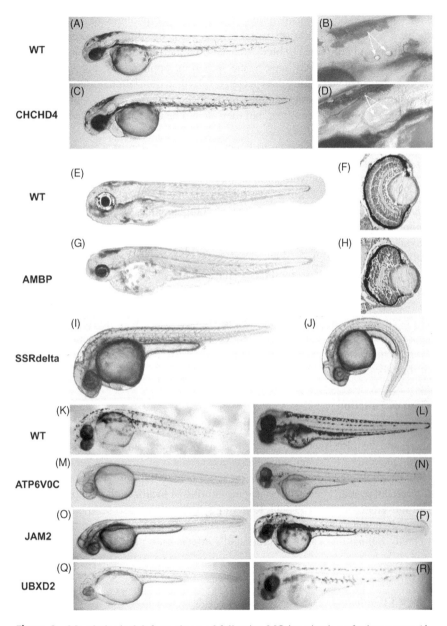

Figure 3 Morphological defects observed following MO inactivation of select genes. (**A**, **B**) Normal otolith morphology observed in 2 dpf untreated embryos. (**C**, **D**) Absence of otoliths, in otherwise normal 2 dpf embryo, following injection of MO targeting CHCHD4. (**B**, **D**) Enlarged view of otic capsules; arrows denote normally formed (**B**) or absent (**D**) otoliths, respectively. (**E**, **F**) Eye morphology in 3 dpf embryos. (**G**, **H**) Abnormally small eyes observed in 3 dpf embryo following injection of MO targeting AMBP. (*Continued*)

knockdown. Another consideration is that some morpholinos have been found to have sequence-specific off-target effects (68). In addition, morpholinos are most commonly designed to block translation initiation sites. Use of morpholinos to block splice sites may have some important applications, but they have a greater chance of generating unexpected results than translation initiation blocking morpholinos and are more complex to use. The potential benefits of splice site targeting morpholinos are great, but they should be carefully designed and tested to insure the intended effect.

Using RNA interference to knock down expression of a target gene with short interfering RNAs (siRNA) is an increasingly popular method for studying gene functions and involvement in biological phenomena in mammalian systems. siRNAs bind to other cellular proteins to form the RNA-induced silencing complex, which in turn blocks translation of complementary RNA sequences. This technology, however, has been less successful in zebrafish research, where it has shown off-target effects including induction of apoptosis that may be p53 mediated (68).

A liability of these methods is the potential for off-target morphological effects. Interactions may occur between the oligonucleotide and the extracellular, cell-surface, and intracellular proteins. Similar to the findings with siRNA, it has been found that some of the off-target effects seen with morpholinos in the zebrafish are p53 dependent and may be ameliorated with coinjection of a p53 morpholino (68). p53 morpholinos alone appear to have no effect on development of the zebrafish (68).

Gain of Function

In gain-of-function studies, zebrafish can be injected with mRNA at the one-cell stage, and the mRNA will then be expressed in every cell during development. This technique allows for organism-wide overexpression of any mRNA target in vivo (69). An additional injection with a transcriptional or translational fusion of green fluorescent protein (GFP) allows for monitoring of localization of expression and cell movement in vivo (69–71). Reporter zebrafish have also been engineered to study signaling and localization of targets of interest. These fish can then be used to study human disease or a compound's effects on a particular target (70–72).

←——

Figure 3 (*Continued*)   (**F, H**) Enlarged view of histological sections of eye in unaffected (**F**) and affected (**H**) embryos. Note differences in both the size and tissue organization of the affected eye. (**I**) Wild-type morphology of 1 dpf embryo. (**J**) Ventral curvature phenotype observed in 1 dpf embryos injected with MO targeting SSRdelta. (**K, L**) Normal pigmentation observed in untreated 1 and 2 dpf embryos. Reduction in pigment observed in 1 and 2 dpf embryos, respectively, following injection of MO targeting ATP6V0 C (**M, N**), or junction adhesion molecule 2 (JAM2) (**O, P**), or UBX domain containing 2 (UBXD2) (**Q, R**). *Source*: From Ref. 67.

Transgenic fish can be produced with transposons, or transposon like methods. While no transposable element has yet been found in the zebrafish, a method of using a synthetic transposon system has been established (73). One such construct is *Sleeping Beauty*, which can create chromosomal insertions in zebrafish germ cells (74). A vertebrate transposon called *Tol2* was identified in the medaka (a related teleost fish) and can create chromosomal integrations in zebrafish germ cells (75,76). These methods have increased proficiency of germline transmission, facilitating transgenic zebrafish generation.

Microinjection

Many of the methods of genetic manipulation described earlier require microinjection of nucleic acids. Various apparatuses are available for microinjection into the zebrafish embryo. These usually consist of a micromanipulator with a glass needle (a pulled capillary tube) and a mechanism to change the pressure within the needle to deliver the injection solution. mRNA injections can be carried out in the yolk of young embryos since cytoplasmic streaming allows large molecules to be transported into the blastoderm from the yolk (77).

DEVELOPMENTAL TOXICANTS STUDIED IN ZEBRAFISH

A number of mammalian teratogens have been studied in the developing zebrafish, including 2,3,7,8-tetrachlorodibenzo-*p*-dioxin (TCDD), ethanol, retinoic acid, endocrine disrupters, cyclopamine, metals, and pesticides. The literature for TCDD and ethanol are reviewed briefly here.

TCDD

TCDD is probably the developmental toxicant that has been most extensively studied in zebrafish. TCDD belongs to the dibenzo-*p*-dioxin class of halogenated aromatic hydrocarbons (HAHs), which are ubiquitous and persistent environmental pollutants. Fish, birds, and mice have been found to be sensitive to TCDD developmental toxicity, leading to concerns over the impact on wildlife populations and potential adverse effects on human development. Furthermore, there are many similarities in the types of developmental anomalies induced by TCDD and in the results of mechanistic studies in mammals, birds, and fish. Those similarities, combined with the conserved nature of vertebrate development and tissue bioaccumulation of HAHs, suggest that findings with TCDD toxicity in developing zebrafish may have relevance to human health, as well [reviewed by Carney et al. (66)].

Zebrafish exposed to TCDD during embryogenesis developed normally until about 48 to 120 hpf, at which stage the characteristic signs of TCDD developmental toxicity were observed. These include abnormal cardiovascular development and function, craniofacial cartilage anomalies, edema, brain anomalies, growth retardation, blocked swim bladder inflation, impaired erythropoiesis, and reduced posthatching survival [reviewed by Carney et al. (66) and by Goldstone and Stegeman (78)]. The earliest toxic response reported following TCDD exposure in zebrafish was a decrease in cardiomyocyte numbers (79), followed by reduced blood flow (16,80). By 72 hpf, a dramatic defect of heart morphology was evident in TCDD-treated zebrafish. Abnormalities in cardiac looping and common cardinal vein remodeling between 48 and 96 hpf resulted in a significantly elongated heart, and altered cardiac function, including an increased rate of ventricular standstill (blocked ventricular contraction), was observed by 96 hpf (79). In addition, TCDD impairs the switch to definitive erythropoiesis that normally occurs around 48 to 96 hpf (81). In zebrafish treated with TCDD, craniofacial cartilage defects have been consistently observed in the lower jaw, pharyngeal arches, and cranium, with all cartilages present but reduced in size and/or having an abnormal shape (16,82,83). It has also been demonstrated that TCDD induced an increase in apoptosis in the dorsal midbrain of zebrafish embryos at 50 hpf (80,84) and an approximately 30% reduction in brain volume and neuron number by 168 hpf (15).

Like other HAHs, TCDD is an aryl hydrocarbon receptor (AHR) ligand, and the toxicity of HAHs appears to be correlated with AHR binding and activation (85,86). Therefore, work has been done to identify and characterize the proteins involved in the zebrafish AHR pathway, including AHRs themselves and their heterodimerization partners, the aryl hydrocarbon receptor nuclear translocators (ARNTs). Zebrafish have three *AHR* genes: *zfAHR1*, *zfAHR2*, and the newly identified *zfAHR1B* (87–89). Although the zebrafish ortholog to the mammalian *AHR* gene is *zfAHR1* (87), *zfAHR2* appears to be the isoform that mediates TCDD toxicity, as indicated by results of functional genomics studies described later. The role of *zfAHR1B* is less clear since it has not been as fully characterized, but it is also bound with high affinity and activated by TCDD (88). In addition, two genes encoding ARNT proteins (*zfARNT1* and *zfARNT2*) have been identified in the zebrafish genome, with *zfARNT2* having multiple splice variants (90–93).

In order to elucidate the mechanisms of TCDD embryo toxicity in zebrafish, researchers have employed several experimental approaches, including careful morphological observation, MOs, mutant lines, and genomics. To date, the most significant advancements in understanding the TCDD mechanism of zebrafish developmental toxicology have come from functional genomics experiments with morpholinos and mutant zebrafish lines. Zebrafish treated with morpholinos targeted against zfAHR1 (zfAHR1-MO) showed reduced sensitivity to TCDD, including improved cardiovascular development, edema, blood flow, erythrocyte maturation, midbrain apoptosis, and reduced induction of zfCYP1A (14,94–98), as well as partial protection against inhibition of craniofacial cartilage growth (97). In contrast, zfAHR2 morphants (injected with zfAHR2-MO) showed no improvement in endpoints of TCDD toxicity (66). Taken together, these results support the hypothesis that zfAHR2 is required for most TCDD developmental toxicity in zebrafish. Similarly, a zfARNT1 morpholino provided partial to complete protection against TCDD toxicity (14,91), while zfARNT2 morphants and mutant zebrafish did not (Fig. 4). (14,99) This suggests that TCDD developmental toxicity in zebrafish is also mediated by zfARNT1 and that zfARNT1 heterodimerizes with zfAHR2 in response to TCDD.

Ethanol

Ethanol has long been a known human teratogen, and it is among the most studied mammalian teratogens. Zebrafish exposed to ethanol during embryonic development had craniofacial, cardiovascular, and axial abnormalities and developmental delays (100–106). The craniofacial effects are remarkably similar to those described in patients with fetal alcohol syndrome or in animal models of fetal alcohol syndrome (107), making zebrafish a valuable model for studying the pathogenesis and mechanism of ethanol-induced teratogenicity.

Zebrafish embryos readily take ethanol up from the surrounding water or buffer-based medium whether they are maintained within the intact chorion (103) or chorions are removed (108). Such exposures result in adverse morphological and functional effects on development. Treatment of zebrafish embryos with ethanol induced an increase in apoptosis and alterations in the skeletal elements of the head (101). Ethanol also induced cyclopia in zebrafish embryos by preventing the migration of the prechordal plate mesoderm, with the late blastula to early gastrula stages being the most sensitive window of development for this effect (100). Larval visual function was impaired by a teratogenic exposure to ethanol on the first (109) or second (110) day of development. Exposure from 2 to 5 dpf caused hypoplasia of the optic nerve and inhibited photoreceptor development and function (111).

The effects of ethanol exposure on gene expression in the zebrafish embryo have also been explored. Blader et al. found *pax-2* expression in the eyes to be decreased in association with ethanol-induced cyclopia, and changes in expression patterns of markers of ventral neural tube and prechordal plate cells were also observed (100). In another study, embryos treated with ethanol at the dome stage (approximately 4 hpf) had an early decrease in expression of *gli1* indicating inhibition of shh signaling, and expression of the telencephalon marker *six3b* was also markedly reduced within hours of ethanol treatment (112).

Zebrafish embryos were more sensitive to adverse effects of acetaldehyde than ethanol. However, the developmental anomalies induced by acetaldehyde occurred at concentrations quite close to those that caused substantial lethality, whereas there was a several-fold difference in teratogenic and lethal concentrations of ethanol (103), and inhibition of the enzymes that convert ethanol to acetaldehyde enhanced the developmental toxicity of ethanol (105). Furthermore, ethanol caused an increase in cell death in the embryo, and while antioxidant cotreatment did not ameliorate the ethanol-induced increase in cell death, glutathione and lipoic acid reduced the incidence of ethanol-induced malformations (105). The mechanisms of ethanol teratogenicity in mammals and zebrafish are less clearly defined than

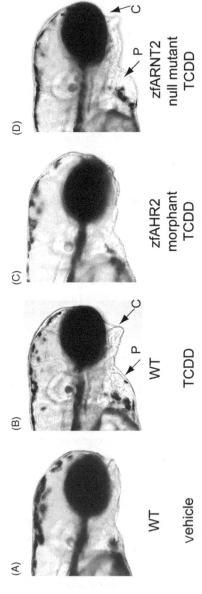

Figure 4 Morpholino targeted against zfAHR2 rescued embryos from TCDD toxicity, whereas an ARNT2 mutant zebrafish shows the typical endpoints of TCDD toxicity. These results helped determine that AHR2 (not AHR1) and ARNT1 (not ARNT2) were the key dimerization partners required for AHR-mediated TCDD developmental toxicity in zebrafish. Wild-type embryos exposed to a vehicle control (**A**) and embryos injected with zfAHR2-MO exposed to 0.4 ng/ml TCDD (**C**) show no endpoints of TCDD toxicity, whereas wild-type (**B**) and ARNT2 mutant embryos (**D**) exposed to 0.4 ng/ml TCDD show typical endpoints of toxicity such as P: pericardial edema and C: craniofacial malformations. Embryos were exposed at 2 hpf for 1 hour to TCDD and then allowed to develop in TCDD-free water prior to observation at 96 hpf. *Source*: From Ref. 1

that of TCDD, but zebrafish appear to be a useful animal model for the continuing exploration of these.

CONCLUDING REMARKS

The zebrafish has become a commonly used model for the study of development (113,114), human disease (115–118), and drug discovery (13,119). The species has also become a very popular emerging model in the field of developmental toxicology. As a model for predictive toxicology or teratology, the zebrafish model allows for the assessment of a large number of compounds in a relatively short time frame and requires less compound for studies than many other animal models. Prescreening can eliminate compounds with obvious adverse effects, saving both time and money. In addition, the model is highly amenable to genetic manipulation, making it an excellent model for genetics research. The combined attributes of the zebrafish make it very attractive for toxicology screening and mechanistic research. For these reasons, the zebrafish is rapidly gaining acceptance as a nonmammalian toxicity model (1,8,120,121).

REFERENCES

1. Hill AJ, Teraoka H, Heideman W, et al. Zebrafish as a model vertebrate for investigating chemical toxicity. Toxicol Sci 2005; 86(1):6–19.
2. Streisinger G, Walker C, Dower N, et al. Production of clones of homozygous diploid zebra fish (Brachydanio rerio). Nature 1981; 291(5813):293–296.
3. Dahm R. Atlas of embryonic stages of development in the zebrafish. In Zebrafish: A Practical Approach, The Practical Approach Series, Volume 261, Nusslein-Volhard C, Dahm R, eds. Oxford University Press, New York, NY, 2002:59–94.
4. Kimmel CB, Ballard WW, Kimmel SR, et al. Stages of embryonic development of the zebrafish. Dev Dyn 1995; 203(3):253–310.
5. Schilling TF. The morphology of larval and adult zebrafish. In Zebrafish: A Practical Approach, The Practical Approach Series, Volume 261, Nusslein-Volhard C, Dahm R, eds. Oxford University Press, New York, NY, 2002:59–94.
6. National Institutes of Health. Trans-NIH zebrafish initiative. [cited 2007, November 26]; Available from: http://www.nih.gov/science/models/zebrafish/.
7. Sanger Institute. The Danio rerio sequencing project. 2007 [cited 2007, November 26]; Available from: http://www.sanger.ac.uk/Projects/D_rerio.
8. Zon LI, Peterson RT. In vivo drug discovery in the zebrafish. Nat Rev Drug Discov 2005; 4(1):35–44.
9. Driever W, Solnica-Krezel L, Schier AF, et al. A genetic screen for mutations affecting embryogenesis in zebrafish. Development 1996; 123:37–46.
10. Haffter P, Granato M, Brand M, et al. The identification of genes with unique and essential functions in the development of the zebrafish, Danio rerio. Development 1996; 123:1–36.
11. Goldsmith P. Zebrafish as a pharmacological tool: The how, why and when. Curr Opin Pharmacol 2004; 4(5):504–512.

12. Parng C. In vivo zebrafish assays for toxicity testing. Curr Opin Drug Discov Devel 2005; 8(1):100–106.

13. Rubinstein AL. Zebrafish assays for drug toxicity screening. Expert Opin Drug Metab Toxicol 2006; 2(2):231–240.

14. Antkiewicz DS, Peterson RE, Heideman W. Blocking expression of AHR2 and ARNT1 in zebrafish larvae protects against cardiac toxicity of 2,3,7,8-tetrachlorodibenzo-p-dioxin. Toxicol Sci 2006; 94(1):175–182.

15. Hill A, Howard CV, Strahle U, et al. Neurodevelopmental defects in zebrafish (Danio rerio) at environmentally relevant dioxin (TCDD) concentrations. Toxicol Sci 2003; 76(2):392–399.

16. Teraoka H, Dong W, Ogawa S, et al. 2,3,7,8-Tetrachlorodibenzo-p-dioxin toxicity in the zebrafish embryo: altered regional blood flow and impaired lower jaw development. Toxicol Sci 2002; 65(2):192–199.

17. Skidmore JF. Resistance to zinc sulphate of the zebrafish (Brachydanio rerio Hamilton-Buchanan) at different phases of its life history. Ann Appl Biol 1965; 56(1):47–53.

18. Roales RR, Perlmutter A. Toxicity of zinc and cygon, applied singly and jointly, to zebrafish embryos. Bull Environ Contam Toxicol 1974; 12(4):475–480.

19. Niimi AJ, LaHam QN. Relative toxicity of organic and inorganic compounds of selenium to newly hatched zebrafish (Brachydanio rerio). Can J Zool 1976; 54(4):501–509.

20. Yosha SF, Cohen GM. Effect of intermittent chlorination of developing zebrafish embryos (Brachydanio rerio). Bull Environ Contam Toxicol 1979; 21(4–5):703–710.

21. Baumann M, Sander K. Bipartite axiation follows incomplete epiboly in zebrafish embryos treated with chemical teratogens. J Exp Zool 1984; 230(3):363–376.

22. Dave G. Effect of pH on pentachlorophenol toxicity to embryos and larvae of zebrafish (Brachydanio rerio). Bull Environ Contam Toxicol 1984; 33(5):621–630.

23. Dave G. The influence of pH on the toxicity of aluminum, cadmium, and iron to eggs and larvae of the zebrafish, Brachydanio rerio. Ecotoxicol Environ Saf 1985; 10(2):253–267.

24. Gorge G, Nagel R. Toxicity of lindane, atrazine, and deltamethrin to early life stages of zebrafish (Brachydanio rerio). Ecotoxicol Environ Saf 1990; 20(3):246–255.

25. Van Leeuwen CJ, Grootelaar EM, Niebeek G. Fish embryos as teratogenicity screens: A comparison of embryotoxicity between fish and birds. Ecotoxicol Environ Saf 1990; 20(1):42–52.

26. Dave G, Xiu RQ. Toxicity of mercury, copper, nickel, lead, and cobalt to embryos and larvae of zebrafish, Brachydanio rerio. Arch Environ Contam Toxicol 1991; 21(1):126–134.

27. Ensenbach U, Nagel R. Toxicokinetics of xenobiotics in zebrafish–comparison between tap and river water. Comp Biochem Physiol C 1991; 100(1–2):49–53.

28. USEPA, Ecological effects test guidelines: OPPTS 850.1400 fish early-life stage toxicity test. Office of Prevention Pesticides and Toxic Substances, U.S. Government Printing Office, Washington, DC, 1996.

29. Bresch H. Early life-stage test in zebrafish versus a growth test in rainbow trout to evaluate toxic effects. Bull Environ Contam Toxicol 1991; 46(5):641–648.

30. Nagel R. DarT: The embryo test with the Zebrafish Danio rerio—a general model in ecotoxicology and toxicology. Altex 2002; 19(Suppl 1):38–48.

31. Parng C, Seng WL, Semino C, et al. Zebrafish: A preclinical model for drug screening. Assay Drug Dev Technol 2002; 1(1 Pt 1):41–48.
32. Jacobs A. Use of nontraditional animals for evaluation of pharmaceutical products. Expert Opin Drug Metab Toxicol 2006; 2(3):345–349.
33. Lillicrap A. Understanding the bioavailibilty of certain classes of chemicals to zebrafish embryos. SETAC Europe Annual Meeting 2007.
34. Adams SL, Zhang T, Rawson DM. The effect of external medium composition on membrane water permeability of zebrafish (Danio rerio) embryos. Theriogenology 2005; 64(7):1591–1602.
35. Sprague J, Bayraktaroglu L, Clements D, et al. The Zebrafish Information Network: the zebrafish model organism database. Nucleic Acids Res 2006; 34(Database issue):D581–D585.
36. Sprague J, Bayraktaroglu L, Bradford Y, et al. The Zebrafish Information Network: the zebrafish model organism database provides expanded support for genotypes and phenotypes. Nucleic Acids Res 2008; 36(Database issue):D768-72.
37. Zebrafish International Resource Center [cited 2007, November 26]; Available from: http://zebrafish.org/zirc/home/guide.php.
38. Solnica-Krezel L, Schier AF, Driever W. Efficient recovery of ENU-induced mutations from the zebrafish germline. Genetics 1994; 136(4):1401–1420.
39. Kwok C, Korn RM, Davis ME, et al. Characterization of whole genome radiation hybrid mapping resources for non-mammalian vertebrates. Nucleic Acids Res 1998; 26(15):3562–3566.
40. Hukriede NA, Joly L, Tsang M, et al. Radiation hybrid mapping of the zebrafish genome. Proc Natl Acad Sci U S A 1999; 96(17):9745–9750.
41. Grunwald DJ, Streisinger G. Induction of recessive lethal and specific locus mutations in the zebrafish with ethyl nitrosourea. Genet Res 1992; 59(2):103–116.
42. Mullins MC, Hammerschmidt M, Haffter P, et al. Large-scale mutagenesis in the zebrafish: in search of genes controlling development in a vertebrate. Curr Biol 1994; 4(3):189–202.
43. Bahary N, Davidson A, Ransom D, et al. The Zon laboratory guide to positional cloning in zebrafish. Methods Cell Biol 2004; 77:305–329.
44. Talbot WS, Schier AF. Positional cloning of mutated zebrafish genes. Methods Cell Biol 1999; 60: 259–286.
45. Henikoff S, Till BJ, Comai L. TILLING. Traditional mutagenesis meets functional genomics. Plant Physiol 2004; 135(2):630–636.
46. Stemple DL. TILLING—a high-throughput harvest for functional genomics. Nat Rev Genet 2004; 5(2):145–150.
47. Wienholds E, Plasterk RH. Target-selected gene inactivation in zebrafish. Methods Cell Biol 2004; 77:69–90.
48. Wienholds E, van Eeden F, Kosters M, et al. Efficient target-selected mutagenesis in zebrafish. Genome Res 2003; 13(12):2700–2707.
49. Pelegri F. Mutagenesis. In Zebrafish: A Practical Approach, The Practical Approach Series, Volume 261, Nusslein-Volhard C, Dahm R, eds. Oxford University Press, New York, NY, 2005:145–174.
50. Walker C. Haploid screens and gamma-ray mutagenesis. Methods Cell Biol 1999; 60:43–70.
51. Ando H, Mishina M. Efficient mutagenesis of zebrafish by a DNA cross-linking agent. Neurosci Lett 1998; 244(2):81–84.

52. Lekven AC, Helde KA, Thorpe CJ, et al. Reverse genetics in zebrafish. Physiol Genomics 2000; 2(2):37–48.
53. Amsterdam A, Burgess S, Golling G, et al. A large-scale insertional mutagenesis screen in zebrafish. Genes Dev 1999; 13(20):2713–2724.
54. Amsterdam A, Hopkins N. Retrovirus-mediated insertional mutagenesis in zebrafish. Methods Cell Biol 1999; 60:87–98.
55. Wang D, Jao LE, Zheng N, et al. Efficient genome-wide mutagenesis of zebrafish genes by retroviral insertions. Proc Natl Acad Sci U S A 2007; 104(30):12428–12433.
56. Gaiano N, Allende M, Amsterdam A, et al. Highly efficient germ-line transmission of proviral insertions in zebrafish. Proc Natl Acad Sci U S A 1996; 93(15):7777–7782.
57. Gaiano N, Amsterdam A, Kawakami K, et al. Insertional mutagenesis and rapid cloning of essential genes in zebrafish. Nature 1996; 383(6603):829–832.
58. Lin S, Gaiano N, Culp P, et al. Integration and germ-line transmission of a pseudo-typed retroviral vector in zebrafish. Science 1994; 265(5172):666–669.
59. Golling G, Amsterdam A, Sun Z, et al. Insertional mutagenesis in zebrafish rapidly identifies genes essential for early vertebrate development. Nat Genet 2002; 31(2):135–140.
60. Wu X, Li Y, Crise B, et al. Transcription start regions in the human genome are favored targets for MLV integration. Science 2003; 300(5626):1749–1751.
61. Stickney HL, Schmutz J, Woods IG, et al. Rapid mapping of zebrafish mutations with SNPs and oligonucleotide microarrays. Genome Res 2002; 12(12):1929–1934.
62. Hudziak RM, Barofsky E, Barofsky DF, et al. Resistance of morpholino phospho-rodiamidate oligomers to enzymatic degradation. Antisense Nucleic Acid Drug Dev 1996; 6(4):267–272.
63. Summerton J. Morpholino antisense oligomers: The case for an RNase H-independent structural type. Biochim Biophys Acta 1999; 1489(1):141–158.
64. Summerton JE. Morpholino, siRNA, and S-DNA compared: Impact of structure and mechanism of action on off-target effects and sequence specificity. Curr Top Med Chem 2007; 7(7):651–660.
65. Chen E, Ekker SC. Zebrafish as a genomics research model. Curr Pharm Biotechnol 2004; 5(5):409–413.
66. Carney SA, Prasch AL, Heideman W, et al. Understanding dioxin developmental toxicity using the zebrafish model. Birth Defects Res A Clin Mol Teratol 2006; 76(1):7–18.
67. Pickart MA, Klee EW, Nielsen AL, et al. Genome-wide reverse genetics framework to identify novel functions of the vertebrate secretome. PLoS ONE 2006; 1:e104.
68. Robu ME, Larson JD, Nasevicius A, et al. p53 activation by knockdown technologies. PLoS Genet 2007; 3(5):e78.
69. Beis D, Stainier DY. In vivo cell biology: Following the zebrafish trend. Trends Cell Biol 2006; 16(2):105–112.
70. Ashworth R, Brennan C. Use of transgenic zebrafish reporter lines to study calcium signalling in development. Brief Funct Genomic Proteomic 2005; 4(2):186–193.
71. Molina GA, Watkins SC, Tsang M. Generation of FGF reporter transgenic zebrafish and their utility in chemical screens. BMC Dev Biol 2007; 7:62.
72. Perz-Edwards A, Hardison NL, Linney E. Retinoic acid-mediated gene expression in transgenic reporter zebrafish. Dev Biol 2001; 229(1):89–101.

73. Kawakami K. Transposon tools and methods in zebrafish. Dev Dyn 2005; 234(2):244–254.

74. Davidson AE, Balciunas D, Mohn D, et al. Efficient gene delivery and gene expression in zebrafish using the Sleeping Beauty transposon. Dev Biol 2003; 263(2):191–202.

75. Balciunas D, Wangensteen KJ, Wilber A, et al. Harnessing a high cargo-capacity transposon for genetic applications in vertebrates. PLoS Genet 2006; 2(11):e169.

76. Kawakami K, Shima A, Kawakami N. Identification of a functional transposase of the Tol2 element, an Ac-like element from the Japanese medaka fish, and its transposition in the zebrafish germ lineage. Proc Natl Acad Sci U S A 2000; 97(21):11403–11408.

77. Gilmour D, Jessen J, Lin S. Manipulating gene expression in the zebrafish. In Zebrafish: A Practical Approach, The Practical Approach Series, Volume 261, Nusslein-Volhard C, Dahm R, eds. Oxford University Press, New York, NY, 2005:121–143.

78. Goldstone HM, Stegeman JJ. Molecular mechanisms of 2,3,7,8-tetrachlorodibenzo-p-dioxin cardiovascular embryotoxicity. Drug Metab Rev 2006; 38(1–2):261–289.

79. Antkiewicz DS, Burns CG, Carney SA, et al. Heart malformation is an early response to TCDD in embryonic zebrafish. Toxicol Sci 2005; 84(2):368–377.

80. Dong W, Teraoka H, Yamazaki K, et al. 2,3,7,8-tetrachlorodibenzo-p-dioxin toxicity in the zebrafish embryo: Local circulation failure in the dorsal midbrain is associated with increased apoptosis. Toxicol Sci 2002; 69(1):191–201.

81. Belair CD, Peterson RE, Heideman W. Disruption of erythropoiesis by dioxin in the zebrafish. Dev Dyn 2001; 222(4):581–594.

82. Henry TR, Spitsbergen JM, Hornung MW, et al. Early life stage toxicity of 2,3,7,8-tetrachlorodibenzo-p-dioxin in zebrafish (Danio rerio). Toxicol Appl Pharmacol 1997; 142(1):56–68.

83. Hill AJ, Bello SM, Prasch AL, et al. Water permeability and TCDD-induced edema in zebrafish early-life stages. Toxicol Sci 2004; 78(1):78–87.

84. Dong W, Teraoka H, Kondo S, et al. 2, 3, 7, 8-tetrachlorodibenzo-p-dioxin induces apoptosis in the dorsal midbrain of zebrafish embryos by activation of arylhydrocarbon receptor. Neurosci Lett 2001; 303(3):169–172.

85. Hankinson O. The aryl hydrocarbon receptor complex. Annu Rev Pharmacol Toxicol 1995; 35:307–340.

86. Heid SE, Walker MK, Swanson HI. Correlation of cardiotoxicity mediated by halogenated aromatic hydrocarbons to aryl hydrocarbon receptor activation. Toxicol Sci 2001; 61(1):187–196.

87. Andreasen EA, Hahn ME, Heideman W, et al. The zebrafish (Danio rerio) aryl hydrocarbon receptor type 1 is a novel vertebrate receptor. Mol Pharmacol 2002; 62(2):234–249.

88. Karchner SI, Franks DG, Hahn ME. AHR1B, a new functional aryl hydrocarbon receptor in zebrafish: Tandem arrangement of ahr1b and ahr2 genes. Biochem J 2005; 392(Pt 1):153–161.

89. Tanguay RL, Abnet CC, Heideman W, et al. Cloning and characterization of the zebrafish (Danio rerio) aryl hydrocarbon receptor. Biochim Biophys Acta 1999; 1444(1):35–48.

90. Hsu HJ, Wang WD, Hu CH. Ectopic expression of negative ARNT2 factor disrupts fish development. Biochem Biophys Res Commun 2001; 282(2):487–492.

91. Prasch AL, Tanguay RL, Mehta V, et al. Identification of zebrafish ARNT1 homologs: 2,3,7,8-tetrachlorodibenzo-p-dioxin toxicity in the developing zebrafish requires ARNT1. Mol Pharmacol 2006; 69(3):776–787.

92. Tanguay RL, Andreasen E, Heideman W, et al. Identification and expression of alternatively spliced aryl hydrocarbon nuclear translocator 2 (ARNT2) cDNAs from zebrafish with distinct functions. *Biochim Biophys Acta* 2000; 1494(1–2):117–128.

93. Wang WD, Wu JC, Hsu HJ, et al. Overexpression of a Zebrafish ARNT2-like Factor Represses CYP1 A Transcription in ZLE Cells. Mar Biotechnol (NY) 2000; 2(4):376–386.

94. Bello SM, Heideman W, Peterson RE. 2,3,7,8-Tetrachlorodibenzo-p-dioxin inhibits regression of the common cardinal vein in developing zebrafish. Toxicol Sci 2004; 78(2):258–266.

95. Carney SA, Peterson RE, Heideman W. 2,3,7,8-Tetrachlorodibenzo-p-dioxin activation of the aryl hydrocarbon receptor/aryl hydrocarbon receptor nuclear translocator pathway causes developmental toxicity through a CYP1 A-independent mechanism in zebrafish. Mol Pharmacol 2004; 66(3):512–521.

96. Dong W, Teraoka H, Tsujimoto Y, et al. Role of aryl hydrocarbon receptor in mesencephalic circulation failure and apoptosis in zebrafish embryos exposed to 2,3,7,8-tetrachlorodibenzo-p-dioxin. Toxicol Sci 2004; 77(1):109–116.

97. Prasch AL, Teraoka H, Carney SA, et al. Aryl hydrocarbon receptor 2 mediates 2,3,7,8-tetrachlorodibenzo-p-dioxin developmental toxicity in zebrafish. Toxicol Sci 2003; 76(1):138–150.

98. Teraoka H, Dong W, Tsujimoto Y, et al. Induction of cytochrome P450 1 A is required for circulation failure and edema by 2,3,7,8-tetrachlorodibenzo-p-dioxin in zebrafish. Biochem Biophys Res Commun 2003; 304(2):223–228.

99. Prasch AL, Heideman W, Peterson RE. ARNT2 is not required for TCDD developmental toxicity in zebrafish. Toxicol Sci 2004; 82(1):250–258.

100. Blader P, Strähle U. Ethanol impairs migration of the prechordal plate in the zebrafish embryo. Dev Biol 1998; 201(2):185–201.

101. Carvan MJ III, Loucks E, Weber DN, et al. Ethanol effects on the developing zebrafish: Neurobehavior and skeletal morphogenesis. Neurotoxicol Teratol 2004; 26(6):757–768.

102. Loucks E, Carvan MJ III. Strain-dependent effects of developmental ethanol exposure in zebrafish. Neurotoxicol Teratol 2004; 26(6):745–755.

103. Reimers MJ, Flockton AR, Tanguay RL. Ethanol- and acetaldehyde-mediated developmental toxicity in zebrafish. Neurotoxicol Teratol 2004; 26(6):769–781.

104. Qian L, Wang Y, Jiang Q, et al. Ethanol disrupts the formation of hypochord and dorsal aorta during the development of embryonic zebrafish. Sci China C Life Sci 2005; 48(6):608–615.

105. Reimers MJ, La Du JK, Periera CB, et al. Ethanol-dependent toxicity in zebrafish is partially attenuated by antioxidants. Neurotoxicol Teratol 2006; 28(4):497–508.

106. Dlugos CA, Rabin RA. Ocular deficits associated with alcohol exposure during zebrafish development. J Comp Neurol 2007; 502(4):497–506.

107. Sulik KK. Genesis of alcohol-induced craniofacial dysmorphism. Exp Biol Med (Maywood) 2005; 230(6):366–375.

108. Bradfield JY, West JR, Maier SE. Uptake and elimination of ethanol by young zebrafish embryos. Neurotoxicol Teratol 2006; 28(5):629–633.

109. Bilotta J, Saszik S, Givin CM, et al. Effects of embryonic exposure to ethanol on zebrafish visual function. Neurotoxicol Teratol 2002; 24(6):759–766.
110. Kashyap B, Frederickson LC, Stenkamp DL. Mechanisms for persistent microphthalmia following ethanol exposure during retinal neurogenesis in zebrafish embryos. Vis Neurosci 2007; 24(3):409–421.
111. Matsui JI, Egana AL, Sponholtz TR, et al. Effects of ethanol on photoreceptors and visual function in developing zebrafish. Invest Ophthalmol Vis Sci 2006; 47(10):4589–4597.
112. Loucks EJ, Schwend T, Ahlgren SC. Molecular changes associated with teratogen-induced cyclopia. Birth Defects Res A Clin Mol Teratol 2007; 79(9):642–651.
113. Berman JN, Kanki JP, Look AT. Zebrafish as a model for myelopoiesis during embryogenesis. Exp Hematol 2005; 33(9):997–1006.
114. Webb AE, Kimelman D. Analysis of early epidermal development in zebrafish. Methods Mol Biol 2005; 289:137–146.
115. Berghmans S, Jette C, Langenau D, et al. Making waves in cancer research: New models in the zebrafish. Biotechniques 2005; 39(2):227–237.
116. Lieschke GJ, Currie PD. Animal models of human disease: Zebrafish swim into view. Nat Rev Genet 2007; 8(5):353–367.
117. McMahon C, Semina EV, Link BA. Using zebrafish to study the complex genetics of glaucoma. Comp Biochem Physiol C Toxicol Pharmacol 2004; 138(3):343–350.
118. Penberthy WT, Shafizadeh E, Lin S. The zebrafish as a model for human disease. Front Biosci 2002; 7:d1439–d1453.
119. Rubinstein AL. Zebrafish: From disease modeling to drug discovery. Curr Opin Drug Discov Devel 2003; 6(2):218–223.
120. Kari G, Rodeck U, Dicker AP. Zebrafish: An emerging model system for human disease and drug discovery. Clin Pharmacol Ther 2007; 82(1):70–80.
121. Spitsbergen JM, Kent ML. The state of the art of the zebrafish model for toxicology and toxicologic pathology research—advantages and current limitations. Toxicol Pathol 2003; 31(Suppl):62–87.

9

Physiologically Based Pharmacokinetic Modeling in the Risk Assessment of Developmental Toxicants

Mathieu Valcke and Kannan Krishnan

Département de Santé Environnementale et Santé au Travail, Faculté de Médecine, Université de Montréal, Montréal, Québec, Canada

INTRODUCTION

The process of risk assessment for developmental toxicants often requires the application of uncertainty factors or extrapolation methods to facilitate the use of animal studies conducted at high-dose levels for deriving acceptable exposure levels for humans (1). Extrapolations of developmental toxicity benchmarks (e.g., NOAEL) from one exposure route to another as well as from animals to humans can only be conducted with the knowledge of appropriate measure of dose. The measure of dose for developmental toxicants ranges from the environmental concentration (potential dose) to the amount delivered to the developing organism (delivered dose) or the amount of putative toxic moiety per unit volume of blood or the target tissue (internal dose).

The dose to the tissues of developing organisms is mainly determined by the extent of maternal exposure, rate of contact (or delivery), and pharmacokinetic determinants. If the tissue is accessible and dose is measurable with currently available methodologies, then the required pharmacokinetic data can be obtained experimentally and used for risk assessment purposes. It is actually reflected by the developmental toxicology study designs that include measurement of the potential toxic moiety or parent chemical in blood, milk, and other biological matrices during specific time frames (2–6). When such data cannot be collected routinely or ethically for the potential toxic moiety (parent chemical or metabolite) associated with each exposure scenario (dose level, route, and windows of susceptibility), then the use of pharmacokinetic models is sought. The pharmacokinetic models are mathematical descriptions of the absorption, distribution, metabolism, and excretion of chemicals in biota, and are often compartmental in nature, even though some models [e.g., physiologically based pharmacokinetic (PBPK) models] characterize the compartments in terms of mechanistic determinants (i.e., physiological, biochemical) while others do not explicitly do so (7,8).

This chapter describes the concepts and tools essential for constructing and applying pharmacokinetic models to evaluate internal dose of chemicals in developing organisms, with particular emphasis on the use of mechanistic PBPK models in the risk assessment of developmental toxicants.

INTERNAL DOSE AND THE RISK ASSESSMENT
OF DEVELOPMENTAL TOXICANTS

For developmental toxicants, the internal dose metric is defined in the context of window of susceptibility associated with prenatal exposures, postnatal exposures, or a combination thereof (Fig. 1). For example, malformations have greater chances of being induced during the window corresponding to 5 to 14 gd in rats (or 18–60 gd in humans) (9). The appropriate internal dose for this susceptibility window might be the area under the chemical concentration versus time curve (AUC), maximal concentration (C_{max}), or other measures (e.g., receptor occupancy, amount metabolized, macromolecular adduct levels). Embryo-lethal effects of valproic acid and caffeine have been correlated with C_{max}; but in the case of retinoids and cyclophosphamide, the incidence of embryo-fetal effects is related to AUC [reviewed in Schwartz (10)]. Generally, when the parent chemical is the toxic moiety, its C_{max} or AUC is often used as the measure of internal dose. On the other hand, if the metabolite is the putative toxic moiety, the rate of its formation or concentration in the target organs is considered as the relevant dose metric. Other measures of tissue exposure may also be appropriate in certain cases depending upon the mode of action (e.g., duration and extent of receptor occupancy, macromolecular adduct formation, or depletion of glutathione) (11).

If the internal dose measures are accessible and measurable in exposed animals and humans during the potential window(s) of susceptibility, then there is no need for pharmacokinetic models. However, one might still need such models to facilitate the computation of internal dose associated with other potential exposure scenarios, routes, and susceptibility windows of interest. In reality, measuring tissue dose in developing organisms during specific window(s) of susceptibility or conducting intentional human exposures to environmental chemicals is not possible or ethical. Furthermore, the available pharmacokinetic data may not correspond to the active toxic moiety, relevant route, or appropriate dose levels. In the absence of experimental data on the biologically active form of a chemical

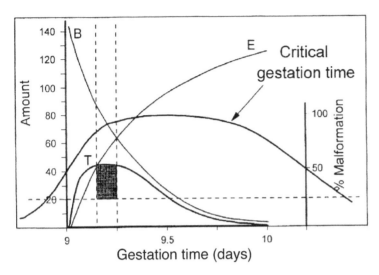

Figure 1 Critical window concept. B: maternal body burden; E: amount eliminated; T: amount in target tissue. *Source*: From Ref. 12.

in target tissues of developing organisms, the limited data on blood concentration of parent chemical, urinary metabolite levels, or fraction absorbed may be used as surrogate of dose metrics. These data are useful for constructing or evaluating pharmacokinetic models that can, in turn, be used to estimate the level of the toxic moiety of interest in the target organ during the window(s) of susceptibility along with attendant characterization of the uncertainty associated with such estimates, as described in the following sections.

TISSUE DOSE AND PHARMACOKINETIC MODELING OF PRENATAL EXPOSURES

Knowledge of the dose to fetus and fetal target organs would enhance the scientific basis of risk assessments. The dose to fetus, depending upon the mode of action, could be maximal concentration, average concentration, concentration at a particular time point, or integrated concentration versus time data over a particular interval of time during development (13). Knowledge of appropriate dose metric is critical since a chemical might cause adverse effects during a certain window during gestation but not during another time frame even if equivalent concentrations are observed at the target organ (12). Since the dose to fetus cannot often be determined directly, measurements of chemical concentrations in maternal plasma may be used, in conjunction with pharmacokinetic models, to calculate the concentration in fetal blood and organs. The measured or simulated tissue dose during prenatal exposures is determined by the maternal physiology and pharmacokinetics, chemical flux between the maternal compartments and fetus/embryo, and the pharmacokinetics within the embryo/fetus (14–16). Essentially then, the pharma-cokinetics of developmental toxicants is determined by the dynamic changes in physiological, biochemical, and physicochemical determinants of toxicant uptake; distribution; and clearance in both the mother and the fetus/embryo.

Toxicant Uptake Associated with Prenatal Exposure

Fetal exposure mainly occurs via transport of chemicals through the placenta (17), although some exposure may occur through the presence of contaminants in the amniotic fluid, particularly until the 20th week of gestation when the fetal skin is quite permeable (18). The transplacental transport generally occurs via passive diffusion, mainly for substances with a molecular weight ≤ 1000 g/mol whereas active transport and facilitated diffusion would appear to play a limited role (19).

Un-ionized lipophilic compounds could traverse the placenta relatively eas-ily because of the lipid content of the placental membranes, whereas the extent of transport of ionized compounds is determined by the difference in maternal and fetal blood pH. Placental perfusion resulting from both maternal and fetal blood flows varies according to the placental form involved (yolk sac during the first trimester, and chorioallantoic placenta afterward), and increases of up to 15% of the maternal cardiac output might occur near term, reaching an average perfusion rate of 500 to 700 ml/min for the whole uterus (20). Along with reduced membrane thickness and increased surface areas, this increase results in greater transfer of xenobiotics from the mother to the fetus/embryo in humans. However, this was not found to be the case with rodents (21).

The breathing rate, food/water consumption rate, and surface area of preg-nant women are known to contribute to increased uptake of chemicals in the environment compared to nonpregnant women (17,22), even though the concen-tration of chemical reaching the fetal compartment would additionally depend upon a number of physiological changes affecting the distribution and clearance.

Toxicant Distribution Following Prenatal Exposure

The physiological volume in which chemicals are distributed increases during pregnancy due to increases in the volumes of plasma, fat, and total body water (22). Similarly, the volume of distribution increases rapidly during fetal growth; from fertilization to birth, the volume of the conceptus increases by several orders of magnitude (Fig. 2). Growth of fetal tissues does not necessarily reflect a linear increase during various gestational stages (Fig. 3). The distribution of toxicants in the fetus is particularly different because of the dynamic changes in its volume as well as the lower lipid content and higher water content compared to adults (24). The body fat content changes from 0.5% at 20 weeks of gestation to about 16% at term compared to $\approx 20\%$ during adulthood (25). Functional differences at the organ level may also influence the distribution. For example, due to the immaturity of the blood–brain barrier, xenobiotics such as methylmercury tend to reach higher concentrations in fetal brain than in adult brain for a similar blood

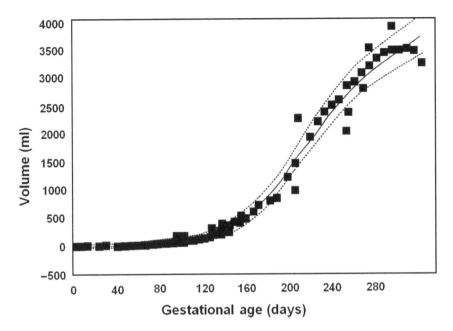

Figure 2   Growth curve of human fetus/embryo. *Source*: From Ref. 12.

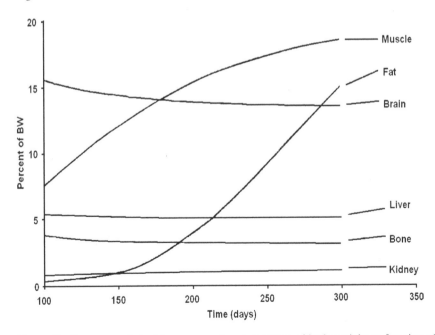

Figure 3   Organ growth in fetus expressed as percentage of body weight as function of gestational age for several tissues. *Source*: From Ref. 23.

concentration (26). However, few data exist on the partitioning of xenobiotics in the fetus in comparison with adult tissues (27,28).

Toxicant Clearance Following Prenatal Exposure

Blood Flow Rate

Blood flow parameters for human fetus are limited but there are ample data in other species such as sheep and goats (29). The cardiac output in the human fetus at term has been estimated to be 1.2 L/min, of which 50% corresponds to umbilical blood flow (20). The proportion of cardiac output flowing through liver in the human fetus/embryo is expected to be high due to umblical venous blood flow. Indeed, in lambs, hepatic blood flow has been estimated to be ≈30% of total cardiac output. However, this figure is reduced significantly at birth due to cessation of umbilical blood flow (30). Among other factors that modulate fetal organ blood flows, fetal hypoxemia and acidosis are important since they have been reported to raise the blood flow to placenta, brain, heart, and adrenals, but decrease the flow to kidneys, lung, and gut (31).

Metabolic Clearance

Human fetal liver can oxidize xenobiotics from the 16th week of gestation (19). The placental contribution to the overall metabolism appears reduced in humans, although it does occur for selected compounds (17). Fetal liver content of phase I enzymes appears to be at about one-third of adult values (during second and third trimesters of gestation). Some isoforms of cytochrome P-450 (CYP) (e.g., 4A, 3E) are found in fetal liver in greater concentrations than in adults, leading to a relatively higher rate of metabolite production in the fetus/embryo as compared to adults for certain toxicants.

Regarding phase II metabolism, acetylase and epoxide hydrolase activities in the human fetus/embryo would appear to be about half of the adult levels (32). Sulfation would appear to be well developed in neonates (33), and glycine conjugation in the near term fetus has been reported to be comparable to that of adults (34). Glutathione S-transferase (GST)π is the main enzyme responsible for conjugation with glutathione in fetal liver as it has been reported to be responsible for about 50% of this activity. Glucuronyl transferase, however, is expressed at very low levels in fetal liver [reviewed in Faustman and Ribeiro (21)].

Toxicants that enter the fetus/embryo via the placenta undergo a fetal hepatic "first pass" effect, since they pass through hepatic parenchyma before reaching the inferior vena cava. However, a percentage of the incoming umbilical blood reaches the vena cava via the ductus venosus, particularly from the 11th week of gestation, and is thus not subject to the first pass effect. This percentage exhibits quite a wide interindividual variability (from 8% to 92%), but a mean estimate of 60% has been proposed (19).

Pharmacokinetic Modeling and Tissue Dosimetry Associated with Prenatal Exposures

The simplest pharmacokinetic models consider the fetus and mother as single homogeneous compartments interconnected by placental transfer of chemicals. The rate of change in the amount of chemical in the mother (dA_{mat}/dt) due to transfer to fetus is computed as follows (35):

$$dA_{mat}/dt = -k A_{mat} \qquad (1)$$

where $-k$ = the rate constant and A_{mat} = the amount of xenobiotic in the maternal compartment.

When the chemical transfer occurs quickly, the fetus:mother concentration ratio is about unity and attained rapidly. Thiopentone, fitting to these criteria, can be modeled using a simple pharmacokinetic description for the mother and fetus (18).

In such models, the clearance between the mother and fetus on both directions is assumed to be equal, as shown below:

$$K_{mf} \times V_m = K_{fm} \times V_f \qquad (2)$$

where K_{mf} is the rate constant for xenobiotic transfer from mother to fetus, K_{fm} is the rate constant for xenobiotic transfer from fetus to mother and V_m and V_f correspond to the volumes of distribution in the maternal and fetal compartments, respectively. The term on the left-hand side of Eqn. 2 represents "efflux clearance for mother" while the right-hand side corresponds to the "efflux clearance for fetus."

A pharmacokinetic model of greater complexity is often warranted; the choice of the type and number of compartments, however, depends largely on the physicochemical properties of the chemical in consideration. For example, for modeling tetracyclines, the mother was represented as a two compartmental system (central and peripheral compartments) linked with a fetal compartment. In this case, the concentration of the chemical in the fetal compartment and the central compartment in mother is identical since the fetal transfer is rapid; conversely, if the transfer to fetus is slow, the ratio is likely to be greater than unity (18).

PBPK models are of greater complexity and flexibility, allowing one to describe the mother as a multicompartmental system and the fetus as a single or a multicompartmental system [Fig. 4 (A)] depending upon the objective and intended use of the resulting model (26). The rate of change in the amount of chemical in the maternal and fetal compartments is based on perfusion-limitation or diffusion-limitation concepts (8). The representation of the placenta in these models can be detailed so that both maternal and fetal circulation can be accounted for in sufficient detail, along with the representation of placenta (the yolk sac or the chorioallantoic placenta) consistent with the gestational period modeled [e.g., Fig. 4 (B)]. In fact, the number of compartments can be as many as 26 for the maternal model and up to 15 for the fetal model (36).

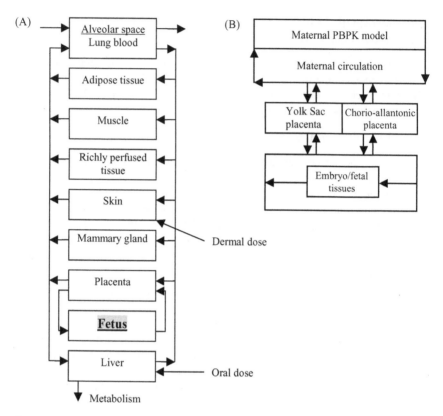

Figure 4 (**A**) Structure of a simple PBPK model facilitating the simulation of chemical concentration in fetus/embryo. (**B**) Structure of a PBPK model that includes the two possible forms of placenta during gestation. *Source*: From Ref. 23.

The rate of transfer of chemicals across the placenta from the mother to the fetus (dA_{pla}/dt) can be described on the basis of Fick's first law and this approach has been used in PBPK modeling (27,37). Accordingly,

$$dA_{pla}/dt = K \times A(C_m - C_f)/D \tag{3}$$

where A = placental surface area available for diffusion, C_m = concentration in maternal blood, C_f = drug concentration in fetal blood, D = thickness of the membrane and K = diffusion constant of the chemical in the membrane.

When blood flow, rather than diffusion, is the limiting factor of chemical transfer, the rate of change in PBPK models is computed on the basis of blood flow rate, concentration gradient, and the corresponding partition coefficients, as follows:

$$dA_{pla}/dt = Q_{pla}(C_a - C_{pla}/P_{pla}) - dA_{fet}/dt \tag{4}$$

and

$$dA_{fet}/dt = Q_{fet}((C_{pla}/P_{pla} \times P_{f:m}) - C_{fet}/P_{fet}) \tag{5}$$

where dA_{pla}/dt = rate of change in the amount of chemical in placenta, dA_{fet}/dt = rate of change in the amount of chemical in fetus, Q_{pla} = blood flow to placenta, Q_{fet} = blood flow to fetus/embryo, C_{pla} = concentration in placenta, C_a = arterial blood concentration in the mother, C_{fet} = concentration in the fetus/embryo, P_{pla} = placenta:blood partition coefficient, P_{fet} = fetal blood:air partition coefficient, and $P_{f:m}$ = fetal blood:maternal blood partition coefficient (38).

Eqns. 4 and 5 can be used along with input parameters corresponding to a particular gestation day; however, in order to be able to implement these models to simulate the kinetics of xenobiotics in the growing fetus during the entire length of gestation, equations to compute physiological parameters in mother as well as fetus need to be integrated within the PBPK model. Compilations and analyses of such physiological data for PBPK modeling are found in Luecke et al. (36,39–41). In this regard, the growth of the fetus can be computed on the basis of its weight (BW) at any time (t) during gestation using the Gompertz equation as follows (23):

$$BW(t) = 0.001374 \times \exp\{(0.19741/0.013063)(1 - \exp[-0.013063 \times d])\}. \tag{6}$$

This equation appears to fit experimental data for day (d) 50 to term well, but overestimates the weight for preceding days. If the fetal submodel is detailed in terms of its description, then the dynamics of growth of the various organs may have to be included. The fetal organ weight (V_t) can be calculated as a function of fetal body weight using the following equation, where a and b are constants whose values are presented in Table 1:

$$V_t = a \times (BW)^b \tag{7}$$

For developing PBPK models, tissue:blood partition coefficients for the fetal organs and maternal organs at various gestational periods are required, and these can be estimated either experimentally or using validated animal-alternative algorithms (8,38).

Table 2 lists chemicals for which PBPK models have been developed in test animals and humans for evaluating internal dose in fetus associated with maternal exposure during pregnancy.

TISSUE DOSE AND PHARMACOKINETIC MODELING OF POSTNATAL EXPOSURES

The target tissue dose, for a given exposure scenario, in developing animals and children might be different or comparable to that of adult animals and humans. Child/adult differences in several pharmacokinetic determinants would appear to decrease with increasing postnatal age, but these factors should still be considered carefully for children of all age groups since the net impact really depends

Table 1 Allometric Parameters for Fetal Organ and
Tissue Growth During Pregnancy

Fetal organ tissues	a	b
Adrenal	0.007467	0.8902
Bone	0.05169	0.9288
Bone marrow	0.01425	0.9943
Brain	0.1871	0.9585
Fat	0.1803	−0.9422
Heart	0.01012	0.9489
Kidney	0.004203	1.255
Liver	0.06050	0.9737
Lung	0.09351	1.552
Pancreas	0.1883	0.3854
Plasma	0.06796	0.9729
Skeletal muscle	0.02668	1.234
Spleen	0.0001302	1.204
Thymus	0.001218	1.093
Thyroid	0.0006470	1.023

Note: The equation $W = a \times (BW)^b$ is used, where W is
the fetal organ weight and BW is the fetal body weight. A
third constant c (= 0.2332, −0.02127, −0.05945, or 0.02909,
respectively) is used in case of fat, kidney, lung, and spleen to
accommodate growth rate differences in these organs and total
weight of human embryo/fetus.
Source: From Ref. 42.

upon the mode of action of chemicals and nature of the toxic moiety (72). The
dose to the tissue during postnatal development is determined by the dynamic
changes in exposure pathways and pharmacokinetic determinants, and it can be
measured experimentally where feasible and ethical or simulated using PBPK
models (72–77). The measured or simulated tissue dose during postnatal exposures is determined by the age-dependent changes in toxicant uptake, clearance,
and distribution, as summarized in Table 3.

Toxicant Uptake Associated with Postnatal Exposure

The intake of toxicants and resulting potential dose (mg/kg/day) associated with
the oral, dermal, and inhalational routes would be greater in younger children compared to adults. The fact that neonates exhibit a greater breathing rate than adults
leads to greater volume of contaminated air passing through the lungs, resulting in a greater amount of chemical inhaled by neonates (mg/kg body weight)
as compared to adults, when they are exposed to the same atmospheric concentration and scenario. The potential dose of a hypothetical air pollutant to
children of several age groups as well as adults is presented in Table 4. The
difference in the calculated dose results from dynamic changes in body weight

Table 2 Chemical-Specific PBPK Models for Simulating Prenatal Dosimetry

Specie	Chemical	Reference
Mouse/rat	Bisphenol A	Kawamoto et al. (43)
	DDE	You et al. (44)
	5,5′-dimethyloxazolidine-2,4-dione	O'Flaherty et al. (45)
	2-Ethoxyethanol and 2-Ethoxyacetic acid	Gargas et al. (46)
	Ethylene glycol	Welsh (47)
	Hydroxyurea	Luecke et al. (39)
	Iodide	Clewell et al. (48, 49)
	Isopropanol	Gentry et al. (50)
	Methadone	Gabrielsson et al. (51) Gabrielsson and Groth (52)
	Methanol	Ward et al. (53)
	2-Methoxyethanol and 2-Methoxyacetic acid	Clarke et al. (54); Terry et al. (55); Welsh et al. (56); O'Flaherty et al. (57); Hays et al. (58); Gargas et al. (59)
	Methylmercury	Gray (26); Faustmann et al. (60)
	Morphine	Gabrielsson and Paazlow (61)
	Perchlorate	Clewell et al. (48,49)
	Pethidine	Gabrielsson et al. (62)
	p-phenylbenzoic acid	Kawahara et al. (63)
	Retinoic acid	Clewell et al. (64)
	TCDD	Emond et al. (65)
	Tetracycline	Olanoff and Anderson (66)
	Theophylline	Gabrielsson et al. (67)
	Trichloroethylene	Fisher et al. (68)
Human	2-Ethoxyethanol and 2-Ethoxyacetic acid	Gargas et al. (58)
	Ethylene glycol	Welsh (47)
	Iodide	Clewell et al. (69)
	Isopropanol	Gentry et al. (50,28)
	Methadone	Gabrielsson et al. (51) Gabrielsson and Groth (52);
	2-Methoxyethanol and 2-Methoxyacetic acid	Welsh et al. (56); Gargas et al. (59)
	Methylene chloride	Gentry et al. (28)
	Methylmercury	Clewell et al. (27)
	Morphine	Gabrielsson and Paazlow (61)
	Nicotine	Gentry et al. (28)
	Perchlorate	Clewell et al. (69)
	Perchloroethylene	Gentry et al. (28)
	Pethidine	Gabrielsson et al. (62)
	Retinoic acid	Clewell et al. (64)
	TCDD	Gentry et al. (28) Maruyama et al. (70)[a]
	Theophylline	Gabrielsson et al. (67)
	Vinyl chloride	Gentry et al. (28)
Others	2,4-Dichlorophenoxyacetic acid [2,4-D]	Kim et al. (71)[b]
	Retinoic acid	Clewell et al. (64)[c]

[a] Fetal compartment was not designed in this study; TCDD concentration in fetus was assumed as being equivalent to the concentration in maternal richly perfused tissues.
[b] Rabbits.
[c] Monkeys.
Source: From Ref. 23.

Table 3 Summary of the Differences in Pharmacokinetic
Determinants in Children/Neonates as Compared to Adults

Pharmacokinetic step	Difference in children/neonate as compared to adults
Absorption	
Oral	
Lipophilic compounds	↑↓
Water soluble compounds	↑↓
Inhalation	
Lipophilic compounds	↑
Water soluble compounds	↑
Particulate	↑
Dermal	
Lipophilic compounds	*I*
Water soluble compounds	*I*
Distribution	
Lipophilic compounds	↑
Water soluble compounds	↑
Protein binding	↓
Metabolism	
Glutathione-S-transferase	↓
Sulfotransferase	↑
Glucuronyl transferase	↓
Cytochrome P450	↓
Carboxyl esterase	↓
Alcohol Dehydrogenase	↓
Excretion	
For lipophilic compounds	
Protein binding	*I*
Glomerular filtration	↓
Tubular secretion	↓
Tubular reabsorption	*I*
For water-soluble compounds	
Protein binding	*I*
Glomerular filtration	↓
Tubular secretion	↓
Tubular reabsorption	*I*

Abbreviations: ↑, Higher than adults; ↓, lower than adults; ↑↓, increase and decrease demonstrated, depending of the substance; I, insufficient data to assess.
Source: From Ref. 78.

during development as well as the breathing rate. Regarding dermal exposures, similar observations can be made. Given that the skin surface area per unit body weight is larger in neonates and young children compared to adults, for an identical exposure scenario, the potential dose received by children would also be

Table 4   Comparison of Exposure Variables in Children of Different Ages with Adults

Age (yrs)	Body weight[a] (kg)	Body height[a] (cm)	Water ingestion rate[b] (l/kg/day)	Food ingestion rate[c] (mg/kg/day)	Inhalation rate[d] (m³/day)	Dermal surface area[d] (cm²)	Dermal dose[e] (mg/kg/day)	Inhaled dose[f] (mg/kg/day)
1	8.6	74.6	0.035	10.6	1.9	4390	4.1	17.0
3	15.0	94.4	0.046	8.1	2.9	6451	3.5	15.0
6	22.2	117.8	0.036	6.7	4.0	8672	4.6	13.7
30	71.8	168.5	0.019	4.1	9.8	18,462	2.1	10.5

[a]Mean values for males and females as described by Haddad et al. (79).
[b]Calculated using the mean water intake (L/day), estimated from the Exposure Factors Handbook (80) divided by the body weight.
[c]Calculated using the values of Exposure Factors Handbook (80) divided by the body weight.
[d]Calculated using the equation described by Haddad et al. (79).
[e]Calculated for a 30 min shower containing 100 μg/L of chloroform, using the equations and data described by Haddad et al. (79).
[f]Calculated with an air concentration of 100 μg/m³ of chloroform, using the equations and data described by Haddad et al. (79).

greater (Table 4). Similarly, the consumption of food and water per unit body weight is greater in neonates compared to adults leading to an increased intake of contaminants in the former group (Table 4) (81). However, the rate and extent of absorption might be lower in newborns due to underdevelopment of some elements of the physiological system, notably the splanchnic blood flow (78,82) but higher due to certain exposure-related activities, for example, ingestion of soil/dusts, hand-to-mouth activity, etc (83,84). Of prime importance in this regard is lactational exposure (85–88), the extent of which depends on several factors including feeding frequency, plasma protein binding in maternal blood, blood flow to breast, physicochemical properties of the substance as well as the composition of milk. The tissue dose resulting from direct and indirect exposurses (e.g., oral, dermal, inhalational, and lactational) of developing organisms is determined by the extent of distribution in the physiological volume as well as clearance processes.

Toxicant Distribution Following Postnatal Exposure

Xenobiotics, following absorption, are distributed in the physiological volume corresponding to the space occupied by water, protein, and fat. The volume of distribution is calculated as the volume of blood plus the product of tissue volume and tissue:blood partition coefficient. Whereas the tissue:blood and blood:air partition coefficients in older children and developing animals are comparable to that of adults (75,89), the absolute volumes of blood and most tissues would appear to increase as a function of age (75,78,82). As shown in Figure 5 (A), the volumes of brain and liver, relative to body weight, are greater in neonates

(A)

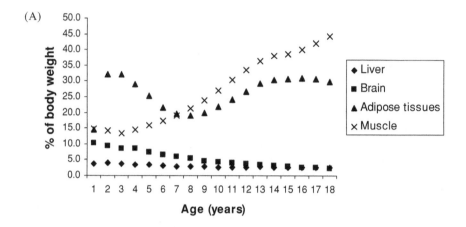

(B)

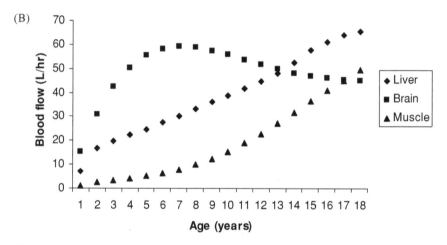

Figure 5 (**A**) Volumes of liver, brain, muscle and adipose tissue volumes as function of age of children; (**B**) Blood flow to liver, brain, and muscle as a function of age in children. *Sources*: From Refs. 73,75,79.

than in adults, whereas the reverse is the case with muscle and adipose tissue. Not only the volumes of tissues but also their composition is somewhat different in neonates compared to adults leading to differences in the distribution and concentration of water-soluble and lipid-soluble chemicals. For example, the lipid content of adipose tissue is greater in neonates compared to adults (55% vs. 25%) (90) whereas the water content of liver, brain, and kidneys is reported to decrease from birth to adulthood by a small percent (5–15%) (91). However, total serum protein concentrations do not seem to change with age, even though the albumin level is observed to be lower in neonates compared to adults (30,78).

Toxicant Clearance Following Postnatal Exposure

Clearance, in the present context, refers to the volume of blood from which toxicant is removed per unit time. Major organs of clearance include liver, kidney, and lungs. Clearance is determined by physiological (e.g., blood flow rate, glomerular filtrate rate) and biochemical factors (enzyme levels). In general, the immaturity of systemic clearance mechanisms in neonates limits their ability to eliminate chemicals effectively, leading to higher internal doses sometimes compared to adults. To reflect the adult/infant differences in metabolic clearance, Alcorn and McNamara (92) developed an Infant Specific Factor that reflects functional maturation relative to adult values of hepatic CYP-mediated clearance and glomerular filtration rate.

Blood Flow Rate

It is well established that the heart rate and cardiac output are greater in newborns compared to older children and adults [e.g., Shock (93), Illiff and Lee (94), Cayler et al. (95), Sholler et al. (96)]. The age-related decrease in cardiac output, during postnatal growth, can be described based on its quantitative relationship to body surface area. The blood flow to individual tissues also changes with postnatal age [Fig. 5 (B)] but it is not always proportional to change in tissue weights (97,98). For example, the average liver blood flow rate in children aged 1, 4, and 12 years, respectively, is 41.1, 43.6, and 44.3 ml/hr/g of liver as compared to 45.6 ml/hr/g of liver in adults. Corresponding data for the brain are 33.1, 45.2, and 47.8 ml/hr/g compared to 33.8 ml/hr/g in adults. Renal blood flow is lower in neonates compared to adults (5–6% vs. 15–25% of cardiac output) (99) whereas blood flow to muscle decreases as a function of postnatal age until 5 years, after which it remains rather fairly constant until adulthood (100–102). Finally, postnatal changes in blood flow to cerebral regions appear to coincide with cognitive development (103).

Metabolic Clearance

Metabolic clearance depends on the affinity and maximal velocity of metabolic reactions as well as the hepatic blood flow. For chemicals exhibiting a high hepatic extraction ratio (e.g., trichloroethylene), the rate of metabolism is essentially limited by hepatic blood flow; whereas, for chemicals exhibiting a low metabolic clearance (e.g., methyl chloroform), the rate limiting factor is the enzyme content and not the liver blood flow. Hepatic transport systems also represent a determinant of hepatic clearance, particularly because of their influence on the effectiveness of tissue efflux and biliary excretion. Limited in vivo evidence shows that hepatic excretory function as well as carrier-mediated hepatocellular uptake might be reduced in infants (30).

    Several phase I enzymes surge within hours after birth whereas others develop gradually during the months following birth (e.g. CYP3A4, CYP2C, and CYP1A2) (104–106). Figure 6 depicts age-dependent changes during postnatal development in CYP450 1A2, 3A4, and 2E1 isoenzymes. Limited data available on the ontogeny of phase II enzymes suggest that GSTπ, dominant fetal form,

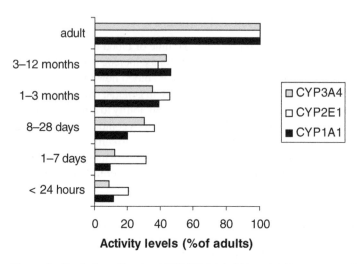

Figure 6   Evolution of levels of CYP450 3A4, 2E1, and 1A1 as compared to adults during postnatal development. *Source*: From Ref. 30.

regresses at birth and is not expressed in adults, whereas other GSTs expressed at low levels in the fetus increase after birth (107,108). Conjugation with glucuronic acid is significantly lower at birth, although the capability for conjugation with sulphate is well developed in neonates (33). The levels of glycine conjugation in newborns are comparable to those of adults (34).

Renal Clearance

Glomerular filtration as well as tubular secretion and resorption are reduced at birth, due to immaturity of the renal system and function (109). Glomerular filtration increases rapidly after birth from 2–4 ml/min (based on creatinine clearance) to 8–20 ml/min at 2 weeks, reaching adult values (127 ml/min) around the age of 6 months (30). Tubular function, on the other hand, increases progressively, reaching adult values at about 1 year of age (82). Reduced renal clearance is also a consequence of poor renal and intrarenal blood flow (110). The magnitude of the reduction in renal clearance has been estimated to be in the order of 30 to 50%, depending upon the lipophilicity of chemicals and mechanism involved (78).

PBPK Modeling and Tissue Dosimetry Associated with Postnatal Exposures

Tissue dose of toxicants associated with postnatal exposures can be simulated using pharmacokinetic models and algorithms (75,77,111). Due to the feasibility of incorporating the dynamic changes in tissue composition, tissue volumes, blood flow rates as well as other physiological determinants of exposure, and pharmacokinetics, PBPK models are increasingly being used to simulate the tissue dose associated with postnatal exposures [e.g. Byczkowski et al. (112), Clewell et al. (113), Rodriguez et al. (114)].

Table 5 Equations Used for Determination of Body and Organ Weights as a Function of Age in Humans, During Prenatal and Postanal Development

Fetal growth	*Fetal weight at a given gestational day (d) [from Corley et al. (23)]* $BW(g) = 0.001374 \times \exp\{(0.19741/0.013063)(1 - \exp[-0.013063 \times d])\}$ *Fetal organs weight in grams (W) for a given body weight (BW) [from Krishnan and Andersen (38)]* Liver: $W = 0.0605 \times BW^{0.9737}$ Brain: $W = 0.1871 \times BW^{0.9585}$ Kidney: $W = 0.004203 \times BW^{1.255} - 0.02127$ Lung: $W = 0.09351 \times BW^{1.552} - 0.05945$
Postnatal growth	*Body weight at a given age (a, years) [from Haddad et al. (73)]* Males: $BW\ (g) = -1.9a^4 + 72.8a^3 - 813.1a^2 + 5535.6\,a + 4453.7$ Females: $BW\ (g) = -2.561a^4 + 85.576a^3 - 855.95a^2 + 5360.6\,a + 4428.5$ *Organs weight in grams (W) for a given age (a, years) [Haddad et al. (73)]* Males: Liver: $W = 0.0072a^5 - 0.3975a^4 + 7.9052a^3 - 65.624a^2 + 262.02a + 157.52$ Brain: $W = 10^4 \times [(a + 0.213)/(6.03 + 6.895a)]$ Kidney: $W = 9.737\text{E-}4a^5 - 0.0561a^4 + 1.1729a^3 - 10.34a^2 + 44.604a + 28.291$ Lung: $W = -0.0346a^4 + 1.5069a^3 - 20.31a^2 + 123.99a + 59.213$ Females: Liver: $W = 0.0057a^5 - 0.3396a^4 + 7.0134a^3 - 59.53a^2 + 251.9a + 139.65$ Brain: $W = 10^4 \times [(a + 0.226)/(6.521 + 7.514a)]$ Kidney: $W = 1.2676\text{E-}3a^5 - 6.6825\text{E-}2a^4 + 1.2345a^3 - 9.4597a^2 + 39.005a + 27.161$ Lung: $W = 6.3\text{E-}3a^5 - 0.3162a^4 + 5.5896a^3 - 42.196a^2 + 160.79a + 50.506$

In general, the structure and equations of the PBPK models for developing organisms are similar to those for adults. In fact, the same general conceptual model as for adults has frequently been used to simulate kinetics of xenobiotics in children. However, differences in exposure pathways should be accounted for along with the substitution of parameters appropriate for the stage of development (i.e., tissue volumes, blood flows, respiratory rate, skin area). Several compilations of physiological parameters for PBPK modeling of postnatal exposures have been published (41,115,116). Table 5 presents some examples of equations that have been used to determine body and several organ weights as a function of age in the growing child. Caution is required in using allometric equations since there may be some discontinuities in slopes for periods of development associated with growth spurts. This has been observed, for example, with glomerular filtration rate and inhalation rate (117).

Table 6 represents the input parameters of a PBPK model used for simulating the inhalation pharmacokinetics of furan in children of various age groups. The

Table 6 Input Parameters for the PBPK Simulation of Furan
Disposition in Adults and Children of Various Ages

Parameters	Adult	Children		
		6 yr	10 yr	14 yr
Alveolar ventilation rate (L/hr)	300	147.03	218.53	290.02
Cardiac output (L/hr)	372	245.23	338.21	403.86
Tissue blood rates (L/hr)				
Liver	96.72	19.5	39.9	54.9
Brain	42.41	59.07	53.73	46.93
Adipose tissue	19.34	12.95	16.73	17.71
Slowly perfused tissues	61.14	7.46	15.09	27.6
Rest of body	152.4	146.25	212.77	256.71
Tissue volume (L)				
Liver	1.8	0.62	0.87	1.26
Brain	1.4	1.31	1.36	1.39
Adipose tissue	14.9	3.68	6.25	11.49
Slowly perfused tissue	35.9	5.71	10.17	18.41
Rest of body	8.17	8.37	11.26	11.64

Source: From Ref. 75.

modeling results indicated that the internal dose, that is, concentration of parent chemical in systemic circulation would decrease with age due to the increase in the rate of hepatic metabolism (Fig. 7). This inhalation PBPK model is essentially identical to the adult model, except for the numerical values of input parameters (75). For other chemicals and scenarios, however, exposure of neonates (pups) via mother's milk could be important [e.g., Byczkowski and Fisher (118), Hallen et al. (119), Hong et al. (120), Hinderliter et al. (121), Marty et al. (122), Chu et al. (123)]. In order to build a PBPK model to simulate infant exposure to chemicals in mother's milk, a conceptual model representing both the exposed mother and the infant is required (Fig. 8). Here, the maternal portion of the PBPK model, similar to an adult model, facilitates the computation of the rate of change in the concentration of chemical in tissues and blood. Additionally, however, it facilitates the simulation of concentration of contaminants in milk as a function of the frequency, duration, route of exposure, and nursing schedule. The rate of change in the amount of chemical in the milk (AML) (dA_{mk}/dt) and mammary gland (dA_{mg}/dt) compartments have been computed as follows (38):

$$dA_{mg}/dt = Q_{mg}(C_a - C_{mlk}/P_{mlk}) - Q_{mlk}C_{mlk} \qquad (8)$$

and

$$dA_{mk}/dt = Q_{mlk}C_{mlk} - Q_{skl}C_{mlk} \qquad (9)$$

where Q_{skl} = suckling rate of the infant, Q_{mlk} = rate of milk production and P_{mlk} = milk:blood partition coefficient of the chemical.

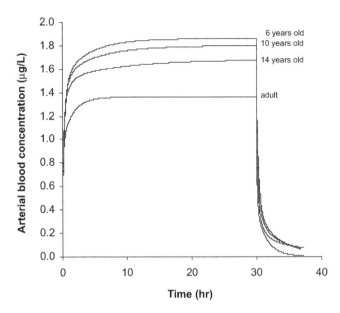

Figure 7 PBPK model simulations of the arterial blood concentration of furan following inhalation exposure. *Source*: From Ref. 75.

A three-compartment rat model representing the dam, milk, and pups proposed by Yoon and Barton (125) describes the rate of change in the amount of chemical in the dam (dA_d/dt) as function of the rate of absorption (RAB_d), rate of elimination ($Ke_d. A_d$), and the secretion into milk ($K_L. A_d$) as follows:

$$dA_d/dt = RAB_d - Ke_d \times A_d - K_L.A_d \tag{10}$$

$$RAB_d = dAB_d/dt = Ka_d \times AG_d \tag{11}$$

$$AG_d = F_d \times dose_d - Ab_d \tag{12}$$

where AG = amount in gut, Ab = amount absorbed, and Ka = oral absorption rate.

The rate of the change in the amount of chemical in the pup (dA_p/dt) was computed as a function of the rate of absorption (RAB_p) minus elimination ($Ke_p. A_p$) (125), as shown below:

$$dA_p/dt = RAB_p - Ke_p \times A_p \tag{13}$$

$$RAB_p = dAB_p/dt = Ka_p \times AG_p \tag{14}$$

$$AG_p = F_p \times [(dose_p)n + AML] - Ab_p \tag{15}$$

The abbreviations in the equations have the same significance as in the dam, with n = the number of pups per litter. However, an additional source of exposure was computed, namely, the exposure via milk ingestion as a function of the AML.

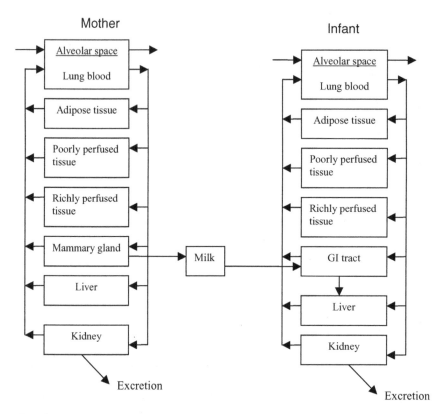

Figure 8   Structure of a PBPK model facilitating the simulation of inhalation and lactational exposures. *Sources*: From Refs. 68,124.

PBPK MODELING AND IMPLICATIONS FOR RISK ASSESSMENT OF DEVELOPMENTAL TOXICANTS

Risk assessments for developmental toxicants based on internal dose improve the scientific basis of the process, compared to the use of potential dose. The internal dose in this context, obtained with the use of PBPK models, refers to the dose metric that is appropriately and closely related to the adverse response, based on current knowledge of mode of action and window of susceptibility. Maximal concentration or average concentration on a particular gestational day, or area under the concentration versus time curve during a particular developmental period or some other parameter, may be the relevant dose metric (13). An example of systematic evaluation to identify relevant dose metrics that facilitated a better understanding of the human risk of malformations based on animal data would be 2-methoxyethanol (54,56).

The identification of the appropriate dose metrics, among the several candidate measures, can be challenging. This is often facilitated by examining the quantitative relationship between the pharmacokinetic dose metrics and developmental toxicity data sets. In simplest scenarios, the relationship can be examined using the conventional dose–response models for specific end points (126). Canonical correlation is a useful procedure when no single biological end point is found to be highly correlated to a single pharmacokinetic parameter. This statistical procedure compares multiple parameters from two sets of variables (X and Y). For example, one might choose to examine the quantitative relationship between the teratogenic parameters, namely, % dead + resorbed fetuses (% D + R), % malformed fetuses (% Malf), and net maternal weight gain (NMWG), as well as uterine weight (UW) and relevant pharmacokinetic variables [e.g., half-life ($t_{1/2}$), the 24-hour blood concentration (24 hour), the peak blood concentration (PkHt), the area under the plasma concentration-time curve (AUC)], as follows (127):

$$X = a(\% \text{ D} + \text{R}) + b(\% \text{ Malf}) + c(\text{NMWG}) + d(\text{UW}) \tag{16}$$

$$Y = e(t_{1/2}) + f(24\text{hr}) + g(\text{PkHt}) + h(\text{AUC}) \tag{17}$$

The magnitude of the coefficients indicates the relative weight that each of the parameters contributed to the correlation, that is, the parameter with the smallest coefficient could probably be dropped from the equation without significant loss of adequacy of the outcome. In this regard, the stepwise multiple regression technique has the advantage of combining terms in a stepwise manner and will only include those terms that actually contribute to the end result (127).

Once these correlations are defined in the test animal system, the challenge would be to extrapolate that relationship to man. In this regard, PBPK models are uniquely useful by facilitating the integration of species-specific physiological parameters associated with specific lifestages to simulate equivalent exposure doses (8,38).

PBPK models also uniquely facilitate the use of in vitro data on certain developmental effects to predict the risk in vivo. In this regard, PBPK models can be used to back-calculate the exposure dose that leads to the relevant plasma concentration, that is, the in vitro effect concentration. Such an in vitro–in vivo extrapolation capability of these simulation models has been successfully validated for the embryotoxicity associated with 2-methoxyethanol, 2-ethoxyethanol, retinoic acid, and methotrexate, but not for 5-fluorouracil (128).

Even though PBPK models facilitate extrapolation of internal dose across the various windows of susceptibility (i.e., the timing of pharmacodynamic events), the timing of birth in rodents and humans differs in relation to the developmental stage of many organ systems and complicates the interpretations (129). A rough approximation is that rodents are born at a point equivalent to the end of the human second trimester, so some developmental events occur postnatally in the rodents that would occur in utero during the third trimester in humans (129). As a result of the timing of birth being different across species in relation to the development of different organs and physiological systems, the tissue dose occurring during the relevant period in each species may not necessarily be concordant. However, with the use of a PBPK modeling framework, equations describing the dynamics of development of various organ systems can be integrated for each species, such that extrapolations of dosimetric parameters can be conducted in a scientifically sound manner.

With regard to postnatal exposures, the goal is not so much an evaluation of the dose–response relationship but rather an assessment of how the exposure level compares with the typical adult. The magnitude of difference in internal dose between the typical adult and children of various groups is assumed to be within a factor of 3.16 (i.e., square root of an order of magnitude, or, 10) (130). PBPK models for inhalation and oral routes have been developed for several contaminants to evaluate the magnitude of adult-child difference in internal dose (72,74–77,131,132). These studies indicate that, with the exception of neonates, the adult:child ratio of internal dose would be within a factor of 2, when both groups are exposed to the same concentration in air or given same oral dose (mg/kg/day). Compared to adult levels, the tissue dose, in terms of unchanged parent chemical, might be greater by as much as a factor of 3 to 4 in newborns [e.g., Nong et al. (77)]. These observations are in accord with results reported for pharmaceuticals (133,134) and have been attributed to both the immaturity of the enzyme systems and the substantial interindividual variability in the expression of CYP isozymes during the early ages. An additional situation of concern relates to those chemicals for which the rate of the toxic metabolite formation is considerably greater than the rate of its clearance. When the latter process depends upon CYP450, glucuronidation, or glomerular filtration, it is likely that the internal dose of metabolite would be greater in neonates compared to adults or older children (135).

The internal dose in neonates associated with inhalation or oral routes should not be viewed in isolation, rather along with the contribution of the lactational exposure pathway. PBPK modeling results indicate that the dose received via lactational exposure might in fact exceed that received from oral exposure to reference dose of certain volatile organic chemicals (87). In this regard, evaluation of the internal dose resulting from aggregate exposures to chemicals would significantly enhance our understanding of tissue dosimetry and risks to children.

CONCLUDING REMARKS

PBPK modeling allows the simulation and description of the dose to target tissue in the context of the risk assessment of developmental toxicants. Relating toxicity outcome to internal dose rather than external dose enhances the scientific basis of assessments and provides a defensible framework for conducting extrapolations across doses, routes, lifestages, and species. This kind of modeling is based on quantitative interrelationships among critical parameters that determine the behavior of the system under study. Unlike empirical models, the PBPK models can be useful in uncovering the biological determinants of tissue dosimetry as well as their relationship to the mode of action of developmental toxicants. These models represent a valuable component of any systematic and comprehensive approach of investigating how chemicals gain entry into, distribute within, and get eliminated from the mother and developing organisms. With regard to developmental toxicants, the use of PBPK models in extrapolating tissue dose for various exposure scenarios and windows of susceptibility is extremely relevant. As information on the developmental profiles of metabolizing enzymes, transporters, serum-binding proteins, and other critical determinants become available for both test animal species and humans, it will be increasingly possible to parameterize PBPK models for predicting target tissue dose essential for the conduct of risk assessment of developmental toxicants.

ACKNOWLEDGMENT

Support from Canadian Institutes of Health Research/Réseau de Recherche en Santé des Populations du Québec Transdisciplinary Training Program In Public and Population Health Research is acknowledged (M. V.).

REFERENCES

1. U.S. Environmental Protection Agency. Guidelines for developmental toxicity risk assessment. Risk Assessment Forum, U.S. Environmental Protection Agency. EPA/600/FR-91/001, Washington DC, 1991.
2. Kimmel CA, Francis EZ. Proceedings of the workshop on the acceptability and interpretation of dermal developmental toxicity studies. Fundam Appl Toxicol 1990; 14:386–398.
3. Mylchreest E, Cattley RC, Foster PM. Male reproductive tract malformations in rats following gestational and lactational exposure to Di(n-butyl) phthalate: An antiandrogenic mechanism? Toxicol Sci 1998; 43:47–60.
4. Parks LG, Ostby JS, Lambright CR, et al. The plasticizer diethylhexyl phthalate induces malformations by decreasing fetal testosterone synthesis during sexual differentiation in the male rat. Toxicol. Sci 2000; 58:339–349.
5. Sturtz N, Bongiovanni B, Rassetto M, et al. Detection of 2,4-dichlorophenoxyacetic acid in rat milk of dams exposed during lactation and milk analysis of their major components. Food Chem Toxicol 2006; 44:8–16.
6. Faber WD, Roberts LS, Stump DG, et al. Inhalation developmental neurotoxicity study of ethylbenzene in Crl-CD rats. Birth Defects Res B 2007; 80:34–48.
7. Renwick AG. Toxicokinetics. In Principles and Methods of Toxicology, Hayes AW. ed. CRC Press, Boca Raton, FL, 2007:179–230.
8. Krishnan K, Andersen ME. Physiologically based pharmacokinetic and toxicokinetic models. In Principles and Methods of Toxicology, Hayes AW. ed. CRC Press, Boca Raton, FL, 2007:231–292.
9. Farrar HC, Blumer JL. Fetal effects of maternal drug exposure. Ann Rev Pharmacol Toxicol 1991; 31:525–547.
10. Schwartz S. Providing toxicokinetic support for reproductive toxicology studies in pharmaceutical development. Arch Toxicol 2001; 75:381–387.

11. Chiu WA, Barton HA, DeWoskin RS, Schlosser P, et al. Evaluation of physiologically based pharmacokinetic models for use in risk assessment. J Appl Toxicol 2007; 27:218–237.
12. Young JF, Branham WS, Sheehan DM, et al. Physiological "constants" for PBPK models for pregnancy. J Toxicol Environ Health 1997; 52:385–401.
13. Mattison DR, Sandler JD. Summary of the workshop on issues in risk assessment: quantitative methods for developmental toxicology. Risk Anal 1994; 14:595–604.
14. Juchau MR. Enzymatic bioactivation and inactivation of chemical teratogens and transplacental carcinogens/mutagens. In The Biochemical Basis of Chemical Teratogenesis, Juchau MR. ed. Elsevier/North Holland, New York, NY, 1981:63–94.
15. Juchau MR, Faustman-Watts E. Pharmacokinetic considerations in the maternal-placental-fetal unit. Clin Obstet Gynecol 1983; 26:379–390.
16. Neubert D. Significance of pharmacokinetic variables in reproductive and developmental toxicity. Xenobiotica 1988; 18(Suppl 1):45–58.
17. Mattison DR. Fetal pharmacokinetic and physiological models. In Developmental Toxicology: Risk Assessment and the Future, Hood RD. ed. Van Nostrand Reinhold, New York, NY, 1990:110–129.
18. Krauer B, Krauer F, Hytten FE. Drug disposition and pharmacokinetics in the maternal-placental-fetal unit. Pharmacol Ther 1998; 10:301–328.
19. Green TP, O'Dea RF, Mirkin BL. Determinants of drug disposition and effect in the fetus. Ann Rev Pharmacol Toxicol 1979; 19:285–322.
20. Finster M, Pedersen H. Placental transfer and fetal uptake of drugs. British J Anaesth. 1979; 51:25S–28S.
21. Faustman EM, Ribeiro PL. Pharmacokinetic considerations in developmental toxicity. In Developmental Toxicology: Risk Assessment and the Future, Hood RD. ed. Van Nostrand Reinhold, New York, NY, 1990:109–135.
22. Mattison DR, Blann E, Malek A. Physiological alterations during pregnancy: Impact on toxicokinetics. Fundam Appl Toxicol 1991; 16:215–218.
23. Corley RA, Mast TJ, Carney EW, et al. Evaluation of physiologically based models of pregnancy and lactation for their application in children's health risk assessments. Crit Rev Toxicol 2003; 33:137–211.
24. Ziegler EE, O'Donnell AM, Nelson SE, et al. Body composition of the reference fetus. Growth 1976; 40:329–341.
25. Stock M, Metcalfe J. Maternal physiology during gestation. In The Physiology of Reproduction, Knobil E, Neill JD, eds. Raven Press, New York, NY, 1994:947–977.
26. Gray DG. A physiologically based pharmacokinetic model for methyl mercury in the pregnant rat and fetus. Toxicol Appl Pharmacol 1995; 132:91–102.
27. Clewell HJ, Gearhart JM, Gentry PR, et al. Evaluation of the uncertainty in an oral reference dose for methylmercury due to interindividual variability in pharmacokinetics. Risk Anal 1999; 19:547–558.
28. Gentry PR, Covington TR, Clewell HJ III. Evaluation of the potential impact of pharmacokinetic differences on tissue dosimetry in offspring during pregnancy and lactation. Regul Toxicol Pharmacol 2003; 38:1–16.
29. Rudolph AM, Heymann MA. The circulation of the fetus in utero. Methods for studying distribution of blood flow, cardiac output and organ blood flow. Circ Res 1967; 21:163–184.

30. Alcorn J, McNamara PJ. Ontogeny of hepatic and renal systemic clearance pathways in infants: Part I. Clin Pharmacokinet 2002; 41:959–998.
31. Welsch F. Placental transfer and fetal uptake of drugs. J Vet Pharmacol Ther 1982; 5:91–104.
32. Cresteil T, Beaune P, Kremers P, et al. Drug-metabolizing enzymes in human foetal liver: Partial resolution of multiple cytochromes P 450. Pediatr Pharmacol 1982; 2:199–207.
33. Levy G, Khanna NN, Soda DM, et al. Pharmacokinetics of acetaminophen in the human neonate: Formation of acetaminophen glucuronide and sulfate in relation to plasma bilirubin concentration and D-glucaric acid excretion. Pediatrics 1975; 55:818–825.
34. Dutton GJ. Developmental aspects of drug conjugation, with special reference to glucuronidation. Ann Rev Pharmacol Toxicol 1978; 18:17–35.
35. Levy G, Hayton WL. Pharmacokinetic aspects of placental drug transfer. In Fetal Pharmacology, Boréus LO. ed. Raven Press, New York, NY, 1973:29–39.
36. Luecke RH, Wosilait WD, Pearce BA. A physiologically based pharmacokinetic computer model for human pregnancy. Teratology 1994; 49:90–103.
37. Finster M, Mark LC. Placental transfer of drugs and their distribution in fetal tissues. In Handbook of Experimental Pharmacology, Brodie BB. Gilette JR. eds. Springer-Verlag, New York, NY, 1971:276–285.
38. Krishnan K, Andersen ME. Physiologically based pharmacokinetic models in the risk assessment of developmental neurotoxicants. In Handbook of Developmental Neurotoxicology, Slikker W, Chang LW, eds. Academic Press, San Diego, CA, 1998:709–725.
39. Luecke RH, Wosilait WD, Young JF. Mathematical analysis for teratogenic sensitivity. Teratology 1997; 55:373–380.
40. Luecke RH, Wosilait WD, Pearce BA, Young JF. A computer model and program for xenobiotic disposition during pregnancy. Comput. Methods Programs Biomed 1997; 53:201–224.
41. Luecke RH, Pearce BA, Wosilait WD, et al. Postnatal growth considerations for PBPK modeling. J Toxicol Environ Health A 2007; 70:1027–1037.
42. Luecke RH, Wosilait WD, Young JF. Mathematical representation of organ growth in the human embryo/fetus. Int J Biomed Comput 1995; 39:337–347.
43. Kawamoto Y, Matsuyama W, Wada M, et al. Development of a physiologically based pharmacokinetic model for bisphenol A in pregnant mice. Toxicol Appl Pharmacol 2007; 224:182–191.
44. You L, Gazi E, Rchibeque-Engle S, et al. Transplacental and lactational transfer of p,p'-DDE in Sprague-Dawley rats. Toxicol Appl Pharmacol 1999; 157:134–144.
45. O'Flaherty EJ, Scott W, Schreiner C, et al. A physiologically based kinetic model of rat and mouse gestation: disposition of a weak acid. Toxicol Appl Pharmacol 1992; 112:245–256.
46. Gargas ML, Tyler TR, Sweeney LM, et al. A toxicokinetic study of inhaled ethylene glycol ethyl ether acetate and validation of a physiologically based pharmacokinetic model for rat and human. Toxicol Appl Pharmacol 2000; 165:63–73.
47. Welsch F. The mechanism of ethylene glycol ether reproductive and developmental toxicity and evidence for adverse effects in humans. Toxicol Lett 2005; 156:13–28.

48. Clewell RA, Merrill EA, Robinson PJ. The use of physiologically based models to integrate diverse data sets and reduce uncertainty in the prediction of perchlorate and iodide kinetics across life stages and species. Toxicol Ind Health 2001; 17:210–222.

49. Clewell RA, Merrill EA, Yu KO, et al. Predicting fetal perchlorate dose and inhibition of iodide kinetics during gestation: A physiologically-based pharmacokinetic analysis of perchlorate and iodide kinetics in the rat. Toxicol Sci 2003; 73:235–255.

50. Gentry PR, Covington TR, Andersen ME, et al. Application of a physiologically based pharmacokinetic model for isopropanol in the derivation of a reference dose and reference concentration. Regul Toxicol Pharmacol 2002; 36:51–68.

51. Gabrielsson JL, Johansson P, Bondesson U, et al. Analysis of methadone disposition in the pregnant rat by means of a physiological flow model. J Pharmacokinet Biopharm 1985; 13:355–372.

52. Gabrielsson JL, Groth T. An extended physiological pharmacokinetic model of methadone disposition in the rat: Validation and sensitivity analysis. J Pharmacokinet Biopharm 1988; 16:183–201.

53. Ward KW, Blumenthal GM, Welsch F, et al. Development of a physiologically based pharmacokinetic model to describe the disposition of methanol in pregnant rats and mice. Toxicol Appl Pharmacol 1997; 145:311–322.

54. Clarke DO, Elswick BA, Welsch F, et al. Pharmacokinetics of 2-methoxyethanol and 2-methoxyacetic acid in the pregnant mouse: A physiologically based mathematical model. Toxicol Appl Pharmacol 1993; 121:239–252.

55. Terry KK, Elswick BA, Welsch F, et al. Development of a physiologically based pharmacokinetic model describing 2-methoxyacetic acid disposition in the pregnant mouse. Toxicol Appl Pharmacol 1995; 132:103–114.

56. Welsch F, Blumenthal GM, Conolly RB. Physiologically based pharmacokinetic models applicable to organogenesis: Extrapolation between species and potential use in prenatal toxicity risk assessments. Toxicol Lett 1995; 82–83:539–547.

57. O'Flaherty EJ, Nau H, McCandless D, et al. Physiologically based pharmacokinetics of methoxyacetic acid: Dose-effect considerations in C57 BL/6 mice. Teratology 1995; 52:78–89.

58. Hays SM, Elswick BA, Blumenthal GM, et al. Development of a physiologically based pharmacokinetic model of 2-methoxyethanol and 2-methoxyacetic acid disposition in pregnant rats. Toxicol Appl Pharmacol 2000; 163:67–74.

59. Gargas ML, Tyler TR, Sweeney LM, et al. A toxicokinetic study of inhaled ethylene glycol monomethyl ether (2-ME) and validation of a physiologically based pharmacokinetic model for the pregnant rat and human. Toxicol Appl Pharmacol 2000; 165:53–62.

60. Faustman EM, Lewandowski TA, Ponce RA, et al. Biologically based dose-response models for developmental toxicants: Lessons from methylmercury. Inhal Toxicol 1999; 11:559–572.

61. Gabrielsson JL, Paalzow LK. A physiological pharmacokinetic model for morphine disposition in the pregnant rat. J Pharmacokinet Biopharm 1983; 11:147–163.

62. Gabrielsson JL, Johansson P, Bondesson U, et al. Analysis of pethidine disposition in the pregnant rat by means of a physiological flow model. J Pharmacokinet Biopharm 1986; 14:381–395.

63. Kawahara M, Nanbo T, Tsuji A. Physiologically based pharmacokinetic prediction of p-phenylbenzoic acid disposition in the pregnant rat. Biopharm Drug Dispos 1998; 19:445–453.

64. Clewell HJ III, Andersen ME, Wills RJ, et al. A physiologically based pharmacokinetic model for retinoic acid and its metabolites. J Am Acad Dermatol 1997; 36:S77–S85.

65. Emond C, Birnbaum LS, DeVito MJ. Physiologically based pharmacokinetic model for developmental exposures to TCDD in the rat. Toxicol Sci 2004; 80:115–133.

66. Olanoff LS, Anderson JM. Controlled release of tetracycline–III: A physiological pharmacokinetic model of the pregnant rat. J Pharmacokinet Biopharm 1980; 8:599–620.

67. Gabrielsson JL, Paalzow LK, Nordstrom L. A physiologically based pharmacokinetic model for theophylline disposition in the pregnant and nonpregnant rat. J Pharmacokinet Biopharm 1984; 12:149–165.

68. Fisher JW, Whittaker TA, Taylor DH, et al. Physiologically based pharmacokinetic modeling of the pregnant rat: a multiroute exposure model for trichloroethylene and its metabolite, trichloroacetic acid. Toxicol Appl Pharmacol 1989; 99:395–414.

69. Clewell RA, Merrill EA, Gearhart JM, et al. Perchlorate and radioiodide kinetics across life stages in the human: Using PBPK models to predict dosimetry and thyroid inhibition and sensitive subpopulations based on developmental stage. J Toxicol Environ Health A 2007; 70:408–428.

70. Maruyama W, Yoshida K, Tanaka T, et al. Simulation of dioxin accumulation in human tissues and analysis of reproductive risk. Chemosphere 2003; 53:301–313.

71. Kim CS, Binienda Z, Sandberg JA. Construction of a physiologically based pharmacokinetic model for 2,4-dichlorophenoxyacetic acid dosimetry in the developing rabbit brain. Toxicol Appl Pharmacol 1996; 136:250–259.

72. Ginsberg G, Hattis D, Miller R, et al. Pediatric pharmacokinetic data: Implications for environmental risk assessment for children. Pediatrics 2004; 113:973–983.

73. Haddad S, Restieri C, Krishnan K. Characterization of age-related changes in body weight and organ weights from birth to adolescence in humans. J Toxicol Environ Health A 2001; 64:453–464.

74. Pelekis M, Gephart LA, Lerman SE. Physiological-model-based derivation of the adult and child pharmacokinetic intraspecies uncertainty factors for volatile organic compounds. Regul Toxicol Pharmacol 2001; 33:12–20.

75. Price K, Haddad S, Krishnan K. Physiological modeling of age-specific changes in the pharmacokinetics of organic chemicals in children. J Toxicol Environ Health A 2003; 66:417–433.

76. Sarangapani R, Gentry PR, Covington TR, et al. Evaluation of the potential impact of age- and gender-specific lung morphology and ventilation rate on the dosimetry of vapors. Inhal Toxicol 2003; 15:987–1016.

77. Nong A, McCarver DG, Hines RN, et al. Modeling interchild differences in pharmacokinetics on the basis of subject-specific data on physiology and hepatic CYP2E1 levels: a case study with toluene. Toxicol Appl Pharmacol 2006; 214:78–87.

78. Clewell HJ, Teeguarden J, McDonald T, et al. Review and evaluation of the potential impact of age- and gender-specific pharmacokinetic differences on tissue dosimetry. Crit Rev Toxicol 2002; 32:329–389.

79. Haddad S, Tardif GC, Tardif R. Development of physiologically based toxicokinetic models for improving the human indoor exposure assessment to water contaminants:

Trichloroethylene and trihalomethanes. J Toxicol Environ Health A 2006; 69:2095–2136.

80. U.S. Environmental Protection Agency. Exposure Factors Handbook. U.S. Environmental Protection Agency, Office of Research and Development National Center for Environmental Assessment Washington Office, Washington, DC, 1997.

81. National Research Council (U.S.). Special characteristics of children. In Pesticides in the Diets of Infants and Children, National Research Council (U.S.) and Committee on Pesticides in the Diets of Infants and Children, eds. National Academy Press, Washington, DC, 1993:23–48.

82. Jacqz-Aigrain E. Pharmacologie du développement. In Médecine et biologie du développement, du gène au nouveau-né, Saliba E, Hamamah S, Gold F, Benhamed M, eds. Masson, Paris, 2001:399.

83. Reed KJ, Jimenez M, Freeman NC, et al. Quantification of children's hand and mouthing activities through a videotaping methodology. J Expo Anal Environ Epidemiol 1999; 9:513–520.

84. Miller MD, Marty MA, Arcus A, et al. Differences between children and adults: Implications for risk assessment at California EPA. Int J Toxicol 2002; 21:403–418.

85. Wilson JT, Brown RD, Cherek DR, et al. Drug excretion in human breast milk: principles, pharmacokinetics and projected consequences. Clin Pharmacokinet 1980; 5:1–66.

86. Jensen AA. Transfert of chemical contaminants into human milk. In Chemical contaminants in human milk, Jensen AA, Slorach SA, eds. CRC Press, Boca Raton, FL, 1991:9–19.

87. Fisher J, Mahle D, Bankston L, et al. Lactational transfer of volatile chemicals in breast milk. Am Ind Hyg Assoc J 1997; 58:425–431.

88. Lee SK, Ou YC, Andersen ME, et al. A physiologically based pharmacokinetic model for lactational transfer of PCB 153 with or without PCB 126 in mice. Arch Toxicol 2007; 81:101–111.

89. Mahle DA, Gearhart JM, Grigsby CC, et al. Age-dependent partition coefficients for a mixture of volatile organic solvents in Sprague-Dawley rats and humans. J Toxicol Environ Health A 2007; 70:1745–1751.

90. Friis-Hansen B. Body composition during growth. In vivo measurements and biochemical data correlated to differential anatomical growth. Pediatrics 1971; 47(Suppl 2):264–274.

91. Dickerson JW, Widdowson EM. Chemical changes in skeletal muscle during development. Biochem J 1960; 74:247–257.

92. Alcorn J, McNamara PJ. Ontogeny of hepatic and renal systemic clearance pathways in infants: Part II. Clin Pharmacokinet 2002; 41:1077–1094.

93. Iliff A, Lee VA. Pulse rate, respiratory rate, and body temperature of children between two months and eighteen years of age. Child Dev 1952; 23:237–245.

94. Shock NW. Basal blood pressure and pulse rate in adolescents. Am J Dis Child 1944; 68:16–22.

95. Cayler GG, Rudolph AM, Nadas AS. Systemic blood flow in infants and children with and without heart disease. Pediatrics 1963; 32:186–201.

96. Sholler GF, Celermajer JM, Whight CM, et al. Echo Doppler assessment of cardiac output and its relation to growth in normal infants. Am J Cardiol 1987; 60:1112–1116.

97. Szantay V, Tamas S, Marian L, et al. Changes of hepatic blood flow in children as a function of age. Rev Roum Med Intern 1974; 11:91–93.
98. Arms AD, Travis CC. Reference physiological parameters in pharmacokinetic modeling. U.S.Environmental Protection Agency. EPA/600/6–88/004, Washington DC, 1988.
99. Hook JB, Baillie MD. Perinatal renal pharmacology. Ann Rev Pharmacol Toxicol 1979; 19:491–509.
100. Lindbjerg IF. Leg muscle blood-flow measured with 133-xenon after ischaemia periods and after muscular exercise performed during ischaemia. Clin Sci 1966; 30:399–408.
101. Amery A, Bossaert H, Verstraete M. Muscle blood flow in normal and hypertensive subjects. Influence of age, exercise, and body position. Am Heart J 1969; 78:211–216.
102. Goetzova J, Skovranek J, Samanek M. Muscle blood flow in children, measured by 133Xe clearance method. Cor Vasa 1977; 19:161–164.
103. Chiron C, Raynaud C, Maziere B, et al. Changes in regional cerebral blood flow during brain maturation in children and adolescents. J Nucl Med 1992; 33:696–703.
104. Hakkola J, Tanaka E, Pelkonen O. Developmental expression of cytochrome P450 enzymes in human liver. Pharmacol Toxicol 1998; 82:209–217.
105. Johnsrud EK, Koukouritaki SB, Divakaran K, et al. Human hepatic CYP2E1 expression during development. J Pharmacol Exp Ther 2003; 307:402–407.
106. WHO (World Health Organization). Principles for evaluating health risks in children associated with exposure to chemicals. Environmental Health Criteria 237. World Health Organization Press, Geneva, Switzerland, 2006.
107. Cresteil T. Onset of xenobiotic metabolism in children: Toxicological implications. Food Addit Contam 1998; 15(Suppl):45–51.
108. Strange RC, Howie AF, Hume R, et al. The development expression of alpha-, mu- and pi-class glutathione S-transferases in human liver. Biochim Biophys Acta 1989; 993:186–190.
109. Renwick AG. Toxicokinetics in infants and children in relation to the ADI and TDI. Food Addit Contam 1998; 15(Suppl):17–35.
110. Kearns GL, Abdel-Rahman SM, Alander SW, et al. Developmental pharmacology—drug disposition, action, and therapy in infants and children. New Engl J Med 2003; 349:1157–1167.
111. Atkinson HC, Begg EJ. Prediction of drug distribution into human milk from physicochemical characteristics. Clin Pharmacokinet 1990; 18:151–167.
112. Byczkowski JZ, Kinkead ER, Leahy HF, et al. Computer simulation of the lactational transfer of tetrachloroethylene in rats using a physiologically based model. Toxicol Appl Pharmacol 1994; 125:228–236.
113. Clewell RA, Merrill EA, Yu KO, et al. Predicting neonatal perchlorate dose and inhibition of iodide uptake in the rat during lactation using physiologically-based pharmacokinetic modeling. Toxicol Sci 2003; 74:416–436.
114. Rodriguez CE, Mahle DA, Gearhart JM, et al. Predicting age-appropriate pharmacokinetics of six volatile organic compounds in the rat utilizing physiologically based pharmacokinetic modeling. Toxicol Sci 2007; 98:43–56.
115. Mirfazaelian A, Fisher JW. Organ growth functions in maturing male Sprague-Dawley rats based on a collective database. J Toxicol Environ Health A 2007; 70:1052–1063.

116.   Timchalk C, Kousba AA, Poet TS. An age-dependent physiologically based pharma-
       codynamic model for the organophosphorus insecticide chlorpyrifos in the prewean-
       ling rat. Toxicol Sci 2007; 98:348–365.
117.   O'Flaherty EJ. Physiologic changes during growth and development. Environ Health
       Perspect 1994; 102(Suppl 11):103–106.
118.   Byczkowski JZ, Fisher JW. Lactational transfer of tetrachloroethylene in rats. Risk
       Anal 1944; 14:339–349.
119.   Hallen IP, Breitholtz-Emanuelsson A, Hult K, et al. Placental and lactational transfer
       of ochratoxin A in rats. Nat Toxins 1998; 6:43–49.
120.   Hong EJ, Choi KC, Jung YW, et al. Transfer of maternally injected endocrine
       disruptors through breast milk during lactation induces neonatal Calbindin-D9k in
       the rat model. Reprod Toxicol 2004; 18:661–668.
121.   Hinderliter PM, Mylchreest E, Gannon SA, et al. Perfluorooctanoate: Placental and
       lactational transport pharmacokinetics in rats. Toxicology 2005; 211:139–148.
122.   Marty MS, Domoradzki JY, Hansen SC, et al. The effect of route, vehicle, and divided
       doses on the pharmacokinetics of chlorpyrifos and its metabolite trichloropyridinol
       in neonatal Sprague-Dawley rats. Toxicol Sci 2007; 100:360–373.
123.   Chu I, Bowers WJ, Caldwell D, et al. Toxicological effects of in utero and lactational
       exposure of rats to a mixture of environmental contaminants detected in Canadian
       Arctic human populations. J Toxicol Environ Health A 2008; 71:93–108.
124.   Shelley ML, Andersen ME, Fisher JW. A risk assessment approach for nursing
       infants exposed to volatile organics through mother's occupational exposure. Appl
       Indus Hygiene 1989; 4:21–26.
125.   Yoon M, Barton HA. Predicting maternal rat and pup exposures: How different are
       they? Toxicol Sci 2008; 102:15–32.
126.   WHO/IPCS (World Health Organization/International Panel on Chemical safety).
       Chemical-specific adjustment factors for interspecies differences and human vari-
       ability: Guidance for use of data in dose/concentration-response assessment. World
       Health Organization Press, Geneva, Switzerland, 2005.
127.   Young JF. Correlation of pharmacokinetic data with endpoints of developmental
       toxicity. Fundam Appl Toxicol 1991; 16;222–224.
128.   Verwei M, van Burgsteden JA, Krul CA, et al. Prediction of in vivo embryotoxic
       effect levels with a combination of in vitro studies and PBPK modelling. Toxicol
       Lett 2006; 165:79–87.
129.   Barton HA. Computational pharmacokinetics during developmental windows of
       susceptibility. J Toxicol Environ Health A 2005; 68:889–900.
130.   Meek B, Renwick A. Guidance for the development of chemical-specific adjustment
       factors: Integration with mode of action framework. In Toxicokinetics and Risk
       Assessment, Lipscomb JC. Ohanian EV. eds. Informa Health Care, New York, NY,
       2007:27–46.
131.   Clewell HJ, Gentry PR, Covington TR, et al. Evaluation of the potential impact of
       age- and gender-specific pharmacokinetic differences on tissue dosimetry. Toxicol
       Sci 2004; 79:381–393.
132.   Walker K, Hattis D, Russ A, et al. Approaches to acrylamide physiologically based
       toxicokinetic modeling for exploring child-adult dosimetry differences. J Toxicol
       Environ Health A 2007; 70:2033–2055.
133.   Hattis D, Ginsberg G, Sonawane B, et al. Differences in pharmacokinetics between
       children and adults–II. Children's variability in drug elimination half-lives and in

some parameters needed for physiologically-based pharmacokinetic modeling. Risk Anal 2003; 23:117–142.

134. Renwick AG, Dorne JL, Walton K. An analysis of the need for an additional uncertainty factor for infants and children. Regul Toxicol Pharmacol 2000; 31:286–296.
135. Ginsberg GL, Foos BP, Firestone MP. Review and analysis of inhalation dosimetry methods for application to children's risk assessment. J Toxicol Environ Health A 2005; 68:573–615.

10

Integration of Whole Animal Developmental Toxicity Data into Risk Assessment

Mark E. Hurtt and Gregg D. Cappon
*Pfizer Global Research and Development,
Groton, Connecticut, U.S.A.*

INTRODUCTION

Risk assessment regardless of the underlying application has evolved as a process and, in fact, it has evolved around a very specific framework (1). Simply put the process involves gathering information, weighing its value, deciding its impact, and then making a decision as to the overall potential risk. This process is very much like the major life decisions we all make. For example, think about the process you go through in selecting a job. Just replace the risk of some unwanted event with the risk of not "liking" or "being happy" with the job. It begins by using the "data collection system" of the yellow legal pad with two columns: pro and con. The list of all the factors scrolls down the left-hand side of the page and comments captured under either or both headers. This method served many well over the years as a means to capture all the important factors that would contribute to the final decision. Those factors included salary, geographical location, benefits, specifics of the role, potential growth and development, reputation of the company, the people you would be working with, your immediate supervisor, and the specific working environment, to name those on the first page. All this information would be collected and then weighted as to its importance. Of course, each factor's importance differs for every individual. For some, salary is an important factor that may receive the top spot, whereas location may be more important for others.

However, I am sure my scientific colleagues are only motivated by the pursuit of knowledge. The process then rests with utilizing some judgment based on previous knowledge and experience and/or input from others. The integration of all this data with the "conditions of satisfaction" then allows an outcome: is the job a good fit? This same thought process applies to many other major decisions we make such as which college to attend (perhaps this decision has more parental influence than the others), buying a house, investing our money, and so on. These applications illustrate a focused and deliberate way of thinking about the question and deriving the best answer. This is very much the risk assessment approach in its simplest form.

This chapter will describe the approach for risk assessment specifically for developmental toxicity. It will focus on the integration of whole animal developmental toxicity data into the risk assessment process. The approach described will come from that employed for a pharmaceutical agent, but will briefly mention approaches used for other types of agents as well as provide appropriate references for more detailed information.

SELECTED TERMINOLOGY

Risk

A concept used to give meaning to things that pose danger to people or what they value. Descriptions of risk are generally stated in terms of a negative, i.e., the likelihood of harm or loss from a hazard.The risk description usually includes identification of what may be harmed or lost, the hazard that may lead to the harm or loss, and a judgment about the likelihood harm or loss will occur.

Risk Analysis

Risk analysis is the application of methods of analysis to matters of risk. Its aim is to increase understanding of the substantive qualities, seriousness, likelihood, and conditions of a hazard or risk and of the options for managing it.

Risk Assessment

The term is used in this chapter to mean the characterization of the potential adverse health effects of human exposures to pharmaceutical agents.

Risk Management

The comprehensive applications of scientifically based methodologies to identify, assess, communicate, and minimize risk throughout the life cycle of a drug so as to maintain a favorable benefit–risk balance in patients.

Developmental Toxicology

Developmental toxicology is defined as the study of adverse effects on the developing organism that may result from exposure prior to conception, during prenatal development, or postnatal to the time of sexual maturity.

Adverse Effect

Any biochemical, physiological, anatomical, pathological, and/or behavioral change that results in functional impairment that may affect the performance of the whole organism or reduce the ability of the organism to respond to additional challenge.

Therapeutic Index

The therapeutic index (TI) (also known as therapeutic ratio or margin of safety) is a comparison of the amount of a therapeutic agent that causes the therapeutic effect to the amount that causes toxic effects.

Risk Communication

An interactive process of sharing knowledge and understanding so as to arrive at well-informed risk management decisions. The goal is a better understanding by experts and nonexperts alike of the actual and perceived risks, the possible solutions, and the related issues and concerns.

FRAMEWORK FOR DEVELOPMENTAL TOXICITY RISK ASSESSMENT

Introduction

The overall approach for risk assessment has generally grown out of the National Research Council (NRC) risk assessment paradigm published in 1983 (1). This publication focused on assessing potential human health risks from exposure to environmental agents in a much more formalized manner than had been done earlier.

The U.S. Environmental Protection Agency (EPA) has used the NRC paradigm as a basis for publication of numerous guidance documents. For developmental toxicity risk assessment, the Guidelines for the Health Assessment of Suspect Developmental Toxicants were published in 1986 (2) and followed in 1991 with an updated version called Guidelines for Developmental Toxicity Risk Assessment (3).The updated document added guidance on the relationship between maternal and developmental toxicity, characterization of the health-related database for developmental toxicity risk assessment, use of the reference dose or reference concentration for developmental toxicity, and the use of the benchmark dose approach. Following the 1991 publication, a number of other risk assessment guidelines were published by the EPA. These include in chronological order, Guidelines for Exposure Assessment (1992) (4), Guidelines for Reproductive Toxicity Risk Assessment (1996) (5), Guidelines for Neurotoxicity Risk Assessment (1998) (6), Guidelines for Ecological Risk Assessment (1998) (7), and Guidelines for Carcinogen Risk Assessment (2005) (8).

The entire premise behind the drive to develop a scientific approach to risk assessment was the need for a consistent approach. A 1994 NRC publication pointed out the need for a risk assessment approach that is more consistent, less fragmented, and more holistic (9). With the exception of an incomplete data set, scientific judgment would then hopefully become the lone significant variable in determining the results of the risk assessment. A 2000 NRC publication (10) concluded that the major advances in developmental biology and genomics can be used to improve qualitative as well as quantitative risk assessments by "integrating toxicological and mechanistic data on a variety of test animals with data on human variability in genes encoding components of developmental processes, genes encoding enzymes involved in metabolism of chemicals, and their metabolites in and out of the cell." The explosion of data collected at the cellular and molecular level will clearly present significant challenges to the application of scientific judgment in the process as our understanding of the end points and relationship to the underlying biologic changes continue to evolve. This scientific expansion will also drive the need for refinements and/or changes in the developmental toxicity risk assessment process.

In contrast to the structured approach utilized by the EPA, the U.S. Food and Drug Administration (FDA) has issued numerous guidance documents for specific study conduct needed to assess potential developmental toxicity; however, the FDA has not finalized a document dealing with the risk assessment for developmental toxicity. The FDA did issue, on November 13, 2001, a draft guideline titled, Integration of Study Results to Assess Concerns about Human Reproduction and Developmental Toxicities (11). The FDA document reflected current Agency thinking at the time but was never finalized and it is still listed as a draft (www.fda.gov/cder/guidance/index.htm). This draft guidance describes a process for estimating the increase in risk for human developmental and reproductive risks as a result of drug exposure when definitive human data are not available. The overall approach integrates nonclinical information from a number of sources (i.e., developmental toxicology, general toxicology, pharmacokinetic and toxicokinetic information) and available clinical data to evaluate a drug's potential to increase the risk of adverse reproductive or developmental outcome in humans. An excellent overview of the draft guidance and its application was recently published (12). Without a formalized guidance, developmental toxicity risk assessments for pharmaceuticals still also follow the NRC paradigm, taking into context the greater scope of information that is typically available for pharmaceuticals as compared to environmental agents in regards to mode of activity, pharmacokinetics in humans and nonclinical models, and extrapolation of nonclinical data to humans.

The four components of the risk assessment process, as originally defined by the NRC publication, are hazard identification, dose–response assessment,

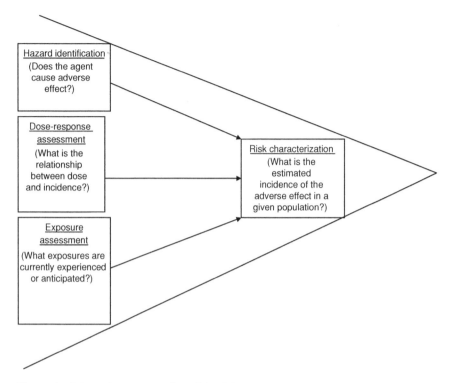

Figure 1 Schematic representation of the risk assessment process.

exposure assessment, and risk characterization (1). The elements of the risk assessment process are schematically represented in Figure 1. While the original paradigm was primarily developed for cancer risk, its foundation has been applied to many noncancer effects. Hazard identification is the process of determining whether exposure to a drug can cause an adverse effect. Over the years, a combined approach to hazard identification and dose–response assessment has evolved (13). This has occurred mainly for three reasons. First is the practical aspect that they are usually completed together and are in many circumstances intertwined and not easily separable. In addition, hazard identification should always be framed in the context of dose. Of course, dose, the specific amount of the drug given, also is described by its route and timing of exposure. Lastly, a dose–response relationship is a hallmark for identification of a drug as causing developmental toxicity (14). Exposure assessment is the process of describing the concentration of the drug that humans are exposed to. The final step in the risk assessment process is risk characterization, which involves the integration of the three earlier components to estimate the overall magnitude of the potential human health risk. A guide for addressing the risk characterization for developmental effects of chemicals has been published earlier (15).

HAZARD IDENTIFICATION

Hazard identification is the most basic part of any risk assessment. Simply put, hazard identification is the process of determining whether exposure to a drug can cause an increase in the potential for developmental toxicity. Ideally, human data as well as animal data should be used for hazard identification. However, since human data are usually not available or are of inadequate quality for assessment, the database available to evaluate developmental toxicity potential is most often totally derived from nonclinical animal studies. Therefore, a default assumption is that if an agent produces an adverse developmental effect in a nonclinical animal study, then it potentially poses a hazard to humans following sufficient exposure during development.

Developmental toxicity is defined as adverse effects on the developing organism and may be manifested as mortality, dysmorphogenesis (structural alterations), alterations to growth, and functional toxicities (3,16). Mortality due to developmental toxicity may occur at any time from early conception to postweaning. Dysmorphogenesis (structural alterations) is generally noted as fetal malformations or variations of the skeleton or soft tissues. Alterations to growth are generally identified by growth retardation, most often indicated by reduced fetal body weight. Functional toxicities include any persistent alteration of normal physiologic or biochemical function, but typically only central nervous system and reproductive function are evaluated following developmental exposure. Other functional toxicities (e.g., toxicity to immune system function) are included for evaluation in specific cases based on information such as mode of action or effects noted in earlier nonclinical studies.

For the risk identification process, it is assumed that all of the four manifestations of developmental toxicity are of concern, as an adverse change in any of the four manifestations is indicative for the potential of a compound to disrupt normal development in humans. However, even though this concept is espoused by developmental toxicologists and regulators, the real-world view is often quite different. This difference was noted in a workshop sponsored by the Teratology Society, where it was shown that clinicians, the primary disseminators of drug safety information to patients, view animal studies very differently than do regulators or developmental toxicologists (17). Clinicians tended to regard specific effects in animal studies as indicative of the same specific effects in humans, whereas developmental toxicologists and regulators regarded adverse outcomes in animal studies as broader signals indicating a general potential to disrupt some developmental processes in humans. The view expressed by clinicians can often result in a tendency to consider only malformations, or malformations and death, as end points of primary concern in nonclinical studies.

The studies most often performed to assess potential developmental toxicity are described in International Conference on Harmonisation Guideline for Detection of Toxicity to Reproduction for Medicinal Products sections 4.1.3 "Study for effects on embryo-fetal development" and 4.1.2. "Study for effects on pre- and postnatal development, including maternal function" (18). Detailed reviews of these study designs have been presented earlier (19,20). Slightly different in design, but conceptually very similar, guidelines for developmental toxicity studies with environmental agents are routinely followed for nonpharmaceutical agents (21,22).

Most new pharmaceutical agents have data available from International Conference on Harmonisation Guideline compliant studies, but little or no reliable human data. These animal studies usually provide sufficient evidence for the scientific determination of the potential for an agent to produce developmental toxicity. However, many environmental agents have incomplete data sets; therefore, for these agents, it is necessary to evaluate the sufficiency of a database for predicting potential developmental toxicity, and if the database is incomplete, make necessary revisions to the risk assessment. A scheme for evaluating the sufficiency of the database for developmental toxicity was prepared by the EPA (Table 1). For some chemicals with insufficient data, the evaluation of structure activity relationships can be used to gain some understanding of the potential hazard for developmental toxicity, and the reliability of structure activity relationships is likely to increase with improved information technology platforms (23).

In addition to having well-conducted laboratory animal studies in two species (most often rat and rabbit) which evaluated end points indicative of the potential manifestations of developmental toxicity, there is often other information that can provide context regarding the relevance of the studies for hazard identification, particularly for pharmaceutical agents. These can range from simple study design features such as consistency in the route of exposure for the nonclinical study and human dosage route to more complex factors such as similarity of metabolic and drug distribution profiles between humans and the nonclinical models or even expression of the target in developing animals and humans (24).

DOSE–RESPONSE ASSESSMENT

Dose–response evaluation involves examining data from nonclinical studies and any available human data on developmental effects with the associated doses, pharmacokinetics, routes, and timing and duration of dose administration. While presented separately, in effect a dose–response relationship assessment is done as part of the hazard identification as the scientific judgment on the relationship between treatment and an adverse developmental outcome is often based on the presence or absence of a dose–response relationship (25).

Dose–response assessment is one area where different approaches are needed for pharmaceuticals and environmental agents. Due to the general lack of pharmacokinetic data, extrapolation from high dose to low dose and extrapolation from animals to humans is required for environmental exposures. The dose–response assessment for environmental agents often makes use of detailed methods to predict effect and no effect levels and to calculate reference dose (15,26). In contrast, maternal exposure data is typically collected in

Table 1 Categorization of the Health-related Database

Sufficient evidence
The Sufficient Evidence category includes data that collectively provide enough
    information to judge whether a reproductive hazard exists within the context of
    effect as well as dose, duration, timing, and route of exposure. This category may
    include both human and experimental animal evidence.

Sufficient of human evidence
This category includes data from epidemiologic studies (e.g., case control and cohort)
    that provide convincing evidence for the scientific community to judge that a causal
    relationship is or is not supported. A case series in conjunction with strong
    supporting evidence may also be used. Supporting animal data may or may not be
    available.

Sufficient experimental animal evidence—limited human data
This category includes data from experimental animal studies and/or limited human
    data that provide convincing evidence for the scientific community to judge if the
    potential for developmental toxicity exists. The minimum evidence necessary to
    judge that a potential hazard exists generally would be data demonstrating an
    adverse developmental effect in a single, appropriate, well-conducted study in a
    single experimental animal species. The minimum evidence needed to judge that a
    potential hazard does not exist would include data from appropriate, well-conducted
    laboratory animal studies in several species (at least two), which evaluated a variety
    of the potential manifestations of developmental toxicity and showed no
    developmental effects at doses that were minimally toxic to the adult.

Insufficient evidence
This category includes situations for which there is less than the minimum sufficient
    evidence necessary for assessing the potential for developmental toxicity, such as
    when no data are available on developmental toxicity; as well as for databases from
    studies in animals or humans that have a limited study design (e.g., small numbers,
    inappropriate dose selection/exposure information, other uncontrolled factors); or
    data from a single species reported to have no adverse developmental effects; or
    databases limited to information on structure/activity relationships, short-term or in
    vitro tests, pharmacokinetics, or metabolic precursors.

Source: From Ref. 3.

nonclinical studies for pharmaceutical agents, thereby alleviating the need for
predictive models.

There are several important considerations for developmental toxicity that
must be factored in when determining a dose–response relationship. First, it is
generally assumed that there is a threshold for the dose response curve below
which a developmental toxicant does not produce adverse effects (27). However, in developmental toxicity studies dose–response relationships can be complicated by the multiple manifestations of developmental toxicity. For example,
an increase in embryo-fetal mortality with increasing exposure can result in an
apparent dose-responsive decrease in malformations with increasing dose (28,29).

Second, because some stages of embryonic development are more vulnerable than others, the developmental stage at which a conceptus is exposed to a developmental toxicant determines both sensitivity to damage and the type of defect (30). Even though regulatory studies typically provide for exposure throughout organogenesis, exposure during a critical time in development can produce an adverse developmental effect, increasing the need for a thorough understanding of the dose–response relationship (31).

While the main value of dose–response relationship is hazard identification and for determination of the no observed adverse effect level (NOAEL) and/or lowest adverse effect level (LOAEL), a thorough understanding of the route, timing, and duration of exposure, species-specific factors, and pharmacokinetics in the test species and humans is necessary for a full assessment of the potential to cause developmental toxicity in humans.

EXPOSURE ASSESSMENT

Exposure assessment is the process of describing the concentration of the agent to which humans are exposed or potentially exposed. The exposure assessment provides an estimate of human exposure levels for particular populations, and must take into account all potential sources of exposure. For pharmaceuticals, the human exposure is obtained from direct measurement under the conditions of intended clinical use. However, for environmental agents, exposure data in humans is either incomplete or unavailable and generally needs to be estimated.

For pharmaceutical agents, the exposure at the maximum recommended human dose is well characterized in the clinic. Also, for most new pharmaceutical agents, a considerable amount of pharmacokinetic data in pregnant animals is available, allowing for an understating of the exposure at the NOAEL and LOAEL. Therefore, by simply dividing the NOAEL exposure by the exposure at the maximum recommended human dose, it is possible to calculate a TI for developmental toxicity.

For environmental agents, exposure assessment describes the composition and size, and presents the types, magnitudes, frequencies, and durations of exposure to an agent. Guidelines for estimating exposures have been developed and published by the EPA (4). Exposure information for environmental agents is usually developed from monitoring data and from estimates based on various scenarios of environmental exposures. This exposure information can then be used to develop a human estimated exposure dose, which with the animal NOAEL can be used to calculate a margin of exposure which is also sometimes referred to as a margin of safety (32). Because of windows of sensitivity to adverse developmental effects during gestation or postnatal development, in addition to simply defining the amount and duration of exposure, in certain cases it may also be necessary to determine exposure during different periods of pregnancy, lactation, or direct exposure to children during postnatal development.

RISK CHARACTERIZATION

Risk characterization is a synthesis and summary of information about a hazard that addresses the needs and interests of the pharmaceutical company, the regulatory agencies, and the physicians and their patients. A 1996 report by the Committee on Risk Characterization of the National Academy of Sciences' National Research Council described risk characterization as a prelude to decision making and that it depends on an iterative, analytic-deliberative process (33). The report established seven principles for risk characterization. The first is that risk characterization should be a decision-driven activity, directed toward informing choices and solving problems. The third principle is that risk characterization is the outcome of an analytic-deliberative process. The report concludes that deliberation frames analysis, analysis informs deliberation, and the feedback between the two benefits the overall process. The report further identified five objectives in the structure of an analytic-deliberative process: getting the science right; getting the right science; getting the right participation; getting the participation right; and developing an accurate, balanced, and informative synthesis.

The risk characterization begins by summarizing the results of the hazard characterization. Does the drug produce a developmental hazard in animals? Does the drug produce a developmental hazard in humans? In addressing these questions, one must consider the confidence in the conclusions, whether the data support alternative conclusions, the data gaps (characterization of the database), and acknowledgement of any major assumptions. The major assumptions generally made in the risk assessment process for developmental toxicity are outlined in Table 2; some of which have already been mentioned. The dose–response assessment, while generally part of the hazard characterization, addresses a hallmark for identifying a drug as causing developmental toxicity (14). For pharmaceuticals, the exposure assessment is generally the easiest part of the overall assessment since exposure is directly measured in humans. Exposure data is generally

Table 2 Assumptions Underlying Developmental Toxicity Risk Assessment

An agent that produces an adverse developmental effect in experimental animal studies will potentially pose a hazard to humans following sufficient exposure during development.

All of the 4 manifestations of developmental toxicity (death, structural abnormalities, growth alterations, and functional deficits) are of concern.

The types of developmental effects seen in animal studies are not necessarily the same as those that may be produced in humans.

The most appropriate species is used to estimate human risk when data are available (e.g., pharmacokinetics). In the absence of such data, the most sensitive species is the most appropriate to use.

There is a threshold for the dose–response curve for agents that produce developmental toxicity.

Source: From Ref. 3.

available from single and multiple dose studies in humans at numerous doses. The exposure characterization also includes a complete pharmacokinetic profile of any major metabolites as well.

The risk characterization allows an overall picture of the developmental risk, based on the hazard, dose–response, and exposure assessments. The risk conclusions generated are determined by a weight-of-evidence approach. All of the earlier mentioned factors (i.e., data quality, human relevancy, assumptions, and scientific judgment) are rolled into the overall conclusion. The expression of risk derived in this final step is used when the health risks are weighed against benefits and other "costs" to determine the appropriate action.

The risk of exposure for developmental toxicity from drugs is expressed as a TI. The TI, therefore, is not a direct measure of risk but a quantitative value used to generate an appropriate level of concern for possible developmental toxicity. It results in indicating the extent to which human exposures are below the observed NOAEL in the study species. A value of 10, for instance, is generally considered as low risk. No acceptable lower limit TI has been established or widely agreed upon since the risk also depends on the benefits of the drug.

BENEFIT–RISK BALANCE

For pharmaceuticals, a significant additional factor is considered in the risk assessment and that is benefit. Exposure to pesticides, many chemicals, and environmental contaminants is generally not intentional and is clearly undesirable. Therefore, the public is not willing to incur a risk even though there are clear benefits. For these areas, a zero-risk outcome is sought. The benefit–risk balance is an important concept within the pharmaceutical industry, its regulators, physicians, and patients. The process of therapeutic intervention by nature will never be "zero risk" and all parties appreciate the importance of maximizing benefit while minimizing risks. The process to identify, assess, communicate, and minimize risk of a drug in order to establish and maintain a favorable benefit–risk balance in patients is termed risk management (34).

Risk Management

While not considered part of the risk assessment process, a few words concerning the risk management process are appropriate based on the benefit–risk assessment performed for pharmaceutical agents. The benefit component for pharmaceuticals is the linchpin that links the risk assessment and risk management. Risk management is the activity which integrates recognition of the risk, risk assessment, developing strategies to manage the risk, and mitigation of the risk. The relationship of the risk assessment and risk management processes are schematically depicted in Figure 2. In general, all techniques to manage risk appear to fall into one or more of the following categories: elimination, mitigation, retention, and transfer (35). For pharmaceuticals, elimination would be to avoid the risk by not

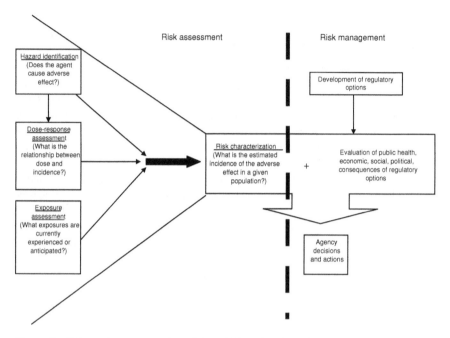

Figure 2 Schematic representation of the relationship of the risk management process to risk assessment.

taking the drug at all, or excluding use during developmental windows of concern for adverse effects. Alternative therapies without the risk may be the favorable approach, if such therapies are available. Additionally, the patient may go without the drug, but this should be done in conjunction with their physician to understand the health risks incurred by not using the therapeutic intervention. For example, many HMG-CoA reductase inhibitors (i.e., statins), which are used to lower cholesterol, have been evaluated for developmental toxicity and do not necessarily produce findings indicative of a high concern for potential human developmental toxicity (36–38). However, because atherosclerosis is a chronic process, discontinuation of lipid-lowering drugs during pregnancy would not be anticipated to have a negative health impact on the mother. Therefore, given negligible benefit during pregnancy, the theoretical risk for fetal harm potentially caused by inhibition of cholesterol synthesis, HMG-CoA reductase inhibitors as a class are contraindicated during pregnancy and labeled as Pregnancy Category X (see Table 3). A mitigation strategy may involve reducing the risk through methods outlined in a risk management plan (RMP). Another option, defined as retention, is to accept the risk as outlined. The patient may conclude the benefit far outweighs the risk and be willing to accept the risk if it were to occur. For example, nicotine replacement therapies (NRT) utilize nicotine to reduce smoking. Even though nicotine has been associated with developmental toxicity and the safety of NRT during pregnancy

Table 3 United States FDA Pharmaceutical Pregnancy Categories

Pregnancy category A	Adequate and well-controlled studies have failed to demonstrate a risk to the fetus in the first trimester of pregnancy (and there is no evidence of risk in later trimesters).
Pregnancy category B	Animal reproduction studies have failed to demonstrate a risk to the fetus and there are no adequate and well-controlled studies in pregnant women OR Animal studies which have shown an adverse effect, but adequate and well-controlled studies in pregnant women have failed to demonstrate a risk to the fetus in any trimester.
Pregnancy category C	Animal reproduction studies have shown an adverse effect on the fetus and there are no adequate and well-controlled studies in humans, but potential benefits may warrant use of the drug in pregnant women despite potential risks.
Pregnancy category D	There is positive evidence of human fetal risk based on adverse reaction data from investigational or marketing experience or studies in humans, but potential benefits may warrant use of the drug in pregnant women despite potential risks.
Pregnancy category X	Studies in animals or humans have demonstrated fetal abnormalities and/or there is positive evidence of human fetal risk based on adverse reaction data from investigational or marketing experience, and the risks involved in use of the drug in pregnant women clearly outweigh potential benefits.

has not been thoroughly evaluated in humans, because the adverse fetal effects of smoking are perceived to be much greater than the potential risk of fetal harm due to NRT, these drugs are routinely prescribed to pregnant women who smoke (39–43). The technique of transferring the risk does not apply to health-based risks from pharmaceuticals.

In 1998, the Council for International Organizations of Medical Sciences published a document intended to provide guidance to industry and regulators in establishing the balance between the benefit and risk of drugs in the postapproval time frame (44). A 1999 report to the FDA Commissioner from a task force on risk management outlined recommendations for improvement in the current system of postmarketing surveillance, which is based around the voluntary spontaneous reporting system of the FDA (45). They proposed a process whereby the FDA would shift from its passive role with action through labeling to a much more proactive role that included the reevaluation of the benefit–risk balance in major postmarketing decision making. Options presented included restrictions in use, restrictions in distribution, mandatory education programs, and slow rollouts of new products and review of other therapeutic alternatives.

On June 12, 2002, the President signed the Public Health Security and Bioterrorism Preparedness and Response Act of 2002, which included the Prescription Drug User Fee Amendments of 2002 (PDUFA III). Following the signing of the act, risk management now has a formal role in the development, review, and approval of new drugs in the U.S. PDUFA III included the development of formal RMPs to be submitted to the FDA during a drug's preapproval period. While the RMPs are voluntary, the recommendation pushes risk management into the preapproval time frame where potential safety signals can be studied, quantified, and followed through the specific product RMP. Recently, Bush et. al. reviewed the current risk management practices and tools, as well as addressed some new approaches (34).

Risk Communication

The goal of risk communication is to present a risk analysis in such a manner that clinicians can aid patients in making well-informed risk management decisions. In 1979, the FDA implemented a pregnancy-labeling format and five-letter classification system for drug use during pregnancy (46). This pregnancy label required that each medication be classified using a five-letter category system, A, B, C, D, or X (Table 3). In addition to the pregnancy category, labels were required to provide information on the potential for the drug to cause reproductive and/or developmental toxicity. This system has been criticized because the information in the pregnancy is often difficult for health-care professionals to interpret and use (47–49). In response to this criticism, the Pregnancy Labeling Taskforce made a recommendation that the current labels should be replaced (U.S. Food and Drug Administration Concept paper on pregnancy labeling, summary of comments from a public hearing and model pregnancy labeling based on recommendations, available at: http://www.fda.gov/ohrms/dockets/ac/99/transcpt/3516r1.doc.). It was recommended that the new label format should be narrative text that is more informative and would include more clinical management advice. The label should also provide a clinically relevant discussion of the available nonclinical and human data that form the basis of the risk assessment. However, at the time of this publication, the revised labeling regulation has not been issued.

SUMMARY REMARKS

The animal data generated in studies exploring potential developmental toxicity is a critical component of the hazard identification and overall risk assessment process for developmental toxicity. The risk assessment process for developmental toxicity, however, is but one component of an overall risk assessment of a given pharmaceutical. It must be integrated with all the other safety data into an overall risk assessment. Most other animal safety issues are supplanted by human data from evaluation in clinical studies. This human data is then integrated into the risk assessment. While human data are the most appropriate for assessing a drug's potential for developmental toxicity, it is generally many years' postmarketing that sufficient data may be available from well-designed epidemiologic studies to inform the risk assessment. Therefore, for developmental toxicity and other end points such as carcinogenicity, the animal data will continue to play a pivotal role in the risk assessment process.

Future improvements in the developmental toxicity risk assessment process will come with additional data. The current explosion of cellular and molecular data will advance our understanding of the mechanisms of normal development and therefore allow new understandings of the causes of abnormal development. All this new data will need to be integrated into the risk assessment process. An example of new data needing to be integrated into the process is in the area of pediatric toxicity. Recent pediatric legislation in the United States and Europe has called for data to inform physicians and patients on the appropriate use of drugs in the pediatric population. In addition, there sometimes is the need to provide nonclinical safety data in young animals for safety effects that cannot be adequately, ethically or safely assessed in pediatric clinical trials. These studies are providing data on postnatal development in our animal models. This data will be a welcome addition to the functional developmental data currently only available from our pre- and postnatal study design.

The FDA has recently published a guidance document addressing the non-clinical safety evaluation of pediatric drug products (50). The guidance indicates that studies in juvenile animals should primarily address the potential effects on growth and development that have not been studied or identified in earlier nonclinical and clinical studies. Therefore, juvenile animal studies of interest are those where identified target organ toxicity in adults is also an organ with significant postnatal development. A series of recently published mini reviews have compared postnatal development in a number of organ systems across species to help determine appropriate animal models (51–59). Additionally, a multistakeholder International Life Sciences Institute sponsored workshop was held to discuss aspects of appropriate study design to address the data needs (60). The data from these juvenile animal studies has already found their way into the risk assessment and informing the risk communication process by appearing in pharmaceutical product labels. The juvenile animal study is just one example of additional data that is being integrated into the developmental risk assessment. This is just the beginning, so do not go and change jobs.

REFERENCES

1. National Research Council. Risk assessment in the federal government: Managing the process. National Academy Press, Washington, DC, 1983:18–49.
2. US EPA. Guidelines for the health assessment of suspect developmental toxicants. Fed Regist 1986; 51:33992–34003.
3. US EPA. Guidelines for developmental toxicology risk assessment. Fed Regist 1991; 56:63798–63826.
4. US EPA. Guidelines for exposure assessment. Fed Regist 1992; 57:22888–22938.
5. US EPA. Guidelines for reproductive toxicity risk assessment. Fed Regist 1996; 61:56274–56322.

6. US EPA. Guidelines for neurotoxicity risk assessment. Fed Regist 1998; 63:26926–26954.
7. US EPA. Guidelines for ecological risk assessment. Fed Regist 1998; 63:26846–26924.
8. US EPA. Guidelines for carcinogen risk assessment. Fed Regist 2005; 70:18865–17817.
9. National Research Council. Science and judgment in risk assessment. National Academy Press, Washington, DC, 1994:16–22.
10. National Research Council. Scientific frontiers in developmental toxicology and risk assessment. National Academy Press, Washington, DC. 2000:1–9.
11. Draft Reviewer Guidance. Integration of study results to assess concerns about human reproductive and developmental toxicities. Fed Regist 2001; 66:56830–56831.
12. Collins TFX, Sprando RL, Schackelford ME, et al. Principles of risk assessment—FDA perspective. In Developmental and Reproductive Toxicology—A Practical Approach, Hood RD, ed. CRC Press, Boca Raton, FL, 2006:877–909.
13. Wilson JG. Current status of teratology: general principles and mechanisms derived from animal studies. In Handbook of Teratology: General Prinicples and Etiology, Wilson JG, Fraser FC, eds. Plenum Press, New York, 1977:47–74.
14. Kimmel CA. Dose response modeling approaches for risk assessment. In Toxicology Forum Proceedings. Aspen, CO, 1991:317–336.
15. Kimmel CA, Kimmel GL, Euling SY. Developmental and reproductive toxicity risk assessment for environmental agents. In Developmental and Reproductive Toxicology: A Practical Approach, Hood RD, ed. 2nd ed. CRC Press, Boca Raton, FL, 2006:841–875.
16. Wilson JG. Environment and birth defects. Academic Press, New York, 1973.
17. Scialli AR, Buelke-Sam JL, Chambers CD, et al. Communicating risks during pregnancy: A workshop on the use of data from animal developmental toxicity studies in pregnancy labels for drugs. Birth Defects Res A 2004; 70:7–12.
18. International Conference on Harmonization (ICH) Harmonized Tripartite Guideline. Detection of toxicity to reproduction for medicinal products. Fed Regist 1994; 59:48746–48752.
19. Tyl RW, Marr ML. Developmental toxicity testing—Methodology. In Developmental and Reproductive Toxicology: A Practical Aproach, Hood RD, ed. 2nd ed. CRC Press, Baco Raton, FL, 2006:201–261.
20. Horimoto M. Standard protocols based on the ICH guidelines for reproductive and developmental toxicity studies: An international implementation. J Tox Sci 1996; 21:451–455.
21. Christian MS, Hoberman AM, Lewis MC. Perspectives on developmental and reproductive toxicity guidelines. In Developmental and Reproductive Toxicology: A Practical Approach, Hood RD, ed. 2nd ed. CRC Press, Boca Raton, FL, 2006:733–798.
22. Francis EZ. Testing of environmental agents for developmental and reproductive toxicity. In Developmental Toxicology, Kimmel CA, Buelke-Sam J, eds. 2nd ed. Raven Press, New York, 1994:403–428.
23. Matthews EJ, Kruhlak NL, Daniel Benz R, et al. A comprehensive model for reproductive and developmental toxicity hazard identification: I. Development of a weight of evidence QSAR database. Regul Toxicol Pharmacol 2007; 47:115–135.

24. Streck RD, Kumpf SW, Ozolins TR, et al. Rat embryos express transcripts for cyclooxygenase-1 and carbonic anhydrase-4, but not for cyclooxygenase-2, during organogenesis. Birth Defects Res B 2003; 68:57–69.
25. Holson JF, Nemec MD, Stump DG, et al. Significance, reliability, and Interpretation of developmental and reproductive toxicity study findings. In Developmental and Reproductive Toxicology: A Practical Approach, Hood RD, ed. 2nd ed. CRC Press, Boca Raton, FL, 2006:329–424.
26. Kimmel CA, Kimmel GL. Risk assessment for developmental toxicity. In Developmental Toxicology, Kimmel CA, Buelke-Sam J, eds. 2nd ed. Raven Press, New York, NY, 1994:429–453.
27. Kimmel CA. Quantitative approaches to human risk assessment for noncancer health effects. Neurotoxicology 1990; 11:189–198.
28. Wilson JG. Mechanisms of teratogenesis. Am J Anat 1973; 136:129–131.
29. Selevan SG, Lemasters GK. The dose-response fallacy in human reproductive studies of toxic exposures. J Occup Med 1987; 29:451–454.
30. Wilson JG. Experimental teratology. Am J Obstet Gynecol 1964; 90(Suppl):1181–1192.
31. Cappon GD, Cook JC, Hurtt ME. The relationship between cyclooxygenase (COX) 1 and 2 selective inhibitors and fetal development when administered to rats and rabbits during the sensitive periods for heart development and midline closure. Birth Defects Res B 2003; 68:47–56.
32. Kimmel CA, Wilson JG, Schumacher HJ. Studies on metabolism and identification of the causative agent in aspirin teratogenesis in rats. Teratology 1971; 4:15–24.
33. National Research Council. Understanding risk: Informing decisions in a democratic society. National Academy Press, Washington, DC, 1996:1–10.
34. Bush JK, Dai WS, Dieck GS, et al. The art and science of risk management: A US research-based industry perspective. Drug Safety 2005; 28:1–18.
35. Dorfman MS. Introduction to risk management and insurance. 9th ed. Prentice Hall Press, New York, 2007.
36. Dostal LA, Schardein JL, Anderson JA. Developmental toxicity of the HMG-CoA reductase inhibitor, atorvastatin, in rats and rabbits. Teratology 1994; 50:387–394.
37. Lankas GR, Cukierski MA, Wise LD. The role of maternal toxicity in lovastatin-induced developmental toxicity. Birth Defects Res B 2004; 71:111–123.
38. von Keutz E, Schluter G. Preclinical safety evaluation of cerivastatin, a novel HMG-CoA reductase inhibitor. Am J Cardiol 1998; 82:11J–17J.
39. Fiore MC. Treating tobacco use and dependence: An introduction to the US Public Health Service Clinical Practice Guideline. Respir Care 2000; 45:1196–1199.
40. Melvin CL, Gaffney CA. Treating nicotine use and dependence of pregnant and parenting smokers: An update. Nicotine Tob Res 2004; 6:107–124.
41. Frishman WH, Mitta W, Kupersmith A, et al. Nicotine and non-nicotine smoking cessation pharmacotherapies. Cardiol Rev 2006; 14:57–73.
42. Oncken CA, Pbert L, Ockene JK, et al. Nicotine replacement prescription practices of obstetric and pediatric clinicians. Obstet Gynecol 2000; 96:261–265.
43. Hudson DB, Timiras PS. Nicotine injection during gestation: Impairment of reproduction, fetal viability, and development. Biol Reprod 1972; 7:247–253.

44. Benefit-Risk Value for Marketed Drugs; Evaluating Safety Signals. Report of CIOMS Working Group IV. In: Council for International Organizations of Medical Sciences (CIOMS). Geneva, 1998.

45. Managing the risks from medical product use; Creating a risk management framework. Report to the FDA Commissioner for the Task Force on Risk Management. (www.fda.gov/oc/tfrm/1999report.html) 1999.

46. United States Food and Drug Administration. Labeling and prescription drug advertising: Content and format for labeling for human prescription drugs. Fed Regist 1979; 44:37434–37467.

47. Friedman JM, Little BB, Brent RL, et al. Potential human teratogenicity of frequently prescribed drugs. Obstet Gynecol 1990; 75:594–599.

48. Doering PL, Boothby LA, Cheok M. Review of pregnancy labeling of prescription drugs: Is the current system adequate to inform of risks? Am J Obstet Gynecol 2002; 187:333–339.

49. Addis A, Sharabi S, Bonati M. Risk classification systems for drug use during pregnancy: Are they a reliable source of information? Drug Saf 2000; 23: 245–253.

50. Food and Drug Administration. Guidance for Industry: Nonclinical Safety Evaluation of Pediatric Drug Products. (http://www.fda.gov/cder/guidance/5671fnl.htm#top) 2006.

51. Watson RE, Desesso JM, Hurtt ME, et al. Postnatal growth and morphological development of the brain: a species comparison. Birth Defects Res B 2006; 77:471–484.

52. Walthall K, Cappon GD, Hurtt ME, et al. Postnatal development of the gastrointestinal system: A species comparison. Birth Defects Res B 2005; 74:132–156.

53. Wood SL, Beyer BK, Cappon GD. Species comparison of postnatal CNS development: Functional measures. Birth Defects Res B 2003; 68:391–407.

54. Zoetis T, Hurtt ME. Species comparison of lung development. Birth Defects Res B 2003; 68:121–124.

55. Zoetis T, Hurtt ME. Species comparison of anatomical and functional renal development. Birth Defects Res B 2003; 68:111–120.

56. Zoetis T, Tassinari MS, Bagi C, et al. Species comparison of postnatal bone growth and development. Birth Defects Res B 2003; 68:86–110.

57. Beckman DA, Feuston M. Landmarks in the development of the female reproductive system. Birth Defects Res B 2003; 68:137–143.

58. Marty MS, Chapin RE, Parks LG, et al. Development and maturation of the male reproductive system. Birth Defects Res B 2003; 68:125–136.

59. Kok Wah Hew KAK. Postnatal anatomical and functional development of the heart: A species comparison. Birth Defects Res B 2003; 68:309–320.

60. Hurtt ME, Daston G, Davis-Bruno K, et al. Juvenile animal studies: testing strategies and design. Birth Defects Res B 2004; 71:281–288.

11

Genomic Approaches in Developmental Toxicology

George P. Daston and Jorge M. Naciff

*Miami Valley Innovation Center, The Procter & Gamble Company,
Cincinnati, Ohio, U.S.A.*

INTRODUCTION

The field of developmental toxicology, like virtually every other life science discipline, is being revolutionized by the advent of genomic tools that allow the investigation of changes in every gene in a biological sample simultaneously. We have long understood that adverse responses to exogenous perturbations, although manifested as changes at the structural or organ level, are attributable to changes that occur at more fundamental levels of biological organization: tissues, cells, and genes. This concept has been at the core of the reductionist, mechanistic approaches to studying abnormal development that have contributed so much to the field over the last four decades. From this research, we have learned that abnormal development can be caused by a number of mechanisms, only some of which are directly attributable to changes in the expression of genes. However, as genomics experiments become more commonplace in toxicology, it is becoming clear that, irrespective of ultimate mechanism, the pathogenesis of virtually all toxic responses involves some change in gene expression (1, 2). The fact that gene expression changes are so universal to toxicity suggests that global analysis of gene expression may be an important tool in understanding toxic mechanisms, predicting toxic response, extrapolating across experimental models and to humans, and addressing many other questions that are important in basic and applied toxicology and risk assessment.

Genome-wide analysis of gene expression became possible with the sequencing of the human genome, which was largely completed in 2000, and that of mice, rats, and other commonly used animal models within a few years thereafter. The information derived from these gene sequencing projects has been used to create high-quality microarrays, which consist of DNA or oligonucleotide probes of multiple genes that have been affixed to a hard surface (and for which the exact location is known). DNA (or cDNA from RNA) from a biological sample is hybridized to these microarrays. It is possible using microarrays to assess the concentration of specific RNAs or DNA sequences for anywhere from hundreds to tens of thousands of genes in a single biological sample. Although there are a number of methods for making microarrays, gene chips that consist of oligonucleotide probe sets tiled onto glass/quartz slides using a photolithographic process (e.g., Affymetrix chips) are generally considered the most powerful and useful in studying global gene expression. Gene chips created in this way can be of very high density (the genome of an entire organism can be represented on a single chip—the size of an old 35-mm photographic negative) and quality control is sufficiently high that experimental replication is good. There are a number of fine reviews of microarray technology (3, 4), others, so the topic will not be belabored here.

Guidance has been published providing minimum standards for the publication of genomics studies in the literature, the so-called MIAME standards, which stands for Minimum Information About a Microarray Experiment (visit www.mged.org for the latest on MIAME standards). The standards are intended to assure a minimum level of quality in microarray experiments, and that an independent laboratory could replicate either the experiments or the data analysis. The most critical elements of MIAME are:

1. The raw data for each hybridization should be available on a publicly accessible website in an accessible format (e.g., CEL or GPR files). There are a number of websites that host data [e.g., the Gene Expression Omnibus from National center for Biotechnology Information (NCBI), the European Bioinformatics Institute from European Molecular Biology Laboratory (EMBL), among others].
2. The final processed (normalized) data for the set of arrays in the experiment, particularly the data used to draw conclusions, should be available.
3. The essential sample annotation including experimental factors and their values (e.g., compound and dose in a dose–response experiment) should be available.
4. The experimental design should be provided, and it should be straightforward to determine which data set is associated with each treatment and end point.
5. Sufficient annotation of the array (e.g., gene identifiers, genomic coordinates, probe oligonucleotide sequences, or reference commercial array catalog number) should be provided to allow others to replicate the experiment or analysis.
6. The essential laboratory and data processing protocols should be published.

In addition to these generic standards, more specific standards have been published for developmental toxicology studies, emanating from an ad hoc working group of the Teratology Society (5). Some of the important considerations for toxicity studies are that it is critical to include either multiple dose levels or multiple time points in an experiment to aid in study interpretation, and for development it is important to have a consistent terminology and annotation of developmental stage. It is also an expectation of microarray experiments that the results be partially verified with a different technique, such as quantitative RT-PCR. This technique does not have the throughput of microarrays, but does allow one to verify the outcome of the microarrays for selected genes.

Protein expression can also be evaluated on a global basis, a procedure known as proteomics. Because proteomics measures proteins, it is one step closer to the biological response to a perturbation. Translation of mRNAs can be selective, so a global analysis of protein expression may present a truer picture of cellular response than an analysis of gene expression. Proteomics also includes techniques that evaluate posttranslational modifications and protein interactions, both of which are outside the scope of DNA or oligonucleotide microarrays. The downside to proteomics is that there are currently no experimental approaches that have the throughput or ease of gene microarrays. The technology still relies on separation of proteins by techniques such as electrophoresis and individual identification. Still, there have been some successful applications of proteomics for developmental toxicology [for a review, see (6)]. However, the number of available proteomics studies in the field is limited; therefore, the rest of this review will concentrate on genomics.

Microarrays can have different purposes. Most of the research in toxicology has been directed toward understanding the effects of exogenous insults on gene expression (i.e., changes in specific mRNA levels) in tissues that are responsive to the insult. Other arrays are designed to evaluate allelic differences in genomic DNA, including single nucleotide polymorphisms (SNPs) that may provide clues to an individual's sensitivity to an insult, among other things.

The potential for microarray technology to improve our understanding of the ways in which exogenous agents cause toxicity, and thereby improve the way that we assess the risk from exposure to these agents, is tremendous. Molecular toxicology in the pregenomics area was restricted largely to evaluating changes in the expression of one or a few genes at a time. Genomics experiments can evaluate the changes in expression in every gene, in the same sample, and do so in a quantitative way, a fact that may be disheartening to many of us who may reflect on this massive gain in efficiency through the prism of having had students (or been students) working for years on dissertation projects whose scope can be exceeded a thousand-fold in a single experiment.

Because patterns of altered gene expression are specific for mode of action, gene expression analysis can be used in screening to predict the effects of an agent in more definitive toxicity studies. Because gene expression is a sensitive indicator of biological response, dose–response assessment, particularly low-dose extrapolation, can be conducted with less uncertainty. Because the sequence and function of developmental genes are highly conserved across species, it may be possible to improve extrapolation across species. Because individual differences in the sequences of critical genes can be identified, it may be possible to uncover the basis for interindividual differences in response. Because microarray experiments are, at their core, hypothesis-generating exercises, they will accelerate progress in identifying mechanisms of action of toxicants.

There are a number of issues in toxicology and risk assessment that microarray experiments have begun to address. These include:

- Predictive toxicology: the ability of relatively small-scale animal or cell-based experiments that predict the outcome of larger-scale, longer duration studies based on presumed mechanism(s) of action.
- Dose–response assessment: the use of the sensitivity of gene expression analysis as a way of elucidating the behavior of the dose–response curve at exposure levels below the no observed adverse effect level (NOAEL).
- Interspecies extrapolation: the use of gene expression data to better understand whether the response seen in an experimental model is relevant to humans, and perhaps to evaluate the relative sensitivity between species.
- Interindividual variability: the use of genetic information to determine the basis for interindividual differences in susceptibility, and to predict the existence of susceptible subpopulations.
- Mechanistic understanding: the use of microarrays to accelerate progress toward understanding the underlying basis for abnormal development.

Each of these aspects of genomics will be discussed in this review.

PREDICTIVE TOXICOLOGY

Virtually all toxic responses (save those that are so acutely toxic that they preclude any subsequent biological response) produce changes in gene expression. Some of these responses are essential and integral to the toxic response, others are secondary and may represent the cell's attempts to deal with the damage done, but regardless, changes in gene expression occur after sufficient exposure to outside agents. On reflection, this is not a surprising observation, in that genes are the repository of the organism's ability to respond to stimuli and to maintain homeostasis, and toxicity can be defined as a stimulus that is sufficient in intensity to perturb homeostasis.

Given that gene expression changes are either the direct response to an exogenous agent or a response to the homeostatic upset produced by an agent, it was surmised and has been subsequently confirmed that changes in gene expression are specific for mode of action. Studies on gene expression response for target organ toxicity, particularly liver and kidney, have identified gene expression profiles that are specific for mode of action. Some of these assessments are massive, evaluating gene expression for several hundred compounds (7). It has been possible to identify limited numbers of genes, which, when coordinately changed in expression level, are diagnostic of a particular mode of action.

Similarly, comprehensive evaluations of gene expression for developmental toxicants have not been done, but there is sufficient information available to support the conclusion that developmental mechanisms are comparably specific. For example, we have identified a gene expression profile for estrogens in the fetal rat uterus (8) and testis (9). Liu et al. (10) have identified a gene expression profile that is specific for those phthalate esters that produce functional and structural teratogenicity in the rat testis. This profile is highly specific; it is not elicited by other phthalates that are not teratogenic (e.g., diethylphthalate) (10), nor is it elicited by some classical antiandrogens (11) that produce some of the same effects on reproductive system development but have different molecular targets.

The specificity of gene expression is being used in the pharmaceutical industry as the basis for preliminary screening for liver and other organ toxicity. Drug discovery, at least in recent times, involves the evaluation of large chemical libraries (often hundreds of thousands of chemicals) for a particular biological activity. This screening is done at a molecular level and usually involves the use of high-throughput bioassays that evaluate the affinity of the chemicals to a specific molecular target and/or the magnitude of response elicited by the ligand when associated with the target. This approach is very useful in identifying large numbers of compounds that have activity toward the molecular target. The downside of the approach is that there were no comparably rapid ways to screen the active subset of the chemical library for agents that may be highly toxic.

Gene expression analysis has begun to address this issue, at least for selected target organs. Compounds can be screened for potential toxicity in cell systems such as hepatocytes. Gene expression analysis adds a level of specificity to the screening that was not previously available. We are unaware of any such screening being carried out on a routine basis for developmental toxicants, but it is theoretically possible to do so, given the compilation of gene expression information on a sufficient number of teratogens and the development of in vitro models that provide a broad enough range of developmental response to be used as the screening platform. Existing models such as rodent whole embryo culture or mouse embryonic stem cells have the potential to become those platforms. Investigations of gene expression in mouse embryonic stem cells suggest that they have the capacity for response to different teratogenic mechanisms (12).

Irrespective of whether this type of preliminary screening has been carried out, toxicity evaluations for new drugs, pesticides, or high production volume chemicals are still carried out using traditional animal tests. Comprehensive toxicity evaluations, including testing for developmental and reproductive toxicity potential, are done as a matter of routine for drugs and pesticides. Commodity chemicals are also tested for toxicity, but the extent of the data set required, and the schedule for conducting the testing, is typically driven primarily by production volume or potential for widespread exposure in the population. Chemical regulations in Europe provide the most explicit guidance on how this is done, with a requirement of a base set of information when a new material is first produced and introduced into commerce, and then increasing numbers of tests when the production of the chemical reaches certain tonnage triggers. The rules in the United States under the Toxic Substances Control Act are somewhat more flexible, but generally speaking, the higher the potential for human exposure, the more likely it is that developmental toxicity testing will be carried out.

While these are good rules, they really only take half of the risk assessment equation into consideration. It would make sense to also consider the potential to produce biological effects in the algorithm that is used to prioritize chemicals for testing. The reason that this has not been done is that screening level tests for toxicity that are of sufficiently high-throughput/short duration to be used for prioritization, and are of sufficiently high predictive quality to support important priority decisions, have not yet been developed and applied for this purpose. (One obvious solution that may occur to some readers is to reapply the screens described earlier for pharmaceuticals. This idea has some merit, but there are some impediments, including the fact that the criteria for acceptable test performance may be far different for drug development. For example, a false positive rate of, say, 80% is not a concern when one can choose from a pool of a thousand potential new drug actives, but would be unacceptable for prioritizing individual compounds.) The first such screens will be coming online with the advent of estrogen/androgen screening. The simplest and most rapid of these screens include reporter gene assays that have high specificity for detecting the binding of ligands to a hormone receptor. While these screens, which analyze a single response (and typically to a gene that is not even native to the cell type), are a far cry from the sophistication envisioned by a genome-wide evaluation of gene expression, they represent a sea change in the way that chemical prioritization will be done in the future. It is now up to the scientific community to provide gene expression–based test methods that can support rapid screening as part of the chemical testing prioritization process.

Another possible use of the power of genomics experiments to identify mode of toxicity is the tailoring of testing based on mode of action. At present, toxicity testing takes a "one-size-fits-all" approach to testing. There is one protocol that has been determined to be the definitive toxicology assessment for a given end point. We typically use only one or two species to test, and rely on one or two strains of those species. Yet we know that these tests have varying levels of reliability depending on mode of toxicity. It will be of interest to see whether gene expression analysis can provide us with enough information about potential toxicity that we can customize testing strategies to pick the experimental model that is most capable of predicting human response, and the optimal testing strategy to most fully evaluate the potential manifestations that may result from the mode(s) of action identified in a screen.

For example, if estrogenic action were identified as the likely mode of action of an untested chemical, one could design a dosing schedule that emphasized exposure during developmental stages that are sensitive to estrogens, such as peri-implantation, the critical period for reproductive organ formation, reproductive tract maturation, and late fetal or early postnatal periods (depending on the experimental model selected) that are critical for brain sex differentiation. One might want to wait until sexual maturity to complete evaluation of reproductive system structure and function. One might also want to select an experimental model that better predicts uterine or mammary tumorigenicity potential than the standard chronic toxicity/cancer bioassay now in use.

Gene expression is of course organ specific, and can also be lifestage specific. We have investigated both of these issues using estrogens as model toxicants. The fetal rat uterus, ovaries, and testes all express estrogen receptors at relatively high levels and are estrogen responsive. Treatment of pregnant rats with estrogens of varying potency (we used ethynyl estradiol, genistein, and bisphenol A to cover a wide range of potency) from the start of the critical period for reproductive system development (gestation day 12) through the late fetal period (gestation day 20) caused a significant number of gene expression changes in both uterus and ovaries (evaluated together) and testes when evaluated using Affymetrix microarrays (8, 9). [Microarray technology has advanced at about the same pace as iPod or flat-screen TV technology, and the earlier experiments we did were with chips that covered about a third of the rat genome (about 8000 genes), while more recent experiments have been run using chips that represent the entire genome. (Coincidentally, retail price for an Affymetrix chip is somewhere between that for an iPod and LCD TV.)] More estrogen-responsive genes happen to be expressed in the female tissues than in the males, but the number of genes for which expression is affected by chemicals with estrogenic activity in either sex is in the hundreds for any given compound. Comparisons across estrogenic compounds for only those genes that are commonly expressed and are significantly dose responsive (in the same direction) decreased the number of gene expression changes to 66 for the female tissues and 44 for the testis. However, there was limited overlap in the transcript profiles for testis and uterus/ovaries. These were the putative transcript profiles for estrogens in these tissues.

The actual number of genes necessary to identify an estrogenic mode of action is far less than this number. In a subsequent study comparing the gene expression from treatment with four estrogens and four nonestrogens, we were able to determine that the number of genes needed to identify an estrogen is on the order of ten or less (Daston et al., unpublished), as long as one chooses judiciously and excludes genes that may be common to more than one type of agent, such as steroid synthesis pathways.

We also assessed gene expression in the uterus and ovaries of rats at two different lifestages: fetal and juvenile (postnatal day 22–25). These reproductive tissues express estrogen receptors at both lifestages. The biological response to estrogen exposure is different, however. The juvenile rat is able to undergo uterotrophic activity in response to an estrogen, a temporary increase in uterine mass and state of differentiation that is comparable to what occurs during the estrus phase of the estrous cycle in the sexually mature rat. The fetal reproductive system is not yet capable of a uterotrophic response; however, we know that it is estrogen-responsive in that we can observe immediate changes in gene expression and potent estrogens have been shown to have latent but persistent effects. We identified characteristic transcript profiles for estrogens at both lifestages (8,13). Again, hundreds of genes were changed in expression at either lifestage. Applying the same stringent criteria of dose responsiveness and statistical significance to the juvenile results as we had to the fetal results, we concluded that 120 genes in the juvenile rat responded to estrogens in a robust manner, versus 66 in the fetus. However, there was not a great deal of overlap in the two sets. Only about half the genes that were changed in the fetal profile were also changed in the juvenile. While most of the genes in common had expression changes in the same direction, some were in opposite directions. These results indicate that lifestage is important in identifying transcript profiles, and suggest that the differences in gene expression underlie the different responses at the tissue and organ levels in the fetal and juvenile rats.

Dose–Response Assessment

Dose–response assessment is the phase of the risk assessment process in which predictions are made about the level at which exposure to a chemical will convey little or no risk of causing an adverse effect. For developmental toxicity, this involves identifying a level in a study that has minimal or no observable adverse effect (NOAEL or benchmark dose), then dividing by uncertainty factors that take into account the possible differences between experimental model and humans in susceptibility, possible variations in susceptibility across the human population (and sometimes other factors). The default values that are used to account for these uncertainties are generally on the order of 100 to 1000 for developmental toxicity. These appear to be protective insofar as we can tell, although given the uncertainty in the magnitude of uncertainty, this conclusion is not universally shared. The basis for arguments around the adequacy of current procedures for setting safe levels is complicated, but much is based on the fact that the resolving power of animal studies is limited and that the shape of the dose–response curve at exposures below those used in the toxicity study cannot be extrapolated with confidence. Animal developmental toxicity studies are conducted with a limited number of animals, usually 20 to 25 per dose group in most regulatory protocols. The dose levels in these studies are exaggerated above human exposure levels, because the intent of the study is to maximize the chance of detecting a hazard. This design is the most practical way to identify a hazard, but it also engenders controversy. First, for chemicals that do present a developmental hazard, there is usually still some residual level of effect at the NOAEL (and there is, by definition, a level of risk at the benchmark dose), but it is below the statistical resolving power of the assay. It is hoped that the application of uncertainty factors results in an exposure level that is below a threshold for toxicity, but since the point at which a threshold is reached is not demonstrable in the study, the threshold level is unknown. The counterbalancing concern is that, because the study is conducted by necessity at exaggerated dosages, the responses observed in the study may be occurring as the result of saturation of homeostatic mechanisms and may not be relevant for predicting risk at ambient exposure levels. It is beyond the scope of this article to provide in-depth discussion of either the threshold issue or the exaggerated dose problem. There are existing reviews on thresholds for developmental toxicants (14). A recent review on dose-dependent transitions in toxicology provides very informative examples of how increases in dose beyond a certain level can sometimes lead to responses that are nonlinear as a result of saturation of homeostasis (15).

Gene expression analysis has the potential to improve the situation by providing information about the nature and magnitude of response below the NOAEL. One of the benefits of gene expression analysis is that it is sensitive: subattomole (10^{-18} M) concentrations of individual messages can be detected, making it possible to track subtle changes. Because we know that gene expression is related to toxicity (sometimes causal, sometimes secondary, but always associated), it is possible to extend observations much farther down the dose–response curve to identify the no-observed effect level for gene expression changes. This level should be much closer to the actual threshold than the NOAEL, and should provide a greater level of confidence that the reference doses for chemicals are, indeed, conservatively chosen.

One interesting approach to applying gene expression data to dose–response assessment has been that of Thomas and coworkers (16). These investigators have evaluated the gene expression changes elicited by respiratory toxicants at a number of dose levels, above and below the NOAEL for toxic effects, and have calculated a benchmark dose for gene expression changes. Rather than calculate a benchmark dose for individual genes, benchmark doses were calculated for groups of genes that were statistically changed at some dose and were in the same gene ontology (GO) category. Genes within the same GO term have similar function; therefore, focusing on groups of genes that are affected coordinately increases the chances that the changes in expression of these genes have biological relevance and is not just coincidental. It is usually possible to detect changes in gene expression at levels below the NOAEL, probably because the gene expression changes are a precursor to the adverse event.

Gene expression analysis has already been used to address one specific dose–response controversy. There are reports in the literature that estrogens given to rodents at doses orders of magnitude below the NOAEL from traditional toxicology studies produce changes in the development of male reproductive organs in rats and mice (17–20). Not only were these effects seen at dose levels were much lower than the NOAEL but also the nature of the effects was different. Numerous attempts have been made to replicate these observations, both in traditional toxicity protocols and in studies designed to reproduce the original study designs as closely as possible. These subsequent studies have not observed the low-dose effects, despite the fact that these studies had greater statistical power (21–23). The U.S. National Toxicology Program convened an expert group to review the collection of studies (24). The working group concluded that the preponderance of evidence indicated no effect at low doses but could not dismiss the results of the studies reporting low-dose effects because of the many factors that may not have been adequately replicated. Furthermore, there is sufficient variability in the responses being measured, such as organ weight, that replicate studies may not be sufficient to resolve the issue.

We used gene expression to address the existence of low-dose effects. Because we knew that the effects of estrogens are mediated by gene expression, and that relatively high doses of a variety of estrogens produced a specific transcript profile in the developing rat reproductive system, we reasoned that there must also be changes in gene expression at low dose in order to produce the structural effects that had been reported. We had no preconceived notion as to which genes might be altered: it could be the same ones that were affected at high dosages, either in the same direction, but at a different level of expression, or in the opposite direction, or a completely different set of genes, or no altered gene expression. Only the latter case would be considered to be a refutation of the existence of the low-dose structural effects. Genomic-scale microarrays make it possible to conduct such unbiased experiments.

We evaluated gene expression over 5 to 6 orders of magnitude of dosage to three estrogens—17α-ethynyl estradiol, bisphenol A, and genistein—in the fetal rat testis. Gene expression was assessed using microarrays that cover more than 15,000 genes in the rat genome. There were no observable changes in gene expression at the low-dose levels (9). The dose–response curves were monotonic for those genes for which expression was changed at the highest dose levels. Gene expression changes were not detectable at levels within 2 to 3 orders of magnitude below the maximally tolerated dose.

This experiment demonstrates the power of gene expression analysis to provide information about biological response at levels far below traditional NOAELs. It provided a novel approach to addressing a controversy that could not have been solved by more replicate studies. This study has been used as support for a risk assessment of bisphenol A that discounts the existence of the purported low-dose effects (25).

INTERSPECIES EXTRAPOLATION

Genomics analyses have the potential to aid the process of interspecies extrapolation by comparing the pattern of gene expression in animal models with that in humans exposed to the compound. The problem with doing this for reproductive or developmental toxicity is that the tissues in humans are inaccessible for sampling. Therefore, there will be a need to establish gene expression biomarkers in accessible tissues such as white blood cells. Only a small amount of work has been done to determine whether it is possible to identify biomarkers of reproductive toxicity in accessible tissues, although there is some indication that estrogens can induce changes in gene expression in white blood cells in rats at dose levels that are comparable to those that produce pharmacological effects in the uterus (26). Other ideas for using surrogate tissues have been presented (27) but we are not aware of any attempts to find gene expression biomarkers for developmental toxicity in easily accessible tissues.

INTERINDIVIDUAL VARIABILITY

It is likely that much of the interindividual variability in response to an environmental insult is attributable to genetic differences. Research in cancer, and to a lesser extent in teratology, has uncovered allelic differences that, by themselves are insufficient to produce an adverse outcome, but when combined with an environmental influence (e.g., cigarette smoking), increase the likelihood of the adverse outcome. There is considerable support for this premise from over a decade of molecular epidemiology research. Progress in this area of research may be accelerated by the advent of microarrays for genomic DNA that evaluate small differences in gene sequence, that is, SNPs. The Environmental Genome Project is in the process of evaluating SNPs for 213 environmentally relevant genes. The human genes included are related to DNA repair, cell cycle control, cell signaling, cell division, homeostasis, and metabolism, and are thought to play a role in susceptibility to environmental exposure (28). The extent of variability in a small population has been determined (29), a necessary first step in the process of using this microarray for association studies. It will be interesting to see how valuable this approach is in understanding developmental susceptibility.

MECHANISTIC UNDERSTANDING

Microarray studies are often considered to be hypothesis generating rather than hypothesis testing. Studies are often designed with the purpose of surveying the changes in gene expression rather than postulating a priori which genes should be affected. The information content of any given microarray study usually surpasses by a wide margin what is already known about the gene expression elicited by a toxicant. Because of the high information content of microarray studies, they will rapidly accelerate the pace at which we understand the mechanisms that underlie abnormal development and developmental disease. Mechanistic information is valuable for risk assessment because it increases confidence that potential hazards have not been missed in the toxicity testing process and in reducing the uncertainty of extrapolating animal results to predict human risks.

Microarray results are put into mechanistic context by computational/bioinformatic analyses that are the subject of another review in this book. There are a number of ways to organize gene expression data sets that help elucidate the ways in which changes in gene expression underlie a biological response. These informatic approaches are essential for organizing data and for generating testable hypotheses regarding mechanism. One useful tool for the sorting of data has been the compilation of a GO database that groups genes according to molecular function, biological process, or cellular component. Within each of these three large categories, genes are grouped in hierarchical clusters that are reminiscent of Linnean taxonomy. For example, nuclear mRNA splicing is a subset of mRNA processing, which is a subset of RNA processing, which is a subset of RNA metabolism, which is a subset of biopolymer metabolism, etc. Grouping of genes in this way makes it possible to determine whether specific biological functions are affected by the treatment being evaluated. There are numerous examples of this kind of clustering in the literature. Figure 1 provides an example of genomic data arranged by GO terms. The intensity of color in the heat map indicates the number of genes that are significantly changed. In this instance, each column represents a different experimental condition and/or statistical rigor.

We have used GO terms to provide us with information that is useful in understanding the gene expression changes that underlie biological responses. The utility can be illustrated by studies that evaluate the time course of gene expression during a short-term biological response, in this case the uterotrophic response to a large dose of a potent estrogen in rats or mice (30–32). The biological response has been well described. Treatment with an estrogen results in a rapid increase in uterine mass that is attributable to glandular epithelial and stromal growth. There is an increase in uterine luminal volume and expansion of the lumen with fluid. The histomorphology of the uterine epithelium changes with increases in epithelial cell height and in the depth and complexity of uterine glands. Then, unless there is further stimulation with estrogens, there is a return to the previous state, with concomitant decreases in uterine mass, luminal fluid volume, cell shape changes, etc.

The gene expression changes that underlie these events are complicated when considered individually, but start to make sense when grouped into higher-order classifications. The genes that are expressed at the earliest time points after estrogen treatment are transcription factors, growth factors, and cell signaling molecules, which are likely to be driving the subsequent growth, changes in differentiation, and changes in tissue architecture that occur a few hours later. Fluid imbibition is one of the earliest noticeable morphological changes, and genes that regulate vascular permeability are also among the most active at early time points. By about 4 hours after treatment, genes for mRNA and protein synthesis become active for a period of a few hours. These gene expression changes are accompanied by changes in the expression of genes for cellular growth and differentiation, as well as a change in the balance of the genes that regulate cell viability in favor or a suppression of apoptosis. DNA replication and cell cycle genes become active next (Fig. 2). In each case, the peaks for gene expression in the various categories precede by a number of hours, the biological events that are noticeable at a tissue and organ level.

After the uterotrophic response reaches its peak, cell division starts to shut down, and the changes in cell cycle machinery tend toward downregulation over time. Changes in the apoptotic pathways tend to promote a decrease in cell number. There is also an increase in the expression of genes associated with the inflammatory response over the latter phases of the uterotrophic response (30–32).

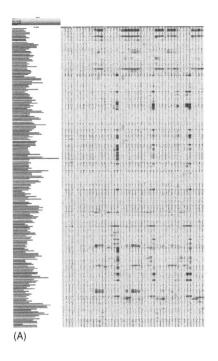

(A)

... GO:0006955 immune response	79	92	116	137	65	86	113	138
.... GO:0001816 cytokine production	9	10	11	14	7	9	14	16
.... GO:0006953 acute-phase response	5	6	8	8	5	5	5	11
.... GO:0006954 inflammatory response	26	32	38	47	27	34	43	51
.... GO:0006959 humoral immune response	27	32	38	40	23	28	35	42
..... GO:0006956 complement activation	15	15	16	16	11	14	15	16
...... GO:0006957 complement activation, alternative pathway	4	4	4	4	3	4	4	4
...... GO:0006958 complement activation, classical pathway	9	9	9	9	7	9	9	9
.... GO:0016064 humoral defense mechanism (sensu Vertebrata)	17	20	24	26	16	19	24	27
...... GO:0019735 antimicrobial humoral response (sensu Vertebrata)	7	8	11	13	6	7	12	14
..... GO:0019730 antimicrobial humoral response	8	9	12	14	7	8	13	16
.... GO:0006968 cellular defense response	13	15	19	23	14	17	21	23
..... GO:0016066 cellular defense response (sensu Vertebrata)	7	8	11	14	7	9	13	15
.... GO:0042087 cell-mediated immune response	6	7	8	10	5	7	10	11
.... GO:0019882 antigen presentation	6	9	10	12	8	8	9	10
..... GO:0019883 antigen presentation, endogenous antigen	3	5	5	7	4	4	5	5
..... GO:0019884 antigen presentation, exogenous antigen	3	4	4	4	4	4	4	4
.... GO:0030333 antigen processing	9	10	10	13	7	8	9	10
..... GO:0019885 antigen processing, endogenous antigen via MHC class I	4	5	5	8	4	4	5	5
..... GO:0019886 antigen processing, exogenous antigen via MHC class II	4	4	4	4	3	4	4	4

(B)

Figure 1 (*See color insert*) Gene expression segregated by GO categories. GO terms are arranged hierarchically into groups according to molecular function, cellular component, or biological process. In this figure, the intensity of color in each grid indicates the number of genes affected in that GO category, and the columns represent a unique experimental condition, such as increasing dose levels, time points, or statistical condition, particularly decreasing stringency for statistical significance from left to right. Figure 1 **A** is a heat map of the entire GO classification for molecular function for an experiment evaluating the effects of an estrogen on fetal uterus. Figure 1 **B** is an enlargement of one part of the heat map in 1 **A**.

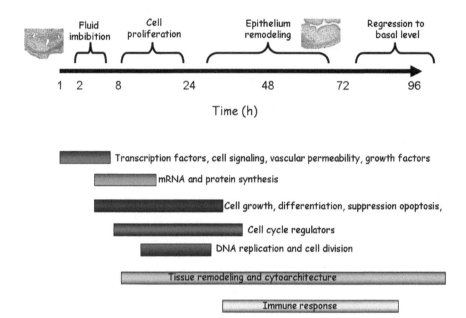

Figure 2 (*See color insert*) Temporal sequence of changes in gene expression as part of the uterotrophic response in rats. Changes in gene expression precede biological response by a few hours, generally. The earliest responses are those that precede tissue growth and differentiation, and include transcription factors and growth factors, as well as genes that regulate fluid imbibition into the uterine lumen, one of the early aspects of the uterotrophic response. These early changes are replaced by upregulation of protein and RNA synthesis machinery, and then cell cycle factors that favor cell division and decrease apoptosis. After the peak uterine growth response, gene expression changes in favor of apoptosis, and genes associated with the inflammatory response also become significantly expressed. The figure was constructed from the results from three labs using rats and mice (30–32) (*Source*: From Refs. 30–32).

It is also possible to evaluate the activity of specific genes within these pathways, or even across pathways, to develop hypotheses about which are critical in particular biological responses. In the case of estrogens, entire reviews have been written on the potential interactions of genes in reproductive system development and function (33) and will not be reproduced here.

Work from Tom Knudsen's laboratory provides another illustration for how gene expression data can be used to develop and test hypotheses. This group has studied the effects of a number of ocular teratogens, particularly those that perturb ocular size. Nemeth et al. (34) have identified a common set of target pathways, including fatty acid metabolism and glycolysis, that are changed in the same direction by a variety of teratogens that affect eye development. The effect appears to involve a perturbation in the transition of energy metabolism in the early embryo from a more glycolytic to a more oxidative state. This hypothesis is supported by observations that modulation of mitochondrial function by a pharmacological agent that acts on the mitochondrial membrane (a peripheral benzodiazepine receptor ligand) can rescue the phenotype (35).

Other teratogenic regimens may target other aspects of intermediary metabolism in ways that were not expected. For example, hyperthermia in mouse embryos was observed to downregulate genes in the cholesterol biosynthesis pathway, in addition to expected effects on heat shock genes (36). Gene expression analysis demonstrated that both hyperthermia and 4-hydroperoxycyclophosphamide had effects on DNA replication and repair, cell cycle, and p53 target genes, which may be the underlying causes of some of the observed similarities in outcome with these two treatments (36).

Efforts to understand the role of folic acid in normal cardiac development have been aided by microarray experiments as well. A comparison of gene expression in mouse embryos null for FolR1, a cellular folate transporter, and wild type revealed that the former were different in several GO groups, including cell migration, cell motility and localization of cells, structural elements of the cytoskeleton, and cell adhesion, among others (37). It is likely that effects on neural crest are at least partially responsible for the adverse cardiac development observed in FolR1 knockout mice (or animals with compromised folate metabolism or nutriture); the changes in gene expression are consistent with changes that would influence a migratory cell type. Rosenquist et al. (38) reported that expression of cell migration and adhesion genes was altered in avian neural crest cells in culture exposed to homocysteine. Homocysteine and folic acid levels are negatively correlated, so this result is supportive of the findings of Zhu and coworkers.

Mechanistic studies like these are critical to our basic understanding of the underlying basis of abnormal development. Understanding the basis for developmental toxicity will provide us with better tools for risk assessment and chemical regulation.

CONCLUSIONS

Genomics technologies have the potential to significantly accelerate progress in understanding developmental toxicity mechanisms and in improving risk assessment. At heart, genomics experiments are hypothesis-generating exercises, allowing the investigator to detect patterns of response across the range of cellular/molecular function in a way that is not possible with other technology. Examples were provided on how microarray experiments can aid in mechanistic understanding and in various aspects of hazard and risk assessment. It is still early days for the technology, so the number of examples is limited. However, the work that has been done supports the strong potential of genomics to address questions in basic and applied developmental toxicology that had earlier been unapproachable.

REFERENCES

1. Nuwaysir EF, Bittner M, Trent J, et al. Microarrays and toxicology: The advent of toxicogenomics. Mol Carcinog 10999; 24:153–159.
2. NRC (National Research Council). Applications of toxicogenomic technologies to predictive toxicology and risk assessment. National Academies Press, Washington, DC, 2007.
3. Lipshutz RJ, Fodor SP, Gingeras TR, et al. High density synthetic oligonucleotide arrays. Nat Genet 1999; 21:20–24.
4. Qian X, Scheithauer BW, Kovacs K, et al. DNA microarrays: Recent developments and applications to the study of pituitary tissues. Endocrine 2005; 28: 49–56.
5. Knudsen TB, Daston GP. MIAME guidelines. Reprod Toxicol 2005; 19:263.
6. Barrier M, Mirkes PE. Proteomics in developmental toxicology. Reprod Toxicol 2005; 19:291–304.
7. Fielden MR, Pearson C, Brennan R, et al. Preclinical drug safety analysis by chemogenomic profiling in the liver. Am J Pharmacogenomics 2005; 5:161–171.
8. Naciff JM, Jump ML, Torontali SM, et al. Gene expression profile induced by 17 α-ethinyl estradiol, bisphenol A, and genistein in the developing female reproductive system of the rat. Toxicol Sci 2002; 68:184–199.
9. Naciff JM, Hess KA, Overmann GJ, et al. Gene expression changes induced in the testis by transplacental exposure to high and low doses of 17-alpha-ethynyl estradiol, genistein, or bisphenol A. Toxicol Sci 2005; 86:396–416.
10. Liu K, Lehmann KP, Sar M, et al. Gene expression profiling following in utero exposure to phthalate esters reveals new gene targets in the etiology of testicular dysgenesis. Biol Reprod 2005; 73:180–192.
11. Shultz VD, Phillips S, Sar M, et al. Altered gene profiles in fetal rat testes after in utero exposure to di(n-butyl) phthalate. Toxicol Sci 2001; 64:233–242.
12. Chapin R, Stedman D, Paquette J, et al. Struggles for equivalence: In vitro developmental toxicity model evolution in pharmaceuticals in 2006. Toxicol In Vitro 2007; 21:1545–1551.
13. Naciff JM, Overmann GJ, Torontali SM, et al. Gene expression profile induced by 17α-ethynyl estradiol in the prepubertal female reproductive system of the rat. Toxicol Sci 2003; 72:314–330.
14. Daston GP. Do thresholds exist for developmental toxicants? A review of the theoretical and experimental evidence. Issues Rev Teratol 1993; 6:169–197.
15. Slikker W Jr, Andersen ME, Bogdanffy MS, et al. Dose-dependent transitions in mechanisms of toxicity. Toxicol Appl Pharmacol 2004; 201:203–225.
16. Thomas RS, Allen BC, Nong A, et al. A method to integrate benchmark dose estimates with genomic data to assess the functional effects of chemical exposure. Toxicol Sci 2007; 98:240–248.

17. Nagel SC, vom Saal FS, Thayer KA, et al. Relative binding affinity-serum modified access (RBA-SMA) assay predicts the relative in vivo bioactivity of the xenoestrogens bisphenol A and octylphenol. Environ Health Perspect 1997; 105:70–76.

18. vom Saal FS, Timms BG, Montano MM, et al. Prostate enlargement in mice due to fetal exposure to low doses of estradiol or diethylstilbestrol and opposite effects at high doses. Proc Natl Acad Sci U S A 1997; 94:2056–2061.

19. Putz O, Schwartz CB, LeBlanc GA, et al. Neonatal low- and high-dose exposure to estradiol benzoate in the male rat: II. Effects on male puberty and the reproductive tract. Biol Reprod 2001; 65:1506–1517.

20. Putz O, Schwartz CB, Kim S, et al. Neonatal low- and high-dose exposure to estradiol benzoate in the male rat: I. Effects on the prostate gland. Biol Reprod 2001; 65:1496–1505.

21. Cagen SZ, Waechter JM Jr, Dimond SS, et al. Normal reproductive organ development in Wistar rats exposed to bisphenol A in the drinking water. Regul Toxicol Pharmacol 1999; 30:130–139.

22. Cagen SZ, Waechter JM Jr, Dimond SS, et al. Normal reproductive organ development in CF-1 mice following prenatal exposure to bisphenol A. Toxicol Sci 1999; 50:36–44.

23. Ashby J, Tinwell H, Haseman J. Lack of effects for low dose levels of bisphenol A and diethylstilbestrol on the prostate gland of CF1 mice exposed in utero. Regul Toxicol Pharmacol 1999; 30:156–166.

24. Melnick R, Lucier G, Wolfe M, et al. Summary of the National Toxicology Program's report of the endocrine disruptors low-dose peer review. Environ Health Perspect 2002; 110:427–431.

25. Willhite CC, Ball GL, McLellan CJ. Derivation of a bisphenol A oral reference dose (RfD) and drinking-water equivalent concentration. J Toxicol Env Health B Crit Rev 2008; 11:69–146.

26. Rockett JC, Kavlock RJ, Lambright CR, et al. DNA arrays to monitor gene expression in rat blood and uterus following 17beta-estradiol exposure: Biomonitoring environmental effects using surrogate tissues. Toxicol Sci 2002; 69:49–59.

27. Rockett JC, Burczynski ME, Fornace AJ, et al. Surrogate tissue analysis: Monitoring toxicant exposure and health status of inaccessible tissues through the analysis of accessible tissues and cells. Toxicol Appl Pharmacol 2004; 194:189–199.

28. Wilson SH, Olden K. The Environmental Genome Project: Phase I and beyond. Mol Interv 2004; 4:147–156.

29. Livingston RJ, von Niederhausem A, Jegga AG, et al. Pattern of sequence variation across 213 environmental response genes. Genome Res 2004; 14:1821–1831.

30. Fertuck KC, Eckel JE, Gennings C, et al. Identification of temporal patterns of gene expression in the uteri of immature, ovariectomized mice following exposure to ethynylestradiol. Physiol Genomics 2003; 15:127–141.

31. Moggs JG, Tinwell H, Spurway T, et al. Phenotypic anchoring of gene expression changes during estrogen-induced uterine growth. Environ Health Perspect 2004; 112:1589–1606.

32. Naciff JM, Overmann GJ, Torontali SM, et al. Uterine temporal response to acute exposure to 17alpha-ethinyl estradiol in the immature rat. Toxicol Sci 2007; 97:467–490.

33. Daston GP, Naciff JM. Gene expression changes related to growth and differentiation in the fetal and juvenile reproductive system of the female rat: Evaluation of microarray results. Reprod Toxicol 2005; 19:381–394.
34. Nemeth KA, Singh AV, Knudsen TB. Searching for biomarkers of developmental toxicity with microarrays: Normal eye morphogenesis in rodent embryos. Toxicol Appl Pharmacol 2005; 206:219–228.
35. O'Hara MF, Nibbio BJ, Craig RC, et al. Mitochondrial benzodiazepine receptors regulate oxygen homeostasis in the early mouse embryo. Reprod Toxicol 2003; 17:365–375.
36. Mikheeva S, Barrier M, Little SA, et al. Alterations in gene expression induced in day-9 mouse embryos exposed to hyperthermia (HS) or 4-hydroperoxycyclophosphamide (4CP): Analysis using cDNA microarrays. Toxicol Sci 2004; 79:345–359.
37. Zhu H, Cabrera RM, Wlodarczyk BJ, et al. Differentially expressed genes in embryonic cardiac tissues of mice lacking Folr1 gene activity. BMC Dev Biol 2007; 7:128.
38. Rosenquist TH, Bennett GD, Bruer PR, et al. Microarray analysis of homocysteine-responsive genes in cardiac neural crest cells in vitro. Dev Dyn 2007; 236:1044–1054.

12

Comparative Bioinformatics and Computational Toxicology

Thomas B. Knudsen and Robert J. Kavlock

National Center for Computational Toxicology (B205–01), Office of Research and Development, U.S. Environmental Protection Agency, Research Triangle Park, North Carolina, U.S.A.

INTRODUCTION

As concern grows that human populations are being exposed to an ever-increasing number of potentially harmful agents, the field of developmental toxicology has become a national research priority (1). Understanding how embryos respond inappropriately to drugs and chemicals remains a central question in the field. Toward the end of the 20th century, it was well demonstrated that chemical agents could perturb transitory genetic signals and responses that direct morphogenesis through direct mechanisms (2,3) or through cellular damage (4,5). Moving into the 21st century, we are gaining wider appreciation of the consequences of teratogenic insult on molecular homeostasis of the embryo (6). DNA microarray analysis, which has become a widely used tool for the generation of gene-expression data on a genomic scale, is a newer technology that is increasingly being applied to experimental teratology for large-scale analysis of gene expression (7,8).

The ultimate goal in large-scale analysis of gene expression is to group genes according to similar expression patterns and functional categories (9–11). An overriding assumption for prenatal developmental toxicity is that genes which follow similar expression across a range of exposure conditions or developmental trajectories are likely to share common molecular regulatory processes and/or participate in related functions (7,9). By partitioning the embryonic transcriptome into expression profiles and compiling a list of statistically enriched biological themes associated with the responsive genes, researchers can use the hierarchical responses of the embryo to help evaluate chemical modes of action. Such information has clear and present impact on understanding the molecular events following the initiating mechanism(s) and leading to an observable phenotype.

A chemical's mode of action includes not only the primary mechanism but the ensuing cascade of heterotypic cellular changes that may include adverse (pathogenic) and adaptive (regulative) responses of the system as a whole (12). As such, gene-expression profiles alone cannot predict developmental toxicity. Biologically meaningful inferences can only be made within the context of cellular consequences and embryological phenotypes. This requires a parallel effort in bioinformatics and computational toxicology to interpret large-scale gene-expression data and predict which alterations are likely to be linked with critical events in teratogenesis. Systems-level integration of high-dimensional data and information generated at all five structural scales (molecular, cellular, tissue, individual, and population) is the mantra of life sciences research in the postgenomics era (13).

This chapter will provide an overview of bioinformatics and computational toxicology in light of our current understanding of developmental systems. It will focus on the practical applications of a life systems approach, and theoretical applications of biocomputing and virtual systems that are driving new attitudes and thinking. The chapter will address logistical questions such as what system models will likely be useful to teratologists, which basic biological units (modules) should be mapped in developmental toxicity, how "control logic" circuits may hold the key to cellular decision making, and how a systems approach may be used to predict and understand teratogenesis. The chapter is not intended to present a detailed review and analysis of the current literature; for the latter, please refer to recent reviews in developmental toxicology (8,14) (and references therein).

COMPARATIVE BIOINFORMATICS

What is bioinformatics? According to the working definition provided in 2000 by the National Institutes of Health (NIH) (http://www.bisti.nih.gov/CompuBioDef. pdf), bioinformatics refers to "research, development, or application of computational tools and approaches for expanding the use of biological, medical, behavioral or health data, including those to acquire, store, organize, analyze, or visualize such data." In practical terms, bioinformatics uses statistical approaches, models, and information management systems to compute biological relationships and discover biological principles. Computing with cellular, biochemical, and molecular information requires sophisticated mathematical models that can then be applied to test formal relationships of features linked together in a network of regulatory and metabolic pathways in development (15). Computational biology is defined by NIH as "the development and application of data-analytical and theoretical methods, mathematical modeling and computational simulation techniques to the study of biological, behavioral, and social systems."

Bioinformatics thus applies principles of information sciences and technologies to make vast, diverse, and complex life sciences data more understandable by humans. But how ready are developmental toxicologists to join the vast bioinformatics nation? Constructing analytical and predictive models for prenatal developmental toxicity is constrained by an incomplete understanding of the fundamental parameters underlying embryonic susceptibility, sensitivity, and vulnerability. To understand a developing embryo from a systems biology perspective, key milestones must be parameterized in terms of structure and dynamics, the critical control processes, and the overall design logic of gene regulatory networks. This understanding (toxicogenomics) is predicated on the availability of high-information content data from studies in developmental biology and experimental teratology, coupled with the availability of bioinformatics resources to help interpret these data in a comparative way (7,8,10–12,14,16–18).

Prenatal Toxicogenomics

A number of studies have addressed global gene-expression profiles in developmental toxicity. The underlying premise has been that regulation of the embryonic transcriptome can be used as a readout to the hidden logic of integrating molecular functions, biological processes, and subcellular responses into higher-order pathways and networks for chemical mode of action and pathogenesis of developmental defects (7). The successful application of toxicogenomics to experimental teratogenesis could lead to rethinking default values and standard protection values currently applied in risk assessments that have traditionally been mired with uncertainties in data or on assumptions about susceptibility across species, low doses, and life stages.

An early application of large-scale gene-expression profiling was from Daston and colleagues who used microarray technology to evaluate low-dose effects of environmental estrogens on female reproductive tract development (19). On the basis of growing concern that exposure to environmental chemicals with estrogenic activity perturbs reproductive tract development, the need arises for sensitive and effective methods of accurately assessing the potential estrogenicity of chemicals on development. Because a direct response on gene expression is anticipated for chemicals that act through estrogen receptors (as with other receptors in the steroid hormone receptor superfamily), Daston and colleagues hypothesized that downstream consequences of direct (genomic) versus indirect (nongenomic) effects of signal transduction pathway(s) could be interpreted for these particular classes of chemicals through classifying the genome-wide response of the system and identifying patterns of transcriptional regulation for specific genes, some of which might have known responsiveness to estrogens, others that might be as yet unidentified targets of estrogen signaling, and others perhaps linked secondarily to the exposure. They tested this hypothesis by exposing pregnant rats to three dose levels of different estrogens, namely, 17α-ethynyl estradiol, bisphenol A, and genistein, from gestation day (GD) 11 to GD 20. RNA from the uterus and ovaries was isolated 2 hours after the last dosing, profiled using Affymetrix rat gene chips, and validated with real-time polymerase chain reaction (RT-PCR). Although each chemical induced its own signature, a "molecular fingerprint" of 66 genes common to all three chemicals could be identified statistically with a similar trend for up-/downregulation, and 55 of these genes (79%) showed a monotonic dose response (19). In follow-up studies, Daston and colleagues applied comparative bioinformatics to gain a deeper understanding of the biological themes associated with estrogenicity's molecular fingerprint on uterine development (8).

Although toxicant-induced alterations in gene expression represent a mechanism of developmental toxicity common to developmental toxicants in general, not all cited toxicants are likely to be direct regulators of gene expression. This raises the question of the extent to which it is possible to broadly understand a chemical's mode of action through genomic studies. Mirkes and colleagues conducted a microarray-based study of mouse embryos exposed to two classical teratogens, hyperthermia (HS) and cyclophosphamide (CP), that have been extensively studied in vivo and in vitro and invoke widespread apoptosis in the neural epithelium (20). The system was GD 8 mouse embryo culture, and the study design was to expose embryos to HS or CP at 24 hours after initiation of the cultures. RNA from the whole embryo proper was isolated 1 hour or 5 hours later, profiled using custom 15 K mouse cDNA arrays, and validated for selected genes. From the multitude of genes that responded to exposure, the authors classified responses as specific to each teratogen and responses that were in common between teratogens. Bioinformatic tools were used to group genes into functional categories. This revealed expected responses such as upregulation of heat shock proteins following HS and of DNA repair enzymes following CP; however, unexpected patterns emerged as well such as HS-induced downregulation of genes for seven enzymes in the sterol biosynthesis pathway. Several p53-responsive genes (e.g., cyclin G1) were induced by both agents suggesting a common link to the p53 pathway.

Use of mouse whole embryo culture as a model for studying teratogenesis raises the question of concordance in the gene expression between in vivo and in vitro embryo development. Hunter and colleagues looked at this question in the craniofacial region of GD 8 mouse embryos using a similar microarray-based study with bioinformatic tools (21). RNA from the craniofacial region rostral to the first branchial arch was isolated at intervals between GD 8 and GD 11, as well as from morphologically normal mouse embryos cultured from GD 8 to GD 9, and profiled using custom 18 K mouse cDNA oligonucleotide arrays. The data sets were analyzed for transcript patterns over the time course as well as to compare the GD 9 in vivo mouse embryo with morphologically similar embryos after 24 hours in culture. From 1240 differentially expressed genes, 63 were clustered into 11 curated pathways in the Kyoto Encyclopedia of Genes and Genomes (KEGG) and/or GenMAPP libraries. The pathways included glycolysis/gluconeogenesis, cholesterol biosynthesis, oxidative phosphorylation, mitogen-activated protein kinase (MAPK) signal transduction, and the Wnt signaling pathway. Due to complex subpatterns described within each pathway, it is difficult to summarize overall patterns; however, the reciprocal effects on oxidative phosphorylation (upregulated) and glycolysis (downregulated) are consistent with the known transition of mouse embryos from anaerobic (glycolytic) to aerobic (oxidative) metabolism. Finally, these authors noted that 329 genes were differentially expressed between in vivo and in vitro developing embryos despite the absence of dysmorphogenesis.

Due to high cellular complexity of the transcriptome and of the cellular heterogeneity in regional bioassays of the embryo, it is worthwhile to profile gene expression in the individual target organ. To explore this issue, Knudsen and colleagues have undertaken experiments in early mouse embryos exposed to various ocular teratogens with the aim of correlating large-scale changes in gene expression to the critical period of eye development (7,22). Using Perkin-Elmer human cDNA arrays (2400 probes, now retired), these authors made a comparison of large-scale changes in gene expression that can be detected in the optic rudiment of the developing mouse and rat embryos across the window of development during which the eye is exceedingly sensitive to teratogen-induced micro-/anophthalmia (e.g., GD 8–10 in mouse). Microarray analysis was performed on RNA from the headfold or ocular region at the optic vesicle and optic cup stages when the ocular primordium is enriched for *Pax6*, a master selector gene during eye morphogenesis. Statistical methods (ANOVA), bioinformatic tools, and pathway analysis programs were applied to identify differentially regulated genes and map them to well-annotated pathways. These authors identified 165 genes with significant differential expression during eye development, including *Pax6* (Fig. 1). Enriched biological themes included fatty acid metabolism (upregulated) and glycolysis (downregulated). Studies such as these that benchmark large-scale gene expression during normal embryonic development are important to identify possible biomarkers that best correlate with species differences and the risks for developmental toxicity in individual target organs.

Gene-expression arrays reveal the potential linkage of altered gene expression in target organs with specific adverse effects leading to disease phenotypes. This hypothesis has been tested in the mouse embryonic forelimb bud following an acute exposure to all-*trans* retinoic acid in vivo on GD 12 (23) or all-*trans* retinol acetate in vitro on GD 11 (24). In the former case, Kochhar and colleagues isolated RNA from the mouse embryonic forelimb bud at 6 hours after maternal exposure to a teratogenic dose of retinoic acid (23). From a microarray-based analysis of 9000 probes from the IMAGE (Integrated Molecular Analysis of Genomes and their Expression) Consortium (http://image.llnl.gov/), these authors identified three-fold-induced expression of several insulin-like growth factors (*Igf1 a, Igf1b*) as well as several members of the Pbx family of highly conserved homeodomain proteins. Validation by RT-PCR and western blotting confirmed that retinoic acid induced the molecular abundance profiles of *Pbx1a, Pbx1b, Pbx2, and Pbx3* and their associated proteins within 3 to 6 hours. Furthermore, the expression of three homeobox genes that interact with PBX proteins during nuclear localization, namely, *Meis1, Meis2,*

Upregulated

	Functional category	Fisher exact test
GO biological process	organogenesis	0.00195
GO biological process	lipid metabolism	0.00555
GO biological process	neurogenesis	0.00564
GO biological process	circulation	0.0299
GO biological process	fatty acid metabolism	0.0424
GO biological process	detection of abiotic stimulus	0.049
GO cellular component	nuclear envelope-endoplasmic reticulum network	0.00568
GO cellular component	endomembrane system	0.0203
GO molecular function	transferase activity, transferring acyl groups	0.0328

Downregulated

	Functional category	Fisher exact test
GO biological process	glycolysis ***	2.7E-09
GO cellular component	membrane fraction	0.046
GO molecular function	magnesium ion binding **	0.0015
GO molecular function	isomerase activity	0.00443
GO molecular function	RNA binding	0.0177
GO molecular function	phosphoric monoester hydrolase activity	0.027

Statistically overrepresented functional categories were computed in the 165 gene list using the *EASEonline* tool of DAVID (http://www.DAVID.niaid.nih.gov). Significant (P≤0.05) gene categories were based on Fisher exact test (P≤0.05) at a threshold = 3 list hits. An asterisk (*) designates those functional categories that remained overrepresented after correcting for multiplicity with the bootstrap

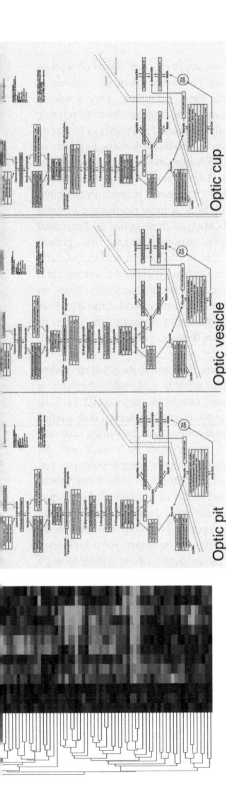

Figure 1 (*See color insert*) Molecular abundance profiles for 165 genes differentially expressed during optic cup morphogenesis in the mouse and rat embryo. Genes are displayed vertically and conditions are displayed horizontally. Expression values are colored to indicate increased (*red*) and decreased (*green*) levels relative to the corresponding reference condition; color intensity indicates magnitude of change on a log2-scale (*black is normal to the reference sample*). Statistically overrepresented functional categories were computed by Gene Ontology (GO) annotation using the Database for Annotation, Visualization, and Integrated Discovery from the National Institute of Allergy and Infectious Diseases (http://www.DAVID.niaid.nih.gov/). Downregulation of the glycolysis/gluconeogenesis pathway was visualized by GenMAPP (http://www.GenMapp.org) and confirmed by PCR (*not shown*). *Source*: From Ref. 22.

and Meis3, were similarly elevated after retinoic acid exposure. Based on a priori knowledge that IGF/PBX/MEIS play roles in early limb outgrowth, it is reasonable to surmise that changes in pathways that use these genes may play a role in the proximodistal limb reduction defects caused by teratogenic doses of retinoic acid. In a similar kind of study, Hales and colleagues used microarrays to profile genome-wide responses of the mouse embryonic limb bud following exposure to all-*trans* retinol acetate in organ culture (24). RNA from the limb bud was isolated at 3 hours of culture, profiled using Clontech Atlas mouse (1176 probes) cDNA arrays, and changes validated by RT-PCR. Most (352) genes responsive to retinol exposure were normally expressed in the limb bud. A large percentage of them were upregulated following exposure. Stratification of these genes by functional categories revealed a broad spectrum of cellular responses ranging from growth factor signaling and transcriptional regulation to cell adhesion and cytoskeletal regulation. This implies an acute effect of retinol excess on general developmental control, thus deregulating genes at multiple levels of signaling networks. Based on bioinformatic tools to construct gene product association networks, these authors proposed that cross talk between signaling cascades may propagate the effects of retinol on the limb and disrupt cellular processes that are critical for normal limb morphogenesis.

It is interesting to note in comparing these two microarray-based studies on retinoid-induced teratogenesis, for example, retinoic acid–induced effects in vivo on GD 11 (23) and retinol acetate-induced effects in vitro on GD 12 (24), that despite very different candidate genes from the microarray screen both studies raised the issue of cross talk between upregulated pathways and resulting network-level changes that may underlie the localized disruption of limb outgrowth. An open question is how network-level responses measured in a specific precursor target cell subpopulation might contribute to localized disruptions of limb outgrowth: are the retinoid-induced effects directly related to the deregulation of pathways driving limb outgrowth, to the replacement of critically damaged proteins, or to a secondary response to compensate for damaged cells? While the large body of data generated from gene-expression profiling can be mined for a high-level perspective of the physiological state of the system, details of the critical biology of the system must be linked to other kinds of assays and experimental designs. For example, Collins and colleagues used whole genome QTL (quantitative trait locus) scans to map the critical chromosomal region for differential susceptibility of two murine strains to 13-*cis* retinoic acid–induced forelimb ectrodactyly: C57 BL/6 N mice are susceptible to this induced phenotype whereas SWV mice are not (25). Results based on 88 *Mit*-markers revealed significant linkages to D11Mit39 and D4Mit170 on chromosomes 11 and 4, respectively. These authors discussed multiple candidate genes for the strain differences, with the strongest arguments for RARα and Wnt3/9b that map to the critical region for differential suscepti-bility on chromosome 11 based on the clinical and developmental information available.

Gene-expression arrays reveal the potential linkage of altered gene expression in precursor target cell populations with specific adverse effects leading to disease phenotypes. But how closely do microarray data reflect early physiological or pharmacological measures that predict toxic event(s)? Fetal Alcohol Syndrome (FAS), a severe consequence of a mother's overindulgence during pregnancy, can cause craniofacial defects, optic defects, mental retardation, and stunted growth. Knudsen and colleagues provide evidence that changes in gene expression within specific molecular pathways are the basis for induced phenotypes (12). They used C57BL/6J and C57BL/6N mice that carry high (B6J) and low (B6N) risk for dysmorphogenesis following maternal exposure to 2.9 g/kg ethanol (two injections spaced 4.0 hours apart on GD 8), and counter-exposure to PK11195, a ligand for the mitochondrial 18 kDa translocator protein (*Bzrp/TspO*) that significantly protected B6J embryos. Microarray analysis of mouse cranial neural folds at 3.0 hours after the first maternal alcohol injection revealed metabolic and cellular reprogramming that was substrain specific and PK11195 dependent. The different responses may be attributed to genetic variation. Mapping the ethanol-responsive KEGG pathways revealed significant upregulation of tight junction, focal adhesion, adherens junction, and regulation of the actin cytoskeleton (and a near-significant upregulation of Wnt signaling and apoptosis pathways) in both substrains. Indeed, additional alcohol-induced changes are found in B6N alone, downregulation of ribosomal proteins and proteasome, and upregulation of glycolysis and pentose phosphate pathways (Fig. 2). Expression networks constructed computationally from these altered genes identified entry points for ethanol at several hubs (*Mapk1, Aldh3a2, CD14, Pfkm, Tnfrsf1 a, Rps6, Igf1, Egfr, and Pten*) and for PK11195 at *Akt1*. These findings are consistent with the growing view that developmental exposure to alcohol alters common signaling pathways linking receptor activation to cytoskeletal reorganization. The programmatic shift in cell motility and metabolic capacity further implies cell signals and responses that are potentially integrated by the mitochondrion.

The ability to determine which gene ontology (GO) terms apply or which biological pathways are enriched among subsets of differentially expressed genes provides an important model to gain understanding of how the embryo reacts as a system to chemical exposure(s). As noted earlier, various methods have been used to rank biological processes predicted from microarray studies. A GO-based algorithm, referred to as "GO-Quant," has been proposed (11). GO-Quant analyzes functional gene categories as they change across dose and time, based on statistics, GO annotation analysis, and Z-score test. The algorithm essentially tests the null hypothesis that all significantly altered genes in randomized variance (ANOVA) are randomly distributed across GO categories (http://depts.washington.edu/irarc/Go-Quant/). This unsupervised approach to defining biological pathways based on the cumulative response of genes in a MAPPFinder pathway for calculating the corresponding ED_{50} for each specific GO term, which is said to be *"important for risk assessment"* (11).

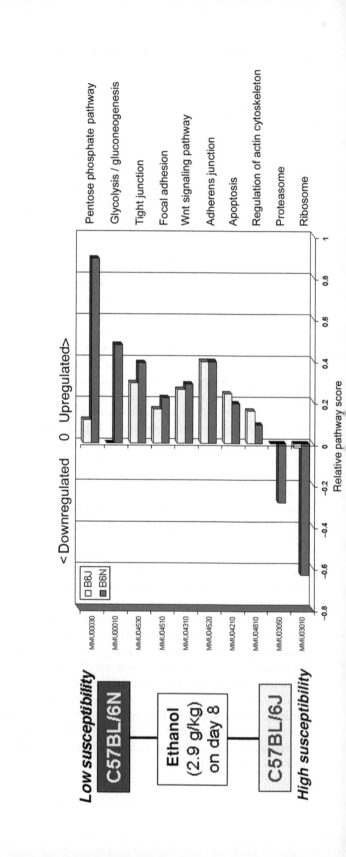

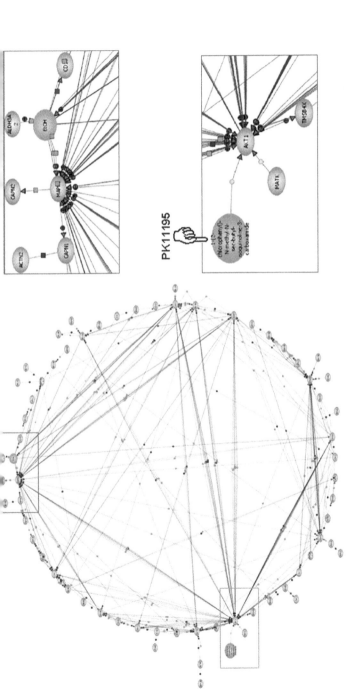

Figure 2 (*See color insert*) Hierarchy of significant KEGG Pathways and predicted computational gene network defining FAS induction. Upper panel: Values above and below zero reflect pathway upregulation and downregulation, respectively, for the response of B6J and B6N cranial neural folds to maternal alcohol treatment (3.0 hr after 1 × 2.9 g/kg on GD 8). Lower panel: Computational gene product association network predicted from genes associated with the KEGG pathways portrayed with Pathway Architect v1.2.0 (Stratagene). Gene products are coded in red and small molecules applied to this study (EtOH, PK11195) in green. Lines indicate curated interactions in the ResNet database for direct binding (*purple*), direct regulation (*blue*), and direct modulation (*grey*). *Source*: From Ref. 12.

Databases to Support Prenatal Toxicogenomics Research

With application of high-throughput screening tools in toxicogenomics to problems in prenatal developmental toxicology researchers are faced with questions about how to handle and interpret complex data sets, particularly with regards to the relative plasticity of precursor target cell populations in the embryo and the realization that morphogenetic processes have critical timing for inductive events and for regulative growth. As demonstrated earlier, research databases and knowledge management systems that hold data and information about molecular biology and conventional toxicology are the cornerstone for a comparative bioinformatics approach and enable the information architecture to support modeling formation of the embryo (*embryo-formatics*) (14).

Biological databases represent a large, organized body of persistent data, usually associated with computerized software designed to update, query, and retrieve components of the data stored within the system. The Human Genome Project raised awareness of the need for extensive cyber-infrastructure and algorithms for the management, organization, and analysis of genomics data (13,26). Cellular, biochemical, and molecular information is increasing at an alarming pace due to genomics data submissions to national gene repositories and a large number of databases derived from primary data sources. With the quantity of genetic and protein data reaching peta bytes (quadrillion bytes), researchers are increasingly dependent on curated databases that adapt to the evolving technologies.

Public databases accessible via the world wide web are tracked annually by the online Molecular Biology Database Collection (http://nar.oxfordjournals.org/). The 2008 update of this collection includes 1078 databases and represents an increase of 110 databases over this time in 2007 (27). A great deal of genomics data is available in the national database repositories. Gene Expression Omnibus (GEO) [http://www.ncbi.nlm.nih.gov/geo/] is the NIH-based repository for high-content transcript data yielded by gene-expression profiling (26). Growth in GEO submissions is evidenced by tenfold increase in records between June 8, 2004, and January 16, 2008. This increase from 18,235 to 194,322 sample records reflects an average rate of 135 new samples entered per day. Many of these samples would be relevant to prenatal developmental toxicologists. Often these data set accession numbers are not reported in the published literature, nor do all references to microarray data in the published literature cycle back to original GEO records. For example, in 2006, we queried PubMed with the keywords "embryo" and "microarray." This query returned 495 records of which 193 actually used the technology to study developing animal systems. The GEO data set (GDS) limiter in PubMed narrowed the list to 47 nonredundant microarray data sets. Including "teratogen" added a few more data sets, for a grand total of 564 public microarray data sets addressing developmental health and disease. Most GDS pertained to mouse embryos and differentiating human cell lines, but a few data sets represented zebrafish, toad, rat, and chick embryos. In the fall of 2006, we projected the growth of public microarray samples to approach 1500 records in embryogenesis and teratogenesis by the year 2010 (14). A list of web-based resources useful for comparative bioinformatics analysis in prenatal developmental toxicology has been recently tabulated (14).

Several open-access and commercial tools are available to derive gene–gene relationships and genetic dependencies based on integrating information obtained on a genomic scale with biological knowledge accumulated through years of research of molecular genetics, biochemistry, and cell physiology (28). Data-driven projects are underway for making large data sets available publicly to be used for studies integrating cellular states with gene-expression signatures following exposure to small molecules (ligands, drugs, and chemicals) (29–31). For example, Lamb et al. (30) created a reference collection of gene-expression profiles from cultured human cells treated with bioactive small molecules and applied data-mining software to establish a "connectivity map" between drugs, genes, and diseases (http://www.broad. mit.edu/cmap). Extensibility of these resources to prenatal developmental toxicology remains to be confirmed; however, at a minimum, this would require network identification from gene-expression data describing normal development, an integration of data for abnormal developmental phenotypes, and the application of scaling laws to move from the embryo to the connectivity map of cells.

Knowledge Management Systems

The amount of information from genome-based discovery efforts has easily outpaced the resources and theory needed to understand it—the problem of "data-overload, information under-load" (32). With so much data in hand, the challenges in data management are shifting from issues of data organization and storage to domain-specific questions such as how to most effectively and completely extract and mine context-based representations. Contextual data integration is a challenge in toxicology where, for example, much of the data are disconnected in operability by loose semantics (metadata) (33). The lack of predictive value in disconnected data gives rise to a "knowledge gap" and in embryology-teratology, just as in other fields, this raises important questions that need to be resolved if the potential for predictive developmental toxicology is to advance (14). Knowledge-base expert systems, data-driven statistical systems, and data-mining algorithms are some of the important techniques that are applied to large databases to identify previously unsuspected relationships or hidden patterns in a data set and to classify or predict behavior. In toxicogenomics, the ability to conduct data meta-analysis on multiple data sets from unrelated studies has been attempted as a means to increase the predictive power of technologies that are data rich but information poor (34).

Information and knowledge management has become a hot research area covering advanced methods for the management of complex biological data sets. Some key topics include knowledge representation and extraction, specialized data structures, modeling of complex biological domains, intelligent methods for information retrieval, and semantic database integration. The application of knowledge management tools opens doors to complex relationships that are not detectable otherwise. Many databases are managed with commercial relational database management systems. Despite their advantages in terms of data management (availability, reliability, scalability, accessibility, and archival), relational database management systems are usually developed with commercial applications in mind.

The demand for a public integrative information management system that may be tailored to specific knowledge domains in developmental toxicology motivated the design, development, and implementation of a web-accessible resource referred to as "Birth Defects Systems Manager" (BDSM) (7,10,14,18,35). BDSM contributes architecture for warehousing preprocessed data and metadata relevant to mouse embryonic development and toxicity as well as tools to model data for interesting patterns across developmental stages, organ systems, and disease phenotypes. Its programming retrieval capability, QueryBDSM (Fig. 3) allows specific queries across experiments to facilitate secondary analysis of developmental genomics data and the applications of comparative bioinformatics across

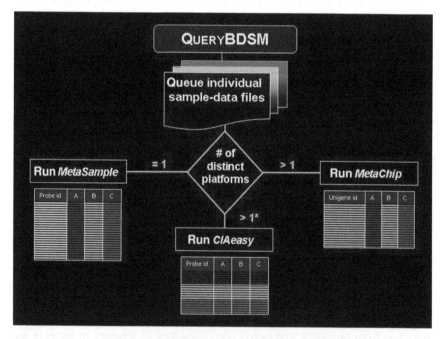

Figure 3 (*See color insert*) Workflow schema for the QueryBDSM module of BDSM. Individual files of normalized microarray data are selected from the GEO library. QueryBDSM determines the number of distinct microarray platforms in the sample queue and merges the data as follows: if all samples come from the same platform, then MetaSample is used; if multiple platforms are represented, then MetaChip is used. CIAeasy compares joint trends in expression data for the same samples run on different platform. *Source*: From Ref. 18.

technology platforms and study types (18). The prototype is currently focused on the mouse embryo but is extensible to all homologene species, including mouse, rat, human, and zebrafish genomes as well.

Formalizing the associative relationships between an anatomical structure and its spatial location, functional system, and chronological stage in the embryo requires hierarchical information. Such detailed information is handled through "ontologies" that link facts as a triad of related terms. Two of the best-known ontology languages, namely, OBO (Open Biomedical Ontology) and OWL (OWL Web Ontology Language), have been used to write developmental ontologies for Theiler Stages [see (36)] and Carnegie Stages [see (37)] in mice and humans, respectively. The OBO and OWL hot links can be found at http://obofoundry.org/ and include web-accessible resources such as CARO (Common Anatomy Reference Ontology), CL (Cell Type), ZFA (Zebrafish Anatomy and Development), and EMAP (Mouse Gross Anatomy and Development) that will be useful for systems-level modeling.

In considering developmental gene expression, the Edinburgh Mouse Atlas Project (EMAP, http://genex.hgu.mrc.ac.uk/) is an authoritative portal for tracking spatiotemporal gene-expression data during mouse embryogenesis (38,39). The core database contains three-dimensional reconstructions of the mouse embryo at various Theiler stages of development (spatial models), a systematic nomenclature of embryo anatomy (anatomical ontology), and embryonic territories (domains) (40–43). As of January 2008, EMAP coverage is 10,020 genes/proteins and 27 Theiler stages. EMAP and BDSM can complement one another as resources for embryo-formatics. Whereas BDSM contributes a low-resolution embryo space/high-content gene space model using microarray data from GEO (10,35), EMAP contributes a high-resolution embryo space/low-content gene space model using experimental data from the Mouse Genome Informatics (http://www.informatics.jax.org/) (41,43).

Expert Systems

Expert systems are computer applications that carry out a degree of logical reasoning similar to those of human beings. As such, they make subject-matter expertise available to nonexperts. The two main components of an expert system are knowledge base (KB) and inference engine (IE) (44): KB holds domain knowledge as a collection of rules, for example, formalized truths extracted from actual experience; and IE draws appropriate deductions by logically compiling the set of rules triggered by an input query. Similarly, Bassan and Worth (45) suggested the two main building blocks as inventories (KB) storing information on chemicals, molecular structures, models, predictions, and experimental data; and tools (IE) for generating estimates, providing information on their reliability, and documenting their use. Although a KB of rules in a developmental toxicology expert system are provided by the experience and expertise of the scientific community, many dozens of rules may be needed to predict teratogenicity. A wide spectrum of informatics resources can be found on the environmental bioinformatics Knowledge Base (ebKB) at http://www.ebkb.org, which is undergoing alpha testing.

Developmental Systems Biology

Although genomics has increased as a research tool in the developmental toxicology community, we are at a relatively early stage in using these data to understand mechanisms of developmental defects. As noted earlier, this requires familiarity with the availability and limitations of numerous biological databases, some of which are domain specific and others that are more generally applied. Because the sensitivity of genomic technologies can reveal more about the potential effects of lower-dose exposure to harmful chemicals than has been possible using traditional techniques, integrating these kinds of data into predictive developmental toxicology can extend conventional testing approaches and fundamentally change the future of risk assessment. However, the myriad of changes in gene expression that accompanies adverse responses to developmental toxicants can get entangled with the adaptive and programmed responses that represent "normal" regulation of the embryonic transcriptome (22). A challenge is to now link changes in genomic/proteomic expression, both adaptive and adverse, to sequential changes in phenotype at a systems level and in a manner that is consistent with the underlying embryological and teratological mechanisms.

What is a System?

The emphasis on "systems" raises the question: what exactly is a system? A basic definition is that a system comprises any group of entities comprising a whole, where each component interacts with at least one other component working toward the same objective. Key properties of many biological systems are similar in concept to mechanical (engineered) systems (Fig. 4).

Thinking about systems has been around for a while. Aristotle (circa 350 BC) spoke about "wholeness" and the concept that one cannot understand the pivotal properties of a complex system solely from the knowledge of its individual parts. He reasoned that behavior of a "system" could only be understood by knowing how the different parts were connected and how these connections influenced collective behavior of its constituent members. Norbert Wiener in 1948 coined the phrase "cybernetics" and showed that in engineered systems the collective behavior can be driven by control processes that are functionally organized and automated by a set of defined rules. Ludwig von Bertalanffy in 1968 addressed "general systems theory" in a physical sense as being dependent upon universal principles of self-organization: nonequilibrium state (energy), hierarchical organization (control), and resilience to perturbation (robustness) (46). Idekar and coworkers defined "systems biology" in 2001 as the integrated study of biological systems at the molecular level, involving perturbation of systems, monitoring molecular expression, integrating response data, and modeling the systems molecular structure and network function (47). Hiroaki Kitano also in 2001 derived a basic paradigm for

Components ←→	Device ←→	Module ←→	System
genes	signals	network	cell
cells	tissue	organ	organism
organisms	family	community	population

Figure 4 Hierarchical organization of systems. The interaction of parts in self-organizing biological systems is similar across scales to mechanical (engineered) systems.

systems biology thinking, using concepts from reverse-engineering: identifying system structure, analyzing system behavior, finding the control points, and testing for optimal system design (48).

Systems Biology

Availability of sophisticated data-mining tools and pattern-recognition software, coupled with knowledge of connectivities from molecular libraries, makes it feasible for developmental biologists and toxicologists to think about biological systems as functional networks that can be associated with the mode of action of chemicals (Fig. 5). Declan Butler, *Nature's* European correspondent, stated that "If biologists do not adapt to the powerful computational tools needed to exploit huge data sets they could find themselves floundering in the wake of advances in genomics" (49). With vast amounts of high-dimensional data now in hand, the knowledge environment has led to an absolute requirement for computerized databases and analysis tools. In the current "decade of informatics," an ability to create mathematical models describing the function of networks of genes, proteins, metabolites, and cells has become just as important as traditional wet laboratory research skills to address these basic and applied research goals (49).

O'Malley and Dupré (51) described two schools of thinking in systems biology: a practical school that focuses thinking on data integration and gene network reconstruction, where the genome is given high informational priority (integrative biology); and a theoretical school that deprioritizes the genome and focuses on network theory and predicting system behavior (biosystems modeling). In reality, the gap between these two schools of thinking has been narrowed with greater recognition of the problem that research to dynamically model cellular responses must employ a variety of experimental conditions to infer biological networks that reflect how the complex system reacts quantitatively to the totality of disease variables. Such a framework must incorporate data from different research platforms and be robust to missing or incomplete data. This presumes reasonable knowledge of biological systems under normal conditions and network-level responses to different stressors (47,52,53).

Of course, a biosystem cannot be broken into component parts and then reassembled blindly. Researchers must use mathematics with some degree of quantification to formalize entailments (implications) that constrain behavior in order to construct models that accurately predict systems-level behavior. Since biological

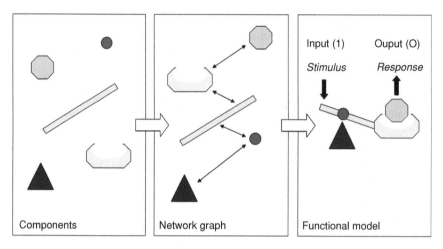

Figure 5 Reconstruction of a functioning system [redrawn after Covert (50), page 194]. Starting with component parts, models are used to build connectivities and test working relationships. In modeling a complex cellular system, statistical methods are used to identify the component parts (genes, proteins, and metabolites) using high-throughput data collection experiments. Bioinformatic tools are then used to reconstruct biological networks. Mathematical methods are used to build models that can simulate input–output relationships. Input–output relationships are tested experimentally and evaluated for how well the model predicts system behavior (e.g., dose–response). Other parts can be added to the model as needed to improve goodness of fit.

processes are inherently nonlinear, many approximations are required to solve these problems numerically and computationally. Therefore, a true systems-based approach to understanding and predicting cellular behaviors can only be accomplished through a comprehensive program that includes statistics, bioinformatics, mathematics, and computer science in addition to the traditional scientific methods (Fig. 6). This requires that biologists and modelers work together to identify critical information and knowledge. *"Molecular biology took Humpty Dumpty apart; mathematical modeling is required to put him back together again . . ."* (54).

Developmental Toxicology in the Era of Systems Biology

One research area using a systems approach to address these kinds of questions is the FAS and Fetal Alcohol Spectrum Defects (FASD). As described earlier (Fig. 2), alcohol exposure in pregnant mice induces network-level changes in the embryonic transcriptome that predicts cellular changes in the receptor-mediated cell adhesion system (12). In contrast, Faustman and colleagues have modeled cellular consequences of maternal alcohol consumption on reduced fetal brain growth (17,55). These latter studies modeled rates of cell growth and cell death in the developing neural epithelium and stage-dependent susceptibility to alcohol using cellular-level data collected from various model systems exposed to

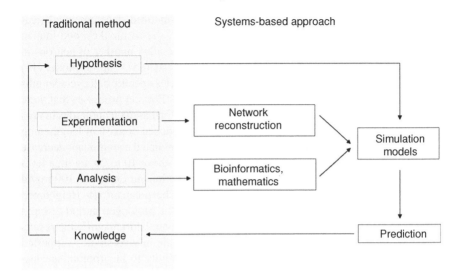

Figure 6 Scientific method for biosystems modeling [redrawn after Covert (50), page 192]. Although several varied definitions have been applied, systems biology can be considered a way of thinking in which computational tools are used to model interactions between and among the basic components, hence connectivities. Both practical (integrative biology) and theoretical (biosystems modeling) schools of thinking are required for a true systems-based approach. These are schools indicated in blue and red, respectively, to expand the traditional scientific method shown in green.

alcohol at different concentrations. The key model parameters therefore included cell cycle rates, differentiation rates, and cell death rates within a critical time period. Progenitor cells making up the periventricular epithelium of the developing rostral neural tube were the target cells, which under normal conditions proliferate and migrate to populate the cortical plate during GD 13 to GD 19 in the rat. Effects on cell death rates during a period of rapid synaptogenesis (postnatal days 0 to 4) were also modeled. Due to the plethora of experimental studies on developmental neurotoxicity, including ones with ethanol but also for other neurotoxicants, these researchers were able to rely extensively on existing data for model parameterization and on outcomes from a number of studies in the published literature on normal brain development and the responses to ethanol. The effort represents a strength of a systems-based approach to dose–response assessment in capitalizing on the diverse types of information already available and integrating this information into a formal model from which inferences can then be drawn. The model was able to demonstrate that although acute ethanol exposure can lead to an intense spike in apoptotic neurons during synaptogenesis, apoptosis did not predict significant long-term loss of neurons. Rather, a relatively small lengthening of the cell cycle at the beginning of neurogenesis was predicted to result in massive neuronal deficits in the mature neocortex.

In the recent extension of work on ethanol-induced developmental neu-rotoxicity, computational models of the rat, mouse, rhesus monkey, and human were developed that described the acquisition of the adult number of neurons in the neocortex during neurogenesis and synaptogenesis (55). The rat model (17) was extended by incorporating time-dependent, species-specific cell cycle lengths and neuronal death rates. All rates were modeled as Poisson processes that were allowed to vary according to an exponential distribution based on numerous in vivo data sets. An assumption was made that ethanol induced consistent lengthening of the cell cycle across species. The human model predicted a significant decrease in neuronal cells at blood concentrations of ethanol above 10 to 20 mg/dl, a level that can be achieved after a single drink. Concentrations in excess of 100 mg/dl were needed to yield similar projected deficits in the pregnant rat. Heightened sensitivity of humans versus rats was attributed to the prolonged period of rapid growth in the former species and thus to a greater potential to magnify the results of small impacts over time. Comparisons to available in vivo studies supported the application of the model across species. The ability to incorporate species-specific dynamic factors demonstrates another utility of systems-level biological models. Through computational tools, experimental results from one species can be placed in context of results expected in another when detailed information about normal biology is available for comparative stages of organ system development as well as pharmacokinetic information to enable a comparison of target organ doses at equivalent stages. If such detailed knowledge were available, it would be possible to test for the ability of an agent to elicit specific malformations based on morphological and physiological similarity across species, especially at the most susceptible early stages of development, when comparative developmental landmarks are most similar (56).

Knudsen and colleagues (14) framed a basic premise in developmental systems biology as: how do cells in a developing embryo integrate complex signals from the genome to make decisions about their behavior or fate, and under what conditions are these decision-making systems susceptible to genetic defects or vulnerable to environmental perturbations? Meta-analysis of microarray-based studies (10) and inferences drawn from gene pathway reconstruction (11) can exploit the extensive genome expression data available from individual studies. The meta-analysis of developing organ systems across species (22), of similar systems across exposure conditions (10,57) or of different systems across development (18), can serve to integrate different kinds of information from multiple nonidentical conditions (7). For example, Rodriguez-Das and colleagues (57) expanded the meta-analysis on microarray data sets from early mouse embryos exposed to several diverse teratogens. They corroborated reports that diverse teratogens share some of the same effects on the embryonic transcriptome while others are unique to the treatment levels or genetic strains (10,12). Shared biological themes uncovered by meta-analysis included glycolysis, cell communication, and proteasome pathways whereas the unique themes included lipoprotein, vitamin, and steroid metabolic pathways (57). In that regard, it will be interesting to learn if the network-level model of FAS (12) can be cobbled with the cell-level model of FASD (55).

Formal models are being constructed in simpler developmental model organisms, such as sea urchin, in order to simulate the hierarchical control within gene regulatory networks that control cellular events in morphogenesis (58–60). Davidson (61) refers to the networks of transcription factors and signaling systems as the "regulatory genome." In developmental toxicology, we might consider a broader collection of system profiles for genetic and cellular networks under all possible gene and environment perturbations as "developmental systeome." Recently, a database of coexpressed gene networks has been published for human and mouse (62). The coexpressed gene database (COXRESdb, http://coxpresdb.hgc.jp) provides networks and gene lists ordered by the strength of coexpression for human and mouse tissues, based on highly coexpressed genes, genes with the same GO annotation, genes expressed in the same tissue, and user-defined gene sets. Extending COXRESdb to include information on the developmental systeome across chemical space, dose responses, time series analysis, stage of development, and test species would give consideration to integrative biological networks.

Biological networks would include not only gene regulatory networks but metabolic networks also. Indeed, cells may be viewed as an integration of these two basic kinds of networks that direct the flow of molecular information in very different ways. Regulatory networks, being signal-flow oriented, serve as information processing systems. They endow the cell with the ability to make adjustments as controllers in response to programmed (genetic) and inducible (environmental) signals. Metabolic networks on the other hand are mass-flow oriented. These relatively fast reactions are driven by flux of metabolites and serve as engines of cellular processes (63). In network theory, "scale" refers to connectivity (64). Highly evolved gene regulatory networks in natural systems tend to show recurrent motifs (65) and scale-free topology (66). To illustrate this concept, consider a sample of 32-node network of genes/proteins (Fig. 7). In a random network, all nodes would have a low-degree of connectivity. The system readily falls apart when one or few nodes (genes) or edges (connections) are disrupted; hence, there is little or no robustness to minor perturbation following toxicant exposure. In contrast, when the same 32 nodes are portrayed as a scale-free topology, most nodes have low connectivity whereas a few nodes (hubs) have a high degree of connectivity. This modular system would display "graceful degradation" following random loss of a node or edge. Hubs are the Achilles' heel of a scale-free network; hence, the system is vulnerable to targeted attack at the hubs. Placing network theory in the context of toxicity pathways may contribute to predictive screening of developmental toxicants and risk assessment based on the premise that "*patterns of altered gene expression are specific for mode of action*" (8). A challenge to the application of computational models in dose–response assessment is resolving the genes/pathways involved in the biological process leading to an adverse response, as well as the quantitative relationship between expression of these genes/pathways and

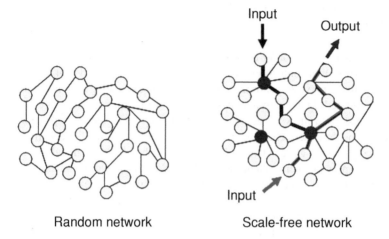

Figure 7 Scale-free network topology [redrawn after Wikepedia commons (67)]. Sample 32 node networks, where nodes represent genes/proteins and edges represent interactions connected randomly or in modular (scale-free) topology. Hubs are colored in black. Signal input and output is shown for several paths in the scale-free network.

changes at higher levels of biological organization at which the adverse effect is manifested.

Risk Assessment Applications of Computational Models

Properly designed and interpreted animal studies have traditionally been the primary tools of safety screening and predictive toxicology (68). The standard procedure for assessing the safety of chemicals to the developing organism involves the use of pregnant laboratory animals (generally rats and/or rabbits), exposed throughout the period of major organogenesis, and subsequent examination of the offspring for growth and morphology. Embryo–fetal effects can be considered as the end point of importance, after excluding that their occurrence is related to maternal toxicity (69,70). EPA's guidelines for developmental toxicity risk assessment (FRL-4038–3, December 5, 1991) are recorded in the Federal Register 56(234):63,798–63,826 and updated in a 1998 workshop (SAP Report No. 99–01 C, January 22, 1999) (71). The guidelines state that when experimental data are looked at "the agents causing human developmental toxicity in almost all cases were found to produce effects in experimental animal studies and, in at least one species tested, types of effects similar to those in humans were generally seen."

In regulatory toxicology, the concept of a "threshold dose" below which no adverse effect is observed provides the basis for recommended maximal exposures that are considered safe. Once a dose–response pattern is characterized and a critical effect is selected in a relevant laboratory animal bioassay, the highest experimental dose level lacking a statistical or biological increase in the critical effect (the NOAEL, or no observed adverse effect level) is extrapolated to humans by dividing by uncertainty factors that take into account the likelihood of humans being more sensitive than the laboratory model and that there are potentially sensitive subpopulations within the general human population. The resulting exposure level, termed a Reference Dose (RfD) by the U.S. Environmental Protection Agency (EPA) or Acceptable Daily Intake by the World Health Organization, is presumed to be without risk over a lifetime of exposure.

In the most basic form of extrapolation of a critical effect in a laboratory animal study to the human situation, little or no information about the metabolism of the chemical or potential modes of action is involved in estimating the safe level of exposure. Because animal models are used, the number of chemicals that can be evaluated is rather limited. For example, more than 100,000 chemical substances are listed in the European Inventory of Existing Chemical Substances (72). The REACH (Registration, Evaluation, and Authorization of Chemicals) legislation (http://ec.europa.eu/enterprise/reach/index_en.htm) passed by the European Parliament in 2006 will require toxicity evaluations of up to 30,000 chemicals, which far exceeds the number of developmental toxicity studies that have been conducted worldwide over the past several decades.

Capacity to accurately predict dose-dependent toxicity in alternate models would greatly aid the risk assessment process through chemical prioritization for animal testing (73); however, alternative testing methods have met with limited success and mixed results in developmental toxicity. The challenge is that such assays must address multiple end points in systems that have relevant developmental characteristics (8). Free living zebrafish embryos are promising for addressing pathways in developmental toxicity (http://www.zygogen. com/) but are limited in pharmacokinetics and human hazard identification. Mouse ES cell (74) and whole embryo culture (http://www.eggcentris. com/) assays have shown promise but are limited in scope for tissue pathways (e.g., cardiomyocytes) and life stage (e.g., early postimplantation). The limitations on the number of chemicals that can be tested using traditional animal bioassay models, and the desire to reduce uncertainties in the extrapolation process of high dose to low dose and animal to humans, is leading to an increased emphasis on the use of computational tools to increase the throughput and to quantitatively integrate various sources of information in the risk assessment process (75).

Computational models of developmental toxicity can facilitate the understanding of how complex factors from the genome and environment interact with the developmental program [reviewed in (16)]. The computational tools can be grouped into two general categories. One category is based primarily on statistical approaches to data and includes benchmark dose (BMD) models and quantitative structure-activity relationship (SAR) models. The former seeks to provide a more uniform departure point for extrapolation of animal testing data while the latter attempts to extend toxicity information across chemicals by building models that link various physicochemical properties with biological responses. The second category of computational tools involves greater consideration of biological processes. Biologically motivated applications involve both physiologically based pharmacokinetic (PBPK) models and mode of action considerations in biologically based dose response (BBDR) models. In the future, computational models of normal embryogenesis will be linked to mode of action information to develop in silico–based predictive models that can be scaled across dose and species (e.g., embryologically based dose response models and virtual embryos).

The following sections summarize the current state of the science in each of these areas of computational developmental toxicology. We will emphasize their potential impact on making better use of existing experimental data, on integrating diverse types of information into better predictive tools across dose and species, and in posing "what if" questions that can be used to explore various facets of concern such as variations in experimental design, chemical form, or greatest sources of uncertainty. The power of computational tools is perhaps best demonstrated where these questions can be explored in silico without involvement and cost of additional experimental information.

BMD Models

The use of the NOAEL-based approach to developmental toxicity risk assessments, with its reliance on pairwise comparisons between a treated and control group, can actually penalize studies with higher statistical power and does not make full use of the dose–response information. To address limitations of the NOAEL, Crump in 1984 proposed the use of mathematical models of the dose response to estimate the lower confidence limit on a predefined risk level (e.g., 5% response) (76). By using all the experimental information, this BMD encourages the use of more powerful experimental designs and makes comparison of potencies across studies more standard. A series of studies evaluating the results of almost 300 standard developmental toxicity studies demonstrated that a variety of mathematical models, including those that incorporate developmental-specific features as litter size and intra-litter correlations, can be readily applied to standard developmental toxicity study designs (77–80). BMDs corresponding to a 5% added risk of effect calculated on quantal end points (e.g., whether an offspring is affected or not) were on average equivalent to traditionally calculated NOAELs. For a continuous variable such as fetal weight, several litter-based benchmark effect levels (a difference in 5% mean fetal weight, a decrease to the 25th percentile mean weight of control litters, a decrease in mean weight by 2 standard errors, and a decrease of 0.5 standard deviation units) result in similar BMD levels and to statistically derived NOAELs. The impact of dose spacing and group size over the NOAEL approach was obvious in both quantal and continuous models. The influence of study design (numbers of dose groups, dose placement, and sample size per group) was further evaluated based on mean squared error of the maximum likelihood estimate of the BMD (81). The best results were obtained when two dose levels had response rates at about the background level, one of which was near the ED05 (e.g., the dose level resulting in a 5% increase in response over background) level.

There was virtually no advantage in increasing the litter size from 10 to 20 litters per group. The poorest results were obtained when only a single dose group with an elevated response was present, and that response rate was much greater than the ED05. Overall, the results indicated that while the BMD approach is readily applicable to standard developmental toxicity designs, minor design modifications would increase the accuracy and precision of the BMD.

The BMD approach has also been applied to compare effective concentrations following exposure of chemicals using whole embryo culture and to in vivo developmental toxicity tests (82). While a clear correlation between in vitro benchmark concentrations and in vivo BMDs was found among the 19 chemicals evaluated, these results showed that predicting in vivo BMDs from in vitro benchmark concentrations would be difficult due to large confidence intervals of the estimate. Presumably, this is largely the result of lack of knowledge of the kinetics of a chemical reaching the target tissue in the two assays. Software to perform BMD calculations is freely available (http://www.epa.gov/ncea/bmds/).

Quantitative SAR Models

With the increasing demands on regulatory agencies throughout the world to assess the hazards of large chemical inventories, the use of SAR models has received considerable attention (83). These models seek to correlate features of chemical structure or chemical properties with some biological activity, and have played an important role in drug discovery where the goal is to find efficacious chemicals against well-characterized targets. SAR models can either be qualitative (does or does not the chemical have the activity) or, if enough potency information is available, they can be used to predict the actual effective dose or concentration of the chemical that would cause some response. Historically, SAR models have proven less useful in defining the hazards of environmental chemicals against apical end points such as developmental toxicity, in large part because of diverse and largely unknown molecular targets and the paucity of carefully designed prospective studies involving sufficient numbers of chemicals (84). These limitations are particularly apparent in the application of SAR models outside carefully defined chemical classes.

To help overcome some of the hurdles to use of SAR models, the Risk Science Institute of the International Life Sciences Institute (ILSI RSI) has been coordinating the efforts of a working group charged with examining the potential of SAR systems for use by regulatory agencies as screening tools for developmental toxicity. ILSI's Developmental Toxicity Database Project has the primary goal of supporting a system to evaluate statistically based SAR prediction models (http://rsi.ilsi.org/devtoxsar.htm) from data and information captured from the published developmental toxicology literature (85). The ILSI working group considered two types of SAR models: rule based and statistically based. Rule-based models tend to require more information on the nature of molecular interactions causing toxicity but have not seen as much effort in developmental toxicology as have the more correlative statistical models that rely on information on the physicochemical properties.

The ILSI working group reviewed the two commercial SAR models for developmental toxicity during the study period of 2003: TOPKAT (86,87) and MultiCase (88,89). Accelrys' Discovery Studio (DS) TOPKAT® model is based on a training set of 234 chemicals tested under similar conditions in the rat (87). MultiCase (89) is based on qualitative determinations of potential for human developmental toxicity in TERIS (The Teratogen Information System) assessments (90) and FDA drug classifications for fetal risk (91). It included 323 chemicals, approximately one-third of which were considered active. In their evaluation, the Working Group noted a number of issues that diminished confidence in the model. One challenge was in the evaluating of data for the presence or absence of developmental effects.

Another is that the framework data for toxicology reside in a highly distributed system of literature, databases, and paper records. As such, developmental toxicology databases are needed to facilitate efforts aimed at determining relationships within classes of chemicals, animal species, and human populations for developmental and reproductive end points. With funding from *Health Canada,* the ILSI database is currently being expanded to capture data using the Developmental and Reproductive (DART) engine of the TOXNET Toxicology Data Network (http://www.nlm.nih.gov/pubs/factsheets/toxnetfs.html) [Beth Julien, personal communication]. TOXNET is a system of databases from the National Library of Medicine that provides information on toxicology, hazardous chemicals, and environmental health including developmental and reproductive studies (92).

For training the models in an objective, rational, reproducible, and transparent fashion, the "activity" being modeled must be clearly defined. For example, does it include all manifestations of developmental toxicity, or is it limited to some particular aspect such as growth retardation or even more specifically to some particular morphological defect? Consequently, another challenge faced by the ILSI working group was how to score the "activity" being modeled given the complexity of the data and the diversity of the data sources (e.g., different strains, species, dosing regimens, examination techniques, presence of potential confounders such as maternal toxicity) and the relative scarcity of studies that limits the chemical features covered in the training set. The ILSI working group recommended that there be a concerted effort to compile in an electronic format the available developmental toxicity data that could be used for model development, examine the use of existing alternative training sets, and develop new tools and approaches. Similar efforts to outline steps needed for regulatory acceptance of SAR models has been provided by the OECD (93).

The DevTox Project (http://www.devtox.org) was initiated by the German Federal Institute for Risk Assessment (BfR, formerly BgVV) and sponsored by the German Federal Ministry of the Environment, Nature Conservation, and Nuclear Safety under the auspices of the WHO's International Programme on Chemical Safety (IPCS). DevTox uses harmonized nomenclature to describe developmental anomalies in laboratory animals, assist in the visual recognition of developmental anomalies with the aid of photographs, and provide a historical control database of developmental effects in laboratory animals. It follows a series of three international terminology workshops held in Berlin that brought together international participants from research, regulatory agencies, and industry. The results of these meetings were compiled and presented in a form useful to professionals working in developmental toxicology (94–96). Although still in planning, a historical control database detailing the naturally occurring incidence of fetal anomalies in laboratory animals will soon become available for regulatory authorities involved in the registration of pesticides, chemicals, biocides, and drugs. Overall, the terminology forms a compendium of malformation terms that could be regarded as potential outcomes from developmental toxicity experiments in animal studies.

With the advent of more mode of action screening as evidenced in the Endocrine Disruptor Screening Program of the U.S. Environmental Protection Agency (EPA) (http://www.epa.gov/scipoly/oscpendo/index.htm) (97) that is examining the potential for chemicals to interact with the estrogen, androgen, and thyroid hormone signaling systems, there is likely to be increased information available in the future to base construction of SAR models that are targeted at lower levels of biological organization than apical measures of developmental toxicity. Already there have been a number of efforts to develop SAR models for binding of environmental chemicals to the estrogen receptor (98–100), which follow on efforts of the pharmaceutical industry to do so for drug discovery.

Even more recent efforts are attempting to use broad scale high-throughput screening assays to analyze hundreds or more molecular or biochemical targets of chemicals in order to develop bioactivity profiles that are characteristic of various toxicological phenotypes including developing systems. The ToxCast™ program at the EPA (31) is using a large number of assays garnered from the drug discovery process (101) to develop prognostic bioactivity profiles for several hundred chemicals for which there is a current wealth of standard toxicological studies including developmental toxicity, multigeneration studies, and subchronic and chronic studies in rodents (102). This approach is not too dissimilar from that recently proposed by the National Research Council in their report on "Toxicity Testing in the Twenty-First Century: A Vision and a Strategy" (73). Like the recommendations for development of improved SAR models, the advent of high-throughput screening with the capacity to test thousands of chemicals in multiple assays will increase the pressure to develop high quality, publicly available and electronically accessible databases of traditional toxicity data and methods to integrate diverse sources of data (103).

Chemical Databases to Support Predictive Screening

Several public and commercial databases support the generalized needs for computational toxicology in predictive screening [reviewed in (34,74,104)]. Toxicity databases were classified by informatics content (literature, data collections, and chemical structure-toxicity), toxicity end points (carcinogenicity, genetic toxicity, target organ toxicity, developmental and reproductive toxicity, and ecological toxicity), and data-mining workflow for common tasks in product development (read-across tables, training sets, and historical control databases) and product safety (submission data and assess risks) (34). For example, in read-across tables, the various entities (chemicals) are ordered by a priori biological measurement data and chemical structure, and the toxicity trends within the same row and column are then evaluated to make predictions. Applying computational (in silico) approaches to the gallows of data for toxicology in general and developmental toxicology in particular raises four fundamental and interrelated problems (104): (i) electronic capture and standardization of data from experimental toxicology, (ii) aggregation and representation of toxicity data into forms that are meaningful and useable by modelers, (iii) interoperability of toxicity "data silos" with the larger world of toxicity data informatics, and (iv) central indexing of chemicals and toxicity data in a web accessible form.

Three public initiatives are directly addressing the aforementioned challenges facing the capture, organization, interoperability, and indexing of toxicity data collections [see (104)]. One is ToxML (Toxicology XML standard). The Extensible Markup Language (XML) is intended for storage and exchange of data or information. As its name implies XML is a markup language, meaning that it is used to convey information about text or other data using embedded codes that are not easily read or deciphered by humans. XML syntax rules can functionally represent data from virtually any subject domain; hence, data can be processed (parsed) using a set of common core software tools that understand universal XML syntax rules (105). ToxML is a publicly available schema of ontology-based fields and controlled vocabulary that includes developmental and reproductive studies (106). Under a cooperative research and development agreement with the U.S. Food and Drug Administration (FDA), Leadscope, Inc. has built databases enriched with food ingredients from the Center for Food Safety and Applied Nutrition (CFSAN) and pharmaceuticals from publicly available data within the FDA's Center for Drug Evaluation and Research (CEDR) [see (104)].

Several other public initiatives come from the EPA. The DSSTox (Distributed Structure-Searchable Toxicity) database network publishes summarized structure-annotated toxicity data files for use in SAR modeling and locator files for public toxicity data inventories (107). Although not focused on developmental toxicity, the DSSTox structure-data files span nearly 7000 chemicals (104). Many of these chemicals have a known hazard to the fetus, including methylmercury, valproic acid, or retinoic acid; however, DSSTox does not explicitly categorize chemicals as developmental toxicants. EPA's new ACToR (Aggregated Computational Toxicology Resource) data integration system, scheduled for release in 2008 (108), is attacking the broader problem of surveying all publicly available chemical toxicity resources and building tools to allow the construction of toxicity data sets for quantitative biological-activity relationship modeling. Currently, the ACToR system contains chemical structure, in vitro bioactivity data, and summary toxicology data for over 500,000 compounds derived from more than 200 sources. In addition, ACToR is serving as the primary data management system for new data being generated in EPA's ToxCast™ research program (31).

Within EPA, the Toxicology Reference Database (ToxRefDB) is being developed to house reference in vivo toxicology data (109). This relational database holds conventional toxicity data from chronic, subchronic, reproductive, and developmental animal studies. Its primary goal is to provide traditional toxicological data in a searchable format, similar to the ILSI effort, but extracting data primarily from EPA Pesticide Data Evaluation Records (DERs) for hundreds of registered pesticidal active ingredients (110). Those records contain data on all toxicologically significant effects from legacy toxicity studies in rats, mice, rabbits, and among other test species.

PBPK Models

Traditional one or two compartment pharmacokinetic models are useful for understanding the kinetics of a particular data set and to estimate parameters such as metabolic half-lives, but have limited ability to provide understanding about other issues such kinetics outside the observed range, the impact of various exposure scenarios, or species differences. PBPK models attempt to overcome these limitations by constructing physiologically relevant, and species and life stage specific, mathematical representation of key events such as blood flows, organ volumes, partition coefficients, metabolic rate constants, and other parameters that are key determinants of absorption, distribution, metabolism, and excretion. PBPK models have proven very useful in describing the time and dose-dependent internal dosimetry from a particular experiment as well as making predictions of target dose in other situations by appropriate adjustments of the model parameters.

Prenatal developmental toxicity poses several challenges for PBPK models. Compartment sizes of maternal and fetal units change over pregnancy, differences in placental physiology must be considered, and exposures can bridge into the perinatal period and thus entail transplacental as well as potential lactational exposures. In addition, the small size of the rodent conceptus during critical periods of development makes experimental validation a challenge. Despite these challenges, considerable progress has been made in the applications of PBPK models. Corley and colleagues reviewed a number of PBPK models that were developed for gestational (14 chemicals) and lactational (3 chemicals) exposures. These authors noted how PBPK models initially developed for adults were adapted for the specific application to developmental toxicants (111). By providing a quantitative framework for simulating target concentrations of a chemical, or of its metabolites, the administered dose can be more rationally coupled to the target tissue response. This can infer clues as to mode of action, point to new experiments that would fill gaps, and promote understanding of the human relevance of exposures in laboratory animal models. Although there were examples in the peer-previewed literature of the application of PBPK in risk assessment (e.g., 2-methoxyethanol) at the time of the review, there were no examples where a regulatory agency used a PBPK model in risk assessment for a developmental or neonatal toxicity—other than to modify uncertainty factors in a draft risk assessment of perchlorate by the U.S. EPA (111). The authors surmised that a lack of apparent use was driven by several factors, including the paucity of cases when developmental toxicity would be the critical effect driving a risk assessment, that positive findings in developmental toxicity studies may lead to product replacement and therefore the avoidance of need for a risk assessment, an insufficient number of scientists in regulatory offices who are trained in development of PBPK models, and the lack of clear guidelines to judge whether a PBPK has been fully validated for regulatory use. Additionally, these authors noted specific research needs such as creation of annotated databases of biological parameters for model construction that are species and strain specific, greater understanding of biochemical mechanisms of placental transport, the impact of uterine position, the metabolic capacity of the dam and fetus during pregnancy, and the use of noninvasive imaging techniques for assessment of tissue function as well as chemical deposition. A number of the criticisms have been addressed since then, including a framework for regulatory acceptance (112) (http://cfpub.epa.gov/ncea/cfm/recordisplay.cfm?deid = 157,668) and the compilation of databases of parameters to use in the development of the models (113) (http://www.epa.gov/comptox/parameters.html).

A recent risk assessment on the fumigant iodomethane by EPA's Office of Pesticide Programs used a PBPK model in conjunction with extensive mechanistic data to model developmental toxicity, respiratory tract lesions, and thyroid hormone perturbations identified as critical effects following inhalation exposure (http://www.epa.gov/opprd001/factsheets/iodomethane.htm). Exposure of pregnant rabbits via inhalation to 0, 2, 10, or 20 ppm iodomethane resulted in increase in fetal loss and decreased fetal weight at 20 ppm in the absence of any adverse maternal effects. The PBPK model was used to identify the most appropriate dose metric and internal dose in calculating a human equivalent concentration (HEC) rather than using the default administered dose and uncertainty adjustments of the reference concentration approach. The model describes the nasal dosimetry and glutathione depletion in the rat to evaluate nasal toxicity. It also describes iodine kinetics in the pregnant rabbit to address developmental toxicity, and the distribution of methyl iodide to the brain to describe the dose metric for neurotoxic effects. The dose metric used for fetal loss was the fetal serum inorganic iodide level after a single day of exposure. Based upon the PBPK model, HECs of 7.4 (nonoccupationally exposed bystander) and 23 ppm (occupational exposure) were derived from the experimental NOAEL of 10 ppm for developmental effects. Taking into consideration the pharmacokinetic differences between species that were accounted for by the PBPK model, the uncertainty factor applied to the HEC was reduced from the default 10× to 3× for pharmacokinetic differences, while the 10× uncertainty factor was retained for the pharmacodynamic extrapolation.

The power of computational models to explore variations in study design was exemplified in a recent study of lactational exposure. Yoon and Barton (114) conducted a simulation study to determine whether using data on adult pharmacokinetics and milk transfer in conjunction with a biologically based model to predict how different pharmacokinetic properties (clearance rates, milk distribution) would affect exposure of the pups. The simulation covered the period from birth to 28 days of life, with the first 21 days including lactational exposure. The model incorporated changes in body weight, milk production and consumption, and food consumption using a model in which the dam and pups were each represented as a one compartment model. Absorption and elimination was via a nonsaturable first-order process. Three exposure scenarios during lactation were studied: chemical delivery in the diet at a fixed concentration, chemical concentration in the diet adjusted for increase in maternal food consumption during lactation, and gavage exposure. These scenarios reflect commonly used study designs in neonatal toxicity studies.

Model output was compared with experimental data for two chemicals (ochratoxin A and 2,4-dichlorophenoxyacetic acid), and then 16 theoretical compounds were studied that varied in elimination rate, volume of distribution, and extent of milk transfer in order to gain insight of experimental design on pup exposures. The results showed that levels of a chemical in the pup frequently do not parallel that in the dam (Fig. 8). Pup levels tended to be lower than in the dam with short half-life compounds whereas the reverse was true for long half-life chemicals.

The PBPK model simulations were able to explore a wide variety of interacting factors; thus, even with a short half-life compound, pup exposure could be similar to that of the dam if there is a smaller volume of distribution or a delay in development of excretion and recirculation of the chemical. In another example, pup exposure was predicted to be very low (less than 1% of the mother) when milk concentrations are much less than maternal blood, or when the volume of distribution was large for a short half-life chemical. The study also pointed out the need for more information on determinants of milk concentration, on dose-related metabolism, and knowledge to craft more physiologically comprehensive models (114). Even with these limitations, however, it is clear that computational models have value not only for integrating present knowledge but also for evaluating study design choices and developing hypothesis to interpret toxicity findings.

Biologically Based Dose–Response Models

Inclusion of mode of action information into computational BBDR models represents the second facet of trying to improve the input of information into the risk assessment process (115). Mode of action refers to the nature of the interaction between a toxicant exposure and the sequence of biological events that ultimately leads to an adverse health outcome. The pioneering effort in BBDR models was the work of Moolgavkar and Luebeck in 1990 in proposing a two-stage model of carcinogenesis (116). The complexities of normal developmental processes, the size of the conceptus during critical developmental periods, the rapid rate of development, and the paucity of detailed information on mechanisms of action

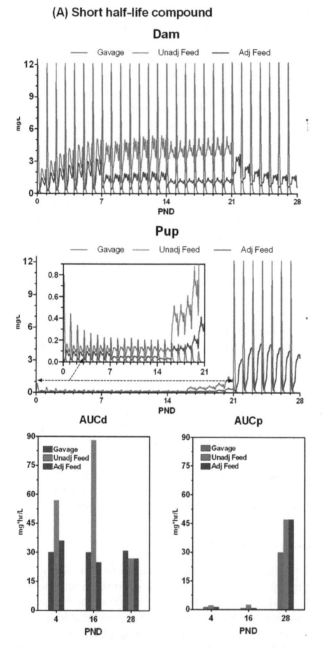

Figure 8 (*See color insert*) Predicted concentrations and area under the curve (AUC) values for base case compounds in the pup versus dam as computed with a PBPK model. Pup levels tended to be lower than in the dam for short half-life compounds whereas the reverse was true for long half-life chemicals. *Source*: From Ref. 114.

(B) Long half-life compound

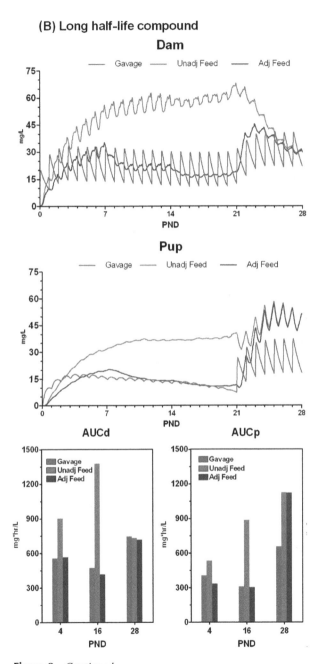

Figure 8 *Continued*

make construction of BBDR models for developmental toxicity resource intensive (117–119).

One of the most comprehensive efforts to develop a BBDR model for developmental toxicity was the work of Lau and colleagues (119–121). These investigators selected the antiproliferative agent 5-fluorouracil (5-FU) as a prototype developmental toxicant by virtue of a relatively well-characterized molecular target, namely, thymidylate synthetase (TS), and severe developmental malformations of the appendicular skeleton following maternal exposure on GD 14. A conceptual model for the proposed mode of action of 5-FU on embryonic development is shown in Figure 9.

5-FU was administered to pregnant rats subcutaneously on GD 14 over a broad dose range (1–40 mg/kg). Drug levels peaked in the maternal serum within 20 minutes and displayed linear pharmacokinetics with a half-life of approximately 14 minutes. Different parameters in the hypothesized toxicity pathway were monitored between 1-hour and 72-hour postexposure. Embryonic TS inhibition was observed within 1-hour postexposure with recovery apparent by 24 hours. Inhibition of TS enzymatic activity followed 5-FU doses as low as 5 mg/kg and peaked at 70 to 80% inhibition with the 40-mg/kg dose level. Therefore, disruption of the molecular function attributed to 5-FU (TS inhibition) was quantitatively related to the applied dose between 5 mg/kg and 40 mg/kg. Significant changes in deoxynucleotide (dNTP) pools followed exposure to the 10-mg/kg dose level and higher. Embryos displayed a rapid depletion of dTTP (expected) and dGTP (unexpected) levels, whereas the other dNTPs remained flat (dATP) or increased (dCTP). Reductions in embryonic DNA and protein contents were significant at the 30-mg/kg dose level and higher. Fetal weight reduction was detected starting on gestation GD 15 and became progressively more severe versus controls as gestation advanced.

TS inhibition drove a model of dNTP pool alterations, which were in turn linked to fetal growth reduction and to the incidence of malformations (Fig. 10). Setzer et al. (120) developed a mathematical model relating maternally administered 5-FU to embryonic TS inhibition. Several outputs were considered: deficit of dTTP alone, deficit of both dTTP and dGTP, and total deviation from control for all four dNTPs. The resulting models predicted that there would be no threshold for dNTP pool alterations. In contrast, the relationship between dNTP pool status and fetal weight reduction suggested a biological threshold downstream of dNTP status. The nature of the critical change was not defined, although it could be due to adverse or adaptive changes in cellular consequences, such as rates of cell growth or death.

Although biologically based and inclusive of a reasonable amount of available data, the linkage model was still considered semiempirical because it relied on experimental data for estimating many of the model parameters and because the mechanistic links between alterations in dNTPs and adverse developmental outcome were not identified. Extension of the model to other exposure conditions, and subsequent experimental confirmation, would be necessary to further enhance the predictive capability of the 5-FU model. Nevertheless, it did

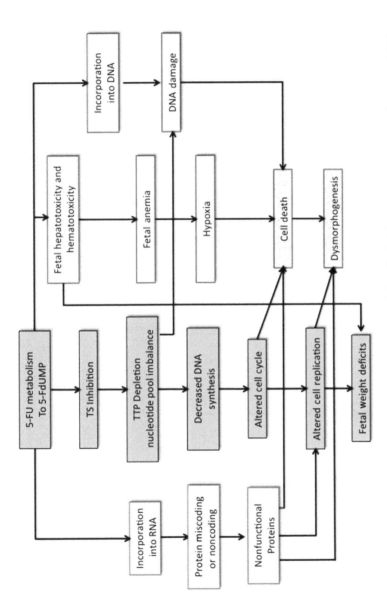

Figure 9 Mechanistic model of 5-FU developmental toxicity. Shaded boxes are parameters relevant to 5-FU's hypothesized toxicity pathway for modeling; other boxes are also possible effects related to 5-FU developmental toxicity. *Source*: From Ref. 121.

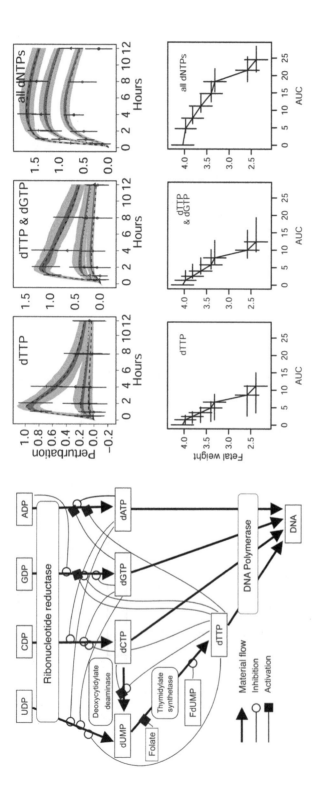

Figure 10 5-FU induced perturbations in embryonic deoxyribonucleotide triphosphate (dNTP) pools and fetal growth. Left panel: Elements of the dNTP biosynthesis pathway included in the BBDR model of 5-FU on rat embryo development. Upper right panel: Time-dependent changes in dNTP pools observed and predicted by the fitted model. Vertical bars represent 95% confidence intervals for the observed values. Gray regions indicate 50% (*dark*) and 95% (*light*) pointwise confidence intervals for the predicted values. 5-FU dosages are 10 mg/kg (*open octagons and short dashes*), 20 mg/kg (*open triangles and medium dashes*), and 40 mg/kg (*pluses and long dashes*). Upper left panel: Relationship of alterations in three measures of dNTP pool perturbations by GD 14 exposure to 5-FU on fetal weight on GD 21. The data suggests a threshold for effect on fetal growth somewhere downstream of the immediate biochemical response on nucleotide pools. *Source:* From Ref. 120.

provide insights into the dose–response relationships and emphasized the need for quantitative information about relationships among various parameters. This information is often lacking in the scientific literature. Furthermore, while it is possible to incorporate mechanistic information into quantitative dose–response models, the process is enormously data intensive and costly. Confidence in the model is ultimately dependent on the level of biological understanding of the system (119). This is only likely to be achieved through a systems approach to confirm linkages between the proximate etiologic events involved (dNTP perturbations), intermediate events accounting for presyptomatic latency (alterations in DNA and RNA synthesis), cellular consequences (cell growth and death), and developmental phenotype (fetal weight reduction or specific malformations). Building empirical linkage models of developmental defects assumes defined pathway(s) that lead to a particular outcome and implies quantitative changes that can be measured in these "toxicity pathways" over dose and time. Homologous mechanisms should be present in human development; therefore, irrespective of whether BBDR models can quantitatively predict human risk, they at least contribute qualitatively to define mode of action, human relevance, and significance of responses. Toward reducing uncertainties in data interpretation, they can also inform the low-dose extrapolation of experimental data (e.g., linear vs. nonlinear). Future BBDR model development is projected to focus on the shape of dose–response relationships for exposure conditions likely to be encountered by humans. Demonstrating the value of BBDR modeling will require close alignment of those working in basic science and science policy.

FUTURE DIRECTIONS

Developmental biology is among the most active areas of modern life sciences research, and a solid foundation of knowledge about the molecular control of development and the nature of interacting systems in development is being laid down (http://www.alttox.org/ttrc/toxicity-tests/repro-dev-tox/way-forward/daston/). As noted earlier, computational models that integrate vast amounts of mechanistic data being generated in developmental biology and genetics would offer a valuable service to the risk assessment community; however, the complexity and diversity of regulatory issues in developmental toxicology make incorporating biological details into risk assessment a nontrivial task.

Computer Simulators of Embryogenesis

In the future, high-tech virtual tissue models of the human embryo may help scientists understand the developmental risks posed by chemicals and other environmental stressors (122). Computer-based tools for modeling morphogenesis based on cellular automata (CA) have attracted recent interest (123–128). So-called "virtual tissues" may someday hold a significant place in the regulatory toxicologist's toolbox to predict potential developmental toxicity.

In late 2007, EPA launched a computational framework for applying these tools to developmental toxicity (129). The goal of this new "Virtual Embryo Project" (*v-Embryo*^{TM}) is to advance EPA's Computational Toxicology program by building capacity for predictive modeling and regulatory analysis of developmental processes and toxicities. *v-Embryo*^{TM} aims to portray critical effects of environmental agents on embryonic tissues as computer simulations that draw from an open-source toolbox addressing knowledge regarding the flow of molecular regulatory information in rudimentary tissues, the cell-autonomous responses to genetic (programmed) and environmental (induced) signals, and the emergent morphogenetic properties associated with collective cellular behavior in any given system.

For proof of principle, *v-Embryo*^{TM} will address early eye development. Motivation for computational modeling of any particular developing system needs a framework simple enough to be computationally tractable and yet a framework complex enough so as to warrant a computational modeling effort. The eye as a morphogenetic system has well-studied anatomical landmarks and developmental phenotypes. It requires precisely timed interactions between surface ectoderm and forebrain neuroepithelium that give rise to the lens and cornea (surface ectoderm) and pigmented retina and sensory retina (neural ectoderm). Many of the key genes and signaling pathways orchestrating these inductive phenomena are conserved across species and systems. As such, eye development is simple enough to capture a great deal of experimental detail for computer simulation and yet complex enough to warrant the engineering of a computational toolbox that can be applied to analyze critical events in silico. Through *v-Embryo*^{TM}, EPA is crafting a model that will indicate how chemical exposures and nutrition and genetic factors interact to affect eye development. The computer animation technology will then help researchers understand what those changes mean for fetal eye structure.

An ideal virtual system would simulate events at one scale that have quantitative and dynamic impact on events at a higher scale (130). Hybrid CA simulators have been used to integrate collective cell behavior into models of spatial patterning and growth during limb morphogenesis (125–128). In hybrid models, the CA is updated using information collected by internal parameters (genetic) and the immediate neighborhood (environment) (124). Cell configuration at a lattice site of the automaton would be specified by complex parameters such as local gene regulatory networks, cellular and extracellular matrix properties, and cell–cell or cell–environment interactions (125–128).

Several open-source software applications are available for CA modeling. Consider, for example, CompuCell3D (CC3D, http://compucell.sourceforge.net). Developed at the University of Notre Dame and Indiana University, this simulation environment CC3D runs three interactive modules to simulate cellular structure, chemical gradients, and cell state-type change. The core modules are all written in C++ and can be downloaded as "source" and as "binary" file directories. An extension, BioLogo, employs XML-based syntax to generate fast C++ code that can be edited by nonprogrammers (127).

v-Embryo™ will apply CC3D and other agent-based models to Bard's developmental ontology (36) that is predicated on four core processes: (i) patterning, which sets up future events in cellular subpopulation; (ii) proliferation and apoptosis as the basis of growth and shaping; (iii) cell differentiation, which refers to changing of cellular phenotype; and (iv) morphogenesis as the generation of spatial organization (e.g., movements). As an example of how these connections can be used, consider these results from a microarray study on the response of the headfold of early mouse embryos to maternal alcohol (12). Part of the response showed differential effects on the input genes into the receptor-mediated focal adhesion pathway, whereby the sensitive strain (B6J) trended upward in expression of these genes versus the refractory strain (B6N) that trended downward under the same conditions. Reprogramming of the receptor-mediated adhesion system at 3-hour postexposure implies differential cell adhesion. Cell autonomous sorting (biological) is modeled by using a cell sorting automaton (computational) in CC3D run through 10,000 flips over several minutes. We can modify cell-type and state parameters in the XML (BioLogo) aspects of CC3D to incorporate predictions from gene-expression networks (Fig. 11).

Projecting Virtual Models of Morphogenesis to Phenotype-Space

Projecting abnormalities from a computer model to a range of abnormal developmental phenotypes such as might be observed in experimental teratology (e.g., "phenotype-space") is a challenge. For example, eye reduction defects (ERDs) may be measured as both continuous (incidence) and quantal (graded) parameters in dose-dependent severity (131). Stochasticity is likely since the outcome is variable in the real world. Therefore, to make a computational experiment more like nonlinear systems in the real world, some degree of stochasticity must be applied either through permutations or random noise. Under those conditions one might predict a slightly different outcome each time a simulation is run. By introducing higher variables such as litter effects or maternal parameters, we might then begin to grow the data in the same way as data might be collected in a traditional test. Applied mathematics can be useful. At the risk of oversimplification we illustrate here, with one example, catastrophe theory.

Catastrophe theory describes sudden shifts in behavior of a nonlinear system (132,133). In catastrophe theory, a perturbation causing changes in the equilibrium of a system can cause a sudden jump to a new state (hysteresis). Most biological systems are assumed to occupy a state between order and chaos. This is referred to as the edge of chaos and is a complex state that provides an optimal balance of flexibility and stability (134). For example, the embryo may exist in stable equilibrium at "State A" or "State B" due to attractor/repeller factors that buffer the edge of chaos. Entering "State B" would take a large or small perturbation to induce transition depending on where in equilibrium space the system resides. The buffer between states may be underdeveloped or compromised under some conditions

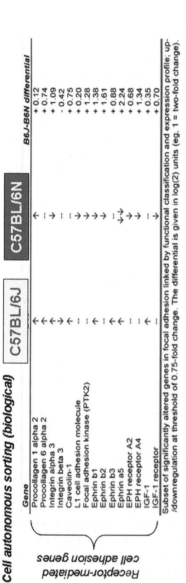

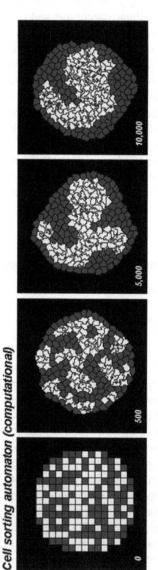

Figure 11 Differential cell adhesion model applied to a simple cell sorting phenotype. Reprogramming of receptor-mediated cell adhesion implied cell adhesivity in the rank order B6 J≥B6 N. The differential adhesion hypothesis prescribes that less adhesive cells sort out from the core. Cell autonomous sorting (biological) is modeled by using a cell sorting automaton (computational) in CC3D run through 10,000 flips over several minutes. We can modify cell-type and state parameters in the XML (BioLogo) aspects of CC3D to incorporate predictions from gene-expression networks (12,127).

and the "critical point" may be where the change required to cause a sudden jump is very small. Catastrophe theory applies to such systems at equilibrium.

Biologically, four levels of codimensionality (x, y, z, and t) can be graphed as a "Zeeman Plot" that divides equilibrium into stable and unstable domains, with the critical point at the interface (Fig. 12). State transitions can be projected onto the graph and used mathematically to predict the range of effects associated with the ERD spectrum, using essential understanding of the system but without detailed knowledge of underlying processes.

Applications to Risk Assessment

Among the short- and long-term goals for *v-Embryo*^{TM}, the question arises as to ultimately if it can be used to test new chemicals for potential development impacts, provide additional data on dose–response at various stages of prenatal development, and how the project may help EPA make informed decisions in the future. It is important to recognize that these models are a long way off in the future for regulatory use. In the shorter term, they can be important for integrating all the various sources of information coming from high-throughput and high-content biological assays with our understanding of normal cell and tissue biology. Industrialized in vitro methods are the focus of new data collection strategies to provide the number of test results needed to build a computational algorithm that has broad coverage of chemical space (73). It is envisaged that once trained and validated *v-Embryo*^{TM} will enable in silico prediction for experimental outcomes on new chemicals. In the longer term, they should be useful for "what-if" kinds of computational experiments to test hypotheses about modes of action, to understand aspects of concentration by time relationships (e.g., what exposure for what length of time will cause the toxicity), and to help predict which populations might be susceptible to prenatal developmental toxicity as we learn more about genetic polymorphisms.

ACKNOWLEDGMENTS

The authors thank members of the Knudsen laboratory (Amar Singh, Yihzi Zhang, Maia Green, and Caleb Bastian) and colleagues in the Center for Environmental Genomics and Integrative Biology (Greg Rempala, Eric Rouchka, Ted Kalbfleisch, Nigel Cooper, Ken Ramos, Susmita Datta, and Rudy Parrish) at the University of Louisville for long-standing intellectual contributions to concepts in this chapter. The authors also thank Imran Shah and Richard Judson of the EPA's National Center for Computational Toxicology for insights into data management and network modeling, and to the editors of this volume, Barbara Abbott and Deborah Hansen, for their technical manuscript review.

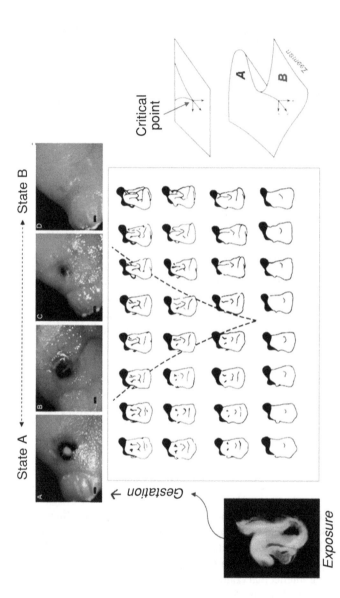

Figure 12 (*See color insert*) Catastrophe theory: projecting output of computational models to phenotype-space. Top panel: A range of malformations can often be detected in a prenatal developmental toxicity study. For example, consider the range of ERDs in mouse fetuses following maternal exposure to ethanol (12). ERD phenotypes range from normophthalmia to different degrees of anophthalmia and complete apparent anophthalmia, representing a transition of the embryo from normal (State A) to abnormal (State B) as development advances from exposure on GD 8 to the fetus at GD 17. The phenotype transition can be modeled through a range of features. Consider, for example, the diagram showing features shifting columnwise from a man's face to a woman (135) [reprinted from (132)]. Each row of images reflects a progression of features that would be drawn and, likewise, would develop over time following maternal alcohol exposure on GD 8, leading to an anatomically complete (**A**) or incomplete (**B**) eye on GD 17. Features in the formal system are added general to specific, much like during natural morphogenesis. Each column represents a possible solution from CC3D; each row represents morphing of features across states A and B. Sweeping different parameters in the model would determine the range of probabilities (%) in each outcome.

DISCLAIMER

This manuscript has ben reviewed by the U.S. Environmental Protection Agency and approved for publication. Approval does not signify that the contents necessarily reflect the views and policies of the Agency, nor does mention of trade names or commercial products constitute endorsement or recommendation for use.

REFERENCES

1. National Research Council. Scientific frontiers in developmental toxicology and risk assessment. National Academy Press, Washington, DC, 2000.
2. Cooper MK, Porter JA, Young KE, et al. Teratogen-mediated inhibition of target tissue response to Shh signaling. Science 1998; 280:1603–1607.
3. Kochhar DM, Jiang H, Penner JD, et al. The use a retinoid receptor antagonist in a new model to study vitamin A-dependent developmental events. Int J Dev Biol 1998; 42:601–608.
4. Fleming A, Copp AJ. Embryonic folate metabolism and mouse neural tube defects. Science 1998; 280:2107–2109.
5. Mirkes PE, Little SA. Teratogen-induced cell death in postimplantation mouse embryos: Differential tissue sensitivity and hallmarks of apoptosis. Cell Death Differ 1998; 5:592–600.
6. Finnell RH, Gelineau-van Waes J, Eudy JD, et al. Molecular basis of environmentally induced birth defects. Ann Rev Pharmacol Toxicol 2002; 42:181–208.
7. Knudsen TB, Charlap JH, Nemeth KR. Microarray applications in developmental toxicology. In Perspectives in Gene Expression, Appasani K, ed. Eaton Publishing/BioTechniques Press, Westboro, MA, 2003:173–194.
8. Daston GP. Genomics and developmental risk assessment. Birth Defects Res A 2007; 79:1–7.
9. Thibault C, Wang L, Zhang L, et al. DNA arrays and functional genomics in neurobiology. Int Rev Neurobiol 2001; 48:219–253.
10. Singh AV, Knudsen KB, Knudsen TB. Computational systems analysis of developmental toxicity: Design, development and implementation of a birth defects systems manager (BDSM). Reprod Toxicol 2005; 19:421–439.
11. Yu X, Griffith WC, Hanspers K, et al. A system-based approach to interpret dose- and time-dependent microarray data: Quantitative integration of gene ontology analysis for risk assessment. Toxicol Sci 2006; 92:560–577.
12. Green ML, Singh AV, Zhang Y, et al. Reprogramming of genetic networks during initiation of the fetal alcohol syndrome. Devel Dynam 2007; 236:613–631.
13. BIOINFOMED Study. Synergy between medical informatics and bioinformatics: Facilitating genomic medicine for future healthcare. European Commission (EC)—Directorate General: Information Society, Brussels. EC-IST 2001–35024, 2003. http://www.infobiomed.org/paginas_en/project.htm
14. Singh AV, Rouhka EC, Rempala GA, et al. Integrative database management for mouse development: systems and concepts. Birth Defects Res C 2007; 81:1–19.
15. Carroll SB. Endless forms most beautiful: The new science of evo devo and the making of the animal kingdom. W. W. Norton & Company, Inc., New York, NY, 2003:350.

16. Cummings A, Kavlock RJ. A systems biology approach to developmental toxicology. Reprod Toxicol 2005; 19:281–290.

17. Gohlke JM, Griffith WC, Faustman EM. A systems-based computational model for dose-response comparisons of two mode of action hypothesis for ethanol-induced neurodevelopmental toxicity. Toxicol Sci 2005; 86:470–484.

18. Singh AV, Knudsen KB, Knudsen TB. Integrative analysis of the mouse embryonic transcriptome. Bioinformation 2007; 1:24–30.

19. Naciff JM, Jump ML, Torontali SM, et al. Gene expression profile induced by 17α-ethynyl estradiol, bisphenol A, and genistein in the developing female reproductive system of the rat. Toxicol Sci 2002; 68:184–199.

20. Mikheeva S, Barrier M, Little SA, et al. Alterations in gene expression induced in day-9 mouse embryos exposed to hyperthermia (HS) or 4-hydroperoxycyclophosphamide (4CP): Analysis using cDNA microarrays. Toxicol Sci 2004; 79:345–359.

21. Karoly ED, Schmid JE, Hunter ES III. Ontogeny of transcription profiles during mouse early craniofacial development. Reprod Toxicol 2005; 19:339–352.

22. Nemeth KA, Singh AV, Knudsen TB. Searching for biomarkers of developmental toxicity with microarrays: Normal eye morphogenesis in rodent embryos. Toxicol Appl Pharmacol 2005; 206:219–228.

23. Qin P, Cimildoro R, Kochhar DM, et al. PBX, MEIS, and IGF-I are potential mediators of retinoic acid-induced proximodistal limb reduction defects. Teratology 2002; 66:224–234.

24. Ali-Khan SE, Hales BF. Novel retinoid targets in the mouse limb during organogenesis. Toxicol Sci 2006; 94:139–152.

25. Lee GS, Cantor RM, Abnoosian A, et al. A gene(s) for all-*trans*-retinoic acid-induced forelimb defects mapped and confirmed to murine chromosome 11. Genetics 2005; 170:345–353.

26. Barrett T, Troup DB, Wilhite SE, et al. NCBI GEO: Mining tens of millions of expression profiles—database and tools update. Nucl Acids Res 2007; 35:D760-D765.

27. Galperin MY. The molecular biology database collection: 2008 update. Nucleic Acids Res 2008; 36 (database issue) D2-D4 (doi:10.1093/nar/gkm1037).

28. Cavalieri D, De Filippo C. Bioinformatic methods for integrating whole-genome expression results into cellular networks. Drug Dis Ther 2005; 10:727–734.

29. Corvi R, Ahr H-J, Albertini S, et al. Meeting report: Validation of toxicogenomics-based test systems: ECVAM-ICCVAM/NICEATM considerations for regulatory use. Env Health Perspect 2006; 114:420–429.

30. Lamb J, Crawford ED, Peck D, et al. The Connectivity Map: Using gene-expression signatures to connect small molecules, genes, and disease. Science 2006; 313:1929–1935.

31. Dix DJ, Houck KA, Martin MT, et al. The ToxCast program for prioritizing toxicity testing of environmental chemicals. Toxicol Sci 2007; 95:5–12.

32. Perrizo W. The role of data mining in turning bio-data into bioinformation. Bioinformation 2007; 1:351–355.

33. Burgoon LD, Boutros PC, Dere E, et al. dbZach: A MIAME-compliant toxicogenomic supportive relational database. Toxicol Sci 2006; 90:558–568.

34. Yang C, Richard AM, Cross KP. The art of data mining the minefields of toxicity databases to link chemistry to biology. Curr Comp Aided Drug Des 2006; 2:135–150.

35. Knudsen KB, Singh AV, Knudsen TB. Data input module for birth defects systems manager. Reprod Toxicol 2005; 20:369–375.

36. Bard J. Systems developmental biology: The use of ontologies in annotating models and in identifying gene function within and across species. Mamm Genome 2007; 18:402–411.

37. Hunter A, Kaufman MH, McKay A, et al. An ontology of human developmental anatomy. J Anat 2007; 203;347–355.

38. Burger A, Davidson D, Baldock R. Formalization of the mouse embryo anatomy. Bioinformatics 2004; 20:259–267.

39. Christiansen JH, Yang Y, Venkataraman S, et al. EMAGE: A spatial database of gene expression patterns during mouse embryo development. Nucl Acids Res 2006; 34:D637–D641.

40. Baldock R, Bard J, Kaufman M, et al. A real mouse for your computer. Bio Essays 1992; 14:501–503.

41. Ringwald M, Baldock R, Bard J, et al. A database for mouse development. Science 1994; 265:2033–2034.

42. Baldock RA, Bard JB, Burger A, et al. EMAP and EMAGE: A framework for understanding spatially organized data. Neuroinformatics 2003; 1:309–325.

43. Hill DP, Begley DA, Finger JH, et al. The mouse Gene Expression Database (GXD): Updates and enhancements. Nucl Acids Res 2004; 32:D568–D571.

44. Gombar VK, Mattioni BE, Zwickl C, et al. Computational approaches for assessment of toxicity: A historical perspective and current status. In Computational Toxicology. Risk Assessment for Pharmaceutical and Environmental Chemicals, Ekins S, ed. John Wiley & Sons, Hoboken, NJ, 2007:815.

45. Bassan A, Worth AP. Computational tools for regulatory needs. In Computational Toxicology. Risk Assessment for Pharmaceutical and Environmental Chemicals, Ekins S, ed. John Wiley & Sons, Hoboken, NJ, 2007:815.

46. Von Bertalanffy L. General systems theory: Foundations, development, applications. George Braziller, New York, 1968:295.

47. Idekar T, Galitski T, Hood L. A new approach to decoding life: Systems biology. Ann Rev Genom Hum Genet 2001; 2:343–372.

48. Kitano H. Foundations of systems biology. The MIT Press, Cambridge, MA, 2001:297.

49. Butler D. Are you ready for the revolution? Nature 2001; 409:758–760.

50. Covert MW. Integrated regulatory and metabolic models. In Computational Systems Biology, Kriete A, Eils R, eds. Elsevier Academic Press, New York, 2006:191–204.

51. O'Malley MA, Dupré J. Fundamental issues in systems biology. Bio Essays 2005; 27:1270–1276.

52. Waters MD, Fostel JM, Wetmore BA, et al. Toxicogenomics and systems toxicology. In Computational Toxicology. Risk Assessment for Pharmaceutical and Environmental Chemicals, Ekins S, ed. John Wiley & Sons, Hoboken, NJ, 2007:815.

53. Ma H-W, Zeng A-P. Reconstruction of metabolic networks from genome information and its structural and functional analysis. In Computational Systems Biology, Kriete A, Eils R, eds. Elsevier Academic Press, New York, 2006:169–189.

54. Schnell S, Grima R, Maini PK. Multiscale modeling in biology. Amer Sci 2007; 95:134–142.
55. Gohlke JM, Griffith WC, Faustman EM. Computational models of ethanol-induced neurodevelopmental toxicity across species: Implications for risk assessment. Birth Defects Res B 2008; 83:1–11.
56. Morford LL, Henck JW, Breslin WJ, et al. Hazard identification and predictability of children's health risk from animal data. Env Health Perspect 2004; 112:266–271.
57. Rodriguez-Zas SL, Ko Y, Adams HA, et al. Advancing understanding of the embryo transcriptome co-regulation using meta-, functional, and gene network analysis tools. Reproduction 2008; 135:213–224.
58. Longabaugh WJR, Davidson EH, Bolouri H. Computational representation of developmental genetic regulatory networks. Devel Biol 2005; 283:1–16.
59. Davidson EH, Erwin DH. Gene regulatory networks and the evolution of animal body plans. Science 2006; 311:796–800.
60. Olson EN. Gene regulatory networks in the evolution and development of the heart. Science 2006; 313:1922–1927.
61. Davidson E. The regulatory genome. gene regulatory networks in development and evolution. Academic Press, London, 2006:289.
62. Obayashi T, Hayashi S, Shibaoka M, et al. COXPRESdb: A database of coexpressed networks in mammals. Nucl Acids Res 2007; 36:D77–D82.
63. Klamt S, Saez-Rodriguez J, Gilles ED. Structural and functional analysis of cellular networks with CellNetAnalyzer. BMC Syst Biol I:2, BioMed Central Ltd., 2007.
64. Alon U. An introduction to systems biology. Design principles of biological circuits. Chapman & Hall/CRC Mathematical and Computational Biology Series, Boca Raton, FL, 2007:301.
65. Shen-Orr SS, Milo R, Mangan S, et al. Network motifs in the transcriptional regulation network of *Escherichia coli*. Nature Gen 2002; 31:64–68.
66. Oikonomou P, Cluzel P. Effects of topology on network evolution. Nature Phys 2006; 2:532–536.
67. Wikepedia contributors. Scale-free network, *Wikipedia, The Free Encyclopedia*, 2 March 2008 (http://en.wikipedia.org/w/index.php?title = Scale-free_network&oldid = 195352029).
68. Combs AB, Acosta DA Jr. An introduction to toxicology and its methodologies. In Computational Toxicology. Risk Assessment for Pharmaceutical and Environmental Chemicals, Ekins S, ed. John Wiley & Sons, Hoboken, NJ, 2007:815.
69. Guittina P, Elefant E, Saint–Salvic B. Hierarchization of animal teratology findings for improving the human risk evaluation of drug. Reprod Toxicol 2000; 14:369–375.
70. Chernoff NA, Rogers EA, Gage MI, et al. The relationship of maternal and fetal toxicity in developmental toxicology bioassays with notes on the biological significance of the "no observed adverse effect level". Reprod Toxicol 2008; 25:192–202.
71. EPA. Guidelines for Developmental Toxicity Risk Assessment; Notice. Fed Reg 56: 63798–63826, 1991 and FQPA 10x Safety Factor Update. FIFRA Scientific Advisory Panel Meeting, SAP Report No. 99–01 C, January 22, 1999. (http://www.epa.gov/scipoly/sap/meetings/1998/december/final3.pdf)
72. Vainio H, Coleman M, Wilburn J. Carcinogenicity evaluations and ongoing studies: The IARC databases. Environ Health Perspect 1991; 96:5–9.

73. National Research Council, Committee on Toxicity and Assessment of Environmental Agents. Testing in the twenty-first century: A vision and a strategy. National Academies Press, Washington, DC, 2007, (http://www.nap.edu/catalog/11970.html).

74. Chapin R, Stedman D, Paquette J, et al. Struggles for equivalence: In vitro developmental toxicity model evolution in pharmaceuticals in 2006. Toxicol In Vitro 2007; 21:1545–1551.

75. Kavlock RJ, Ankley GA, Blancato J, et al. Computational toxicology—a state of the science mini review. Toxicol Sci 2008; 103:14–27.

76. Crump K. A new method for determining allowable intakes. Fund Appl Toxicol 1984; 4:854–871.

77. Faustman EM, Allen BC, Kavlock RJ, et al. Dose-response assessment for developmental toxicity. I. Characterization of the database and determination of the no observed adverse effect level. Fund Appl Toxicol 1994; 23:478–486.

78. Allen BC, Kavlock RJ, Kimmel CA, et al. Dose response assessment for developmental toxicity. II. Comparison of generic benchmark dose estimates with no observed adverse effect levels. Fund Appl Toxicol 1994; 23:487–495.

79. Allen BC, Kavlock RJ, Kimmel CA, et al. Dose response assessment for developmental toxicity. III. Statistical models. Fund Appl Toxicol 1994; 23:496–509.

80. Kavlock RJ, Allen BC, Faustman EM, et al. Dose-response assessment for developmental toxicity. IV. Benchmark doses for fetal weight changes. Fund Appl Toxicol 1995; 26:211–222.

81. Kavlock RJ, Schmid JE, Setzer RW. A simulation study of the influence of study design on the estimation of benchmark doses for developmental toxicity. Risk Anal 1996; 16:399–410.

82. Piersma AH, Janer G, Wolterink G, et al. Quantitative extrapolation of in vitro whole embryo culture embryotoxicity data to developmental toxicity in vivo using the benchmark dose approach. Toxicol Sci 2008; 101:91–100.

83. OECD. Guidance Document on the Validation of (Quantitative) Structure-Activity Relationships [(Q)SAR] Models. OECD Environmental Health and Safety Publications, Series on Testing and Assessment, No. 69 ENV/JM/MONO(2007)2, 2007.

84. Kavlock RJ. Structure-activity approaches in the screening of environmental agents for developmental toxicity. Reprod Toxicol 1993; 7:113–166.

85. Julien E, Willhite CC, Richard AM, et al. Challenges in constructing statistically based structure-activity models for developmental toxicity. Birth Defects Res A 2004; 70:902–911.

86. Accelrys. http://www.accelrys. com/products/topkat/index.html. Accessed Jan 15 2008, 2004.

87. Gombar VK, Enslein K, Blake BW. Assessment of developmental toxicity potential of chemicals by quantitative structure-activity relationship models. Chemosphere 1995; 31:2499–2510.

88. MultiCase. http://wwwmulticasecom/ 2004. Accessed Jan 15 2008.

89. Ghanooni M, Mattison DR, Zhang YP, et al. Structural determinants associated with the risk of human developmental toxicity. Am J Obstet Gynecol 1997; 176: 799–805.

90. Friedman JM, Little BB, Brent RL, et al. Potential human teratogenicity of frequently prescribed drugs. Obstet Gynecol 1990; 75:594–599.

91. Briggs GG, Freeman RK, Yaffee SJ. Drugs in pregnancy and lactation: A reference guide to fetal and neonatal risk. 5th ed. Williams and Wilkins, Baltimore, MD, 2005:1888.

92. NLM TOXNET. National Institutes of Health, National Library of Medicine (NLM), Toxicology Data Network (TOXNET). (http://toxnet.nlm.nih.gov), 2007.

93. OECD. Report on Regulatory Uses and Applications in OECD Member Countries of (Quantitative) Structure-Activity Relationships [(Q)SAR] Models in the Assessment of New and Existing Chemicals. OECD Environmental Health and Safety Publications, Series on Testing and Assessment, No. 58 ENV/JM/MONO(2006)25, 2007.

94. Wise LD, Beck SL, Beltrame D, et al. Terminology of developmental abnormalities in common laboratory mammals (version 1). Teratology 1997; 55:249–292.

95. Chahoud I, Buschmann J, Clark R, et al. Classification terms in developmental toxicology: need for harmonization. Report of the second workshop on the terminology in developmental toxicology Berlin, 27–28 August 1998. Reprod Toxicol 1999; 13:77–82.

96. Solecki R, Bürgin H, Buschmann J, et al. Harmonisation of rat fetal skeletal terminology and classification. Report of the third workshop on the terminology in developmental toxicology*1 Berlin, 14–16 September 2000. Reprod Toxicol 2001; 15:713–721.

97. Foster PMD, Gray G. Toxic Responses of the Reproductive System. In Casarette and Doull's Toxicology: The Basic Science of Poisons, 7th ed. McGraw Hill, New York, 2007:1309.

98. Tong W, Xie Q, Hong H, et al. Assessment of prediction confidence and domain extrapolation of two structure-activity relationship models for predicting estrogen receptor binding activity. Environ Health Perspect 2004; 112:1249–1254.

99. Lui H, Papa E, Grammatica P. QSAR prediction of estrogen activity for a large set of diverse chemicals under the guidance of OECD principles. Chem Res Toxicol 2006; 19;1540–1548.

100. Shi LM, Fang H, Tong W, et al. QSAR models using a large diverse set of estrogens. J Chem Inf Comput Sci 2001; 41:186–195.

101. Houck KA, Kavlock RJ. Understanding mechanisms of toxicity: Insights from drug discovery research. Toxicol Appl Pharm 2008; 227:163–178.

102. EPA ToxCast. U. S. Environmental Protection Agency's National Center for Computational Toxicology Tox CastTM Program, 2007. (http://www.epa.gov/ncct/toxcast/)

103. Richard A. Future of toxicology–predictive toxicology: An expanded view of "chemical toxicity". Chem Res Toxicol 2006; 19:1258–1262.

104. Richard A, Yang C, Judson R. Toxicity data informatics: Supporting a new paradigm for toxicity prediction. Toxicol Meth Mech 2008; 18:103–118.

105. Watt AH. Sams teach uourself XML in 10 minutes. Sams Publishing, Indianapolis, IN, 2003:292.

106. Tox ML. A Publicly Available Toxicity XML Standard and Controlled Vocabulary for Representing Toxicity Data. Database Schema available at Leadscope® Internet site: http://www.leadscope. com/toxml.php, 2007.

107. EPA DSSTox. U.S. Environmental Protection Agency's Distributed Structure-Searchable Toxicity (DSSTox) Database Network, 2007. (http://www.epa.gov/ncct/dsstox/index.html)

108. Judson R, Richard A, Dix D, et al. ACToR—aggregated computational toxicology resource. Toxicol Appl Pharmacol 2008; in press (online, doi: 10.1016/j.taap.2007.12.037).
109. Martin MT, Houck KA, McLaurin K, et al. Linking regulatory toxicological information on environmental chemicals with high-throughput screening (HTS) and genomic data. Toxicol CD–J Soc Toxicol 2007; 96:219–220.
110. EPA Regulating Pesticides. U.S. Environmental Protection Agency's Office of Pesticide Programs, Regulating Pesticides, 2007. (http://www.epa.gov/pesticides/regulating)
111. Corley RA, Mast TJ, Carney EW, et al. Evaluation of physiologically based models of pregnancy and lactation for their application in children's health assessments. Crit Rev Toxicol 2003; 33:137–211.
112. Clark LH, Setzer RW, Barton HA. Framework for evaluation of physiologically based pharmacokinetic models for use in safety and risk assessment. Risk Analysis 2004; 24:1697–1717.
113. Clewell HJ, Tueguarden J, McDonald T, et al. Review and evaluation of the potential impact of age- and gender-specific pharmacokinetic differences in tissue dosimetry. Crit Rev Toxicol 2002; 32:329–390.
114. Yoon M, Barton HA. Predicting maternal and pup exposures: How different are they? Toxicol Sci 2008; 102:15–32.
115. Andersen ME, Dennison JE. Mode of action and tissue dosimetry in current and future risk assessments. Sci Total Environ 2001; 274:3–14.
116. Moolgavkar SH, Luebeck G. Two-event model for carcinogenesis: Biological, mathematical and statistical considerations. Risk Anal 1990; 10:323–341.
117. Leroux B, Leisenring W, Moolgavkar S, et al. A biologically based dose-response model for developmental toxicology. Risk Anal 1996; 16:449–458.
118. Faustman EM, Lewandowski T, Ponce R, et al. Biologically-based dose-response models for developmental toxicology: Lessons from methylmercury. Inhal Toxicol 1999; 11:559–572.
119. Lau C, Andersen ME, Crawford-Brown D, et al. Evaluation of biologically based dose-response modeling for developmental toxicity: A workshop report. Regul Toxicol Pharmacol 2000; 31:190–199.
120. Setzer RW, Lau C, Mole ML, et al. Toward a biologically based dose-response model for developmental toxicity of 5-fluorouracil: a mathematical construct. Toxicol Sci 2001; 59:49–58.
121. Lau C, Mole ML, Copeland F, et al. Toward a biologically based dose-response model for developmental toxicity of 5-fluorouracil: Acquisition of experimental data. Toxicol Sci 2001; 59:37–48.
122. Kavlock RJ, Setzer RW. The road to embryologically based dose-response models. Environ Health Perspect 1996; 104(Suppl 1):107–121.
123. Graner F, Glazier JA. Simulation of biological cell sorting using a two-dimensional extended Potts model. Phys Rev Lett 1992; 69:2013–2016.
124. Dormann S, Deutsch A. Modeling of self-organized avascular tumor growth with a hybrid cellular automaton. In Silico Biol 2002; 2:0035.
125. Izaguirre JA, Chaturvedi R, Huang C, et al. CompuCell, a multi-model framework for simulation of morphogenesis. Bioinformatics 2004; 20:1129–1137.

126. Cickovski TM, Huang C, Chaturvedi R, et al. A framework for three-dimensional simulation of morphogenesis. IEEE/ACM Trans Comput Biol Bioinform 2005; 2: 1–15.

127. Cickovski TM, Izaguirre JA. Biologo, a domain-specific language for morphogenesis. ACM Trans. Programming Languages and Systems, 2005. http://www.nd.edu/tcickovs/acmtr2e.pdf

128. Robertson SH, Smith CK, Langhans AL, et al. Multiscale computational analysis of Xenopus laevis morphogenesis reveals key insights of systems-level behavior. BMC Syst Biol 2007; 1:46, doi:10.1186/1752–0509–1–46.

129. Risk Policy Report. EPA High-Tech "Virtual Embryo Project" Will Target Developmental Risk. vol 15, no. 2, January 8, 2008. http://www.InsideEPA. com, 2008.

130. Krietes A, Eils R. Computational systems biology. Elsevier Academic Press, London, 2006:409.

131. Wubah JA, Setzer RW, Lau C, et al. Exposure-disease continuum for 2-chloro-2'-deoxyadenosine, a prototype ocular teratogen. I. Dose-response analysis. Teratology 2001; 64:154–169.

132. Saunders PT. An introduction to catastrophe theory. Cambridge University Press, New York, 1980:144.

133. Gilmore R. Catastrophe theory for scientists and engineers. General Publishing Company, Ltd., Toronto, 1981:666.

134. Konopka AK. Systems biology. Principles, methods, and concepts. CRC Press, Taylor & Francis Group, Boca Raton, FL, 2007:244.

135. Fisher GH. Preparation of ambiguous stimulus materials. Perc Psychophys 1967; 2:421–422.

13

Investigating Drug Effects in Human Pregnancy

Christina D. Chambers

Departments of Pediatrics and Family and Preventive Medicine, University of California, San Diego, La Jolla, California, U.S.A.

INTRODUCTION

Regulatory agencies and a variety of public health entities face the responsibility of evaluating the safety of maternal use of medications with respect to the developing fetus. This is a challenging task for a variety of methodological reasons, but is also a monumental undertaking because of the large number of medications to which pregnant women are likely to be exposed and for whom this information is critically needed. As congenital anomalies are the leading cause of infant mortality and number of years of potential life lost in the United States, well-designed and well-conducted human studies that can identify potentially teratogenic medications are needed to help prevent teratogen-induced malformations. Of similar importance, well-designed human studies are also needed to provide reassurance regarding the absence of substantial increased risks for birth defects so that clinicians and patients can appropriately and confidently use necessary medications during pregnancy (1,2).

FREQUENCY AND VARIETY OF MEDICATION USE AMONG PREGNANT WOMEN

Numerous studies have demonstrated that pregnant women commonly use several medications over the course of gestation. For example, a 2004 study involving

361

review of prescription records for over 150,000 pregnant women in eight health maintenance organizations in the United States found that 64% of these women were prescribed at least one drug other than a vitamin or mineral sometime during pregnancy. Moreover, 39% of all women in the sample received at least one prescription during the first trimester. On average, women received 2.7 drug dispensings and 1.7 different chemical entities over the course of pregnancy (3). Recent evidence also suggests that over-the-counter medications are used even more commonly by pregnant women. Using two large case-control data sets, a 2005 report estimated that acetaminophen, ibuprofen, and pseudoephedrine were used by at least 65%, 18%, and 15% of pregnant women, respectively (4).

Inadvertent first-trimester exposures to medications are a frequent occurrence due to the high incidence of unplanned pregnancies in which women may take drugs well into the first 4 to 6 weeks postconception prior to pregnancy recognition (5). In addition, many maternal conditions, both acute and chronic, require intentional treatment after pregnancy is confirmed. For example, relatively common chronic conditions that can necessitate treatment in women of childbearing age include clinical depression with an estimated prevalence between 8.0% and 20.0% (6), asthma 3.7 to 8.4% (7), epilepsy 0.4 to 1.0% (8,9), and rheumatoid arthritis and other autoimmune disorders 1.0 to 2.0% (10). Furthermore, for some of these maternal conditions, a decision not to treat (or to undertreat) could lead to events, such as uncontrolled seizure activity or psychiatric episodes, which could be detrimental to the woman, the pregnancy, and/or the fetus itself (11–13).

Thus, the development of adequate information on drug safety in pregnancy is essential not only for the identification of potentially harmful exposures that might be avoided or managed, but also for the establishment of acceptable margins of safety for drugs that offer potential benefit to women during their pregnancies.

SOURCES OF PREGNANCY SAFETY DATA

Clinical trials are considered the gold standard for evaluating drug safety. For ethical reasons, pregnant women typically are not recruited for trials during any phase of drug development. If and when unintended pregnancies occur during the course of a trial or postmarketing safety study, these subjects can provide useful preliminary information regarding exposure and outcome (14). However, these data usually involve a very small number of subjects and have not been collected with the intention of evaluating pregnancy outcome, and therefore have limited value.

Once a medication is marketed, there are a number of resources that can provide observational data regarding drug safety in pregnancy. These include the following:

Isolated Clinical Case Reports and Case Series

Individual case reports describing a maternal medication exposure in association with a malformation phenotype in the infant can be one of the first methods of generating the hypothesis that a specific agent is teratogenic. However, these reports must be initiated spontaneously, and therefore may involve investigator as well as publication bias. Furthermore, without a known denominator of exposed pregnancies that do or do not result in infants with that specific malformation or pattern of malformations, it is difficult to determine if the reported defect(s) represent an increase over baseline. If the phenotype is sufficiently unique, for example, the isotretinoin embryopathy (15), then a series of case reports can strongly suggest a hypothesis that can be confirmed using other methods. In fact, affected cases that have been recognized and evaluated by astute clinicians have been the key method whereby most known human teratogens have been identified, dating back to Gregg and the rubella embryopathy (16).

Systematic Adverse Event Reporting

Formal mechanisms whereby drug-related adverse pregnancy outcomes are collected can provide a systematic method for accumulation of case reports from a variety of resources. For example, under the U.S. FDA's Adverse Event Reporting System (AERS), manufacturers and distributors of FDA-approved pharmaceuticals are mandated to report events such as congenital anomalies that occur following exposure to their product as they become aware of such events. Recently, this effort has been expanded to include over-the-counter medications and dietary supplements. The FDA also receives reports through the MedWatch program, an educational and promotional effort which facilitates spontaneous reporting from health-care providers (17,18). Finally, consumers may provide pregnancy-related adverse event information directly to the FDA.

One advantage of such systems is that reports can be accumulated from a variety of resources relatively soon after a new drug is marketed. Although these systems have typically not been productive in identifying new human teratogens, they have been useful resources for exploring the specific characteristics surrounding teratogens identified through other methods. For example, the angiotensin II converting enzyme inhibitor (ACE inhibitor) fetopathy was first reported by a clinician (19). However, the frequency of similar or related abnormalities in relation to gestational timing of exposure and dose of the drug was identifiable through review of a series of 110 ACE inhibitor adverse event reports submitted to the FDA through 1999 (20). Similarly, case reports and cohort studies that identified the increased risk for a variety of neonatal complications with late pregnancy exposure to selective serotonin reuptake inhibitors (21,22) have been confirmed and classified into possible pathogenetic subtypes using adverse event reporting data (23,24).

The primary limitations of such systems are similar to those of case reports appearing in the medical literature. Reports initially rely on the motivation of the clinician to contribute to the system and therefore may involve bias in the types and number of actual events that are reported. Spontaneous reporting systems rely on the clinician to make a link between medication exposure and pregnancy outcome, a link perhaps more likely for outcomes normally rare and extremely severe, and less likely for outcomes considered relatively common or with subtle presentation. In addition, adverse event reports do not provide denominator information on the number of exposed, affected, or unaffected pregnancies that could be used to develop a birth prevalence rate for purposes of comparison to baseline rates for a specified outcome in the general population. Finally, due to the variety of reporters and sources of information contributing to such databases, the completeness and validity of the data may vary.

Pregnancy Drug Exposure Registries

Dating back to the acyclovir registry established by the drug manufacturer (25,26), traditional pregnancy registries represent one method of evaluating drug safety in pregnancy that is increasingly being used. A current listing is available on the U.S. FDA's Office of Women's Health website (http://www.fda.gov/womens/registries/registries.html). All traditional pregnancy registries involve spontaneous reporting of exposed pregnancies. The collection of exposure and outcome data is usually accomplished through the healthcare provider and/or the pregnant woman who initiate contact with the registry. Although pregnancy outcome reports can be collected retrospectively, most current drug registries also identify and follow exposed pregnancies prospectively, that is, ascertain pregnant women sometime before delivery, and collect exposure and other information prior to the known outcome of that pregnancy.

The traditional pregnancy exposure registry method has a number of advantages including the accumulation of information at one centralized location on pregnancy exposures to a specific drug, which can yield early information for newly marketed medications. Particularly, if the exposure is rare, this may be the most efficient method for collecting pregnancy outcome data as quickly as possible. Furthermore, industry-sponsored registries can use their existing infrastructure both nationally and internationally to more efficiently identify potential registry participants (27). The registry approach when used to accumulate prospective reports can provide good quality information about the temporal association between exposure and outcome. In addition, prospective registry designs, unlike adverse event reporting systems and case reports, provide a defined denominator of exposed women who do or do not have adverse outcomes. This facilitates comparisons of congenital anomaly rates to those of a reference group.

With relatively small samples sizes typically numbering well under 1000 exposed pregnancies, these registries may have the ability to detect two-to-five-fold significant increases in the overall frequency of major congenital anomalies that are evident at birth relative to the overall birth prevalence of major congenital anomalies in the general population (27–29). Especially for high-risk teratogens such as isotretinoin or thalidomide, such an approach is arguably the most efficient, cost-effective, and timely method for identifying such agents quickly. For high-risk teratogens like isotretinoin that are associated with a characteristic and frequently occurring pattern of major congenital anomalies recognizable at birth, only a small number of exposed pregnancies are necessary to infer potential teratogenicity (29).

However, there are several limitations of the traditional pregnancy registry approach. Spontaneous reporting of exposed pregnancies may involve selection bias so that pregnancies that are enrolled in the registry are not representative of all pregnant women who have taken the drug. It is also difficult to project sample sizes. Even with successful identification and recruitment of a high proportion of all exposed pregnancies occurring in the population, the absolute number of exposed pregnancies in the registry is likely to be low and therefore power to detect increased risks is low.

To further complicate the situation, enrolled pregnancies may include wide variability in the characteristics of exposures in terms of the maternal disease being treated, gestational windows of exposure, and/or dose of the drug, thereby further reducing the power to detect increased risks in a particular susceptible subset of pregnancies.

The power issue is also relevant in the typical study design with a goal of detection of an overall increased risk in all major birth defects. Detection of an overall increased risk in all defects combined is not consistent with what is known about most known human teratogens which are typically associated with increased risks for specific patterns of birth defects and other adverse outcomes, rather than an increase in all birth defects across the spectrum. The low frequency of any specific congenital anomaly makes it unlikely that a pregnancy registry will have sufficient power to detect any but the most dramatic effects for a specific outcome. Thus, an important function of a pregnancy registry is to generate hypotheses on the basis of "signal" detection when higher than expected numbers of specific malformations are reported, with additional studies required to confirm or refute the signal (30).

Other limitations of traditional pregnancy registries include the difficulty in identifying an appropriate comparison group. Many registry designs do not include a concurrently enrolled comparison group. Instead, outcomes in exposed pregnancies are frequently compared to externally derived reference rates for some other population. External reference rates, without the ability to adjust for possible confounding factors that differ between exposed women and the reference population, may not represent the most appropriate comparison. Some registry designs do involve recruitment of an internal comparison group with collection of information on potential confounders so that comparisons can adjust for differences between groups (31). Disease-based registry designs, such as the Antiepileptic Drugs in Pregnancy Registry, can address this problem in part by comparing pregnancies with a specific drug exposure to pregnancies with exposure to other medications used for the same disease (8,31), or to pregnancies exposed to the drug only in a later gestational window (32).

Finally, as traditional registries rely on a wide variety of clinicians and/or mothers to report pregnancy outcome, there is potential for misclassification of outcomes such as major congenital anomalies with respect to accurate and complete diagnosis and/or suspected etiology (33). Furthermore, subtle or less easily recognizable teratogenic effects, such as the fetal alcohol syndrome or the minor structural abnormalities that comprise the anticonvulsant embryopathy, are unlikely to be identified or described accurately by the obstetrician or general pediatrician who is reporting outcomes to a registry. In addition, in some registries, a substantial proportion of pregnancy exposure reports are lost to follow-up with no documented outcome information, thereby potentially biasing conclusions that can be drawn from these data.

To help standardize as well as improve the reliability and validity of pregnancy registry data, the U.S. FDA has produced a guidance document (34) that establishes some principles for the design and conduct of pregnancy registries. A second guidance document sets standards for reviewers who are evaluating human data on the effects of in utero drug exposure on the developing fetus (35). Taken together, these guidelines provide the nucleus of a set of standards for the collection and interpretation of pregnancy exposure and outcome data that can contribute to consistency and improved quality in the collection and evaluation of safety data generated through pregnancy registries.

Case-Control Studies

In the study of teratogenic causes of birth defects, case-control studies can be classified into one of two approaches.

In the first approach, a specific hypothesis is tested using cases and controls to measure the association between a medication exposure and a specified birth defect or group of defects. This approach requires that *a priori* decisions be made regarding the research questions, selection of the appropriate control group, and adequate power and sample size. For example, based on concerns raised in the literature, this design was successfully used to document a statistically significant association between congenital facial nerve paralysis, or Möebius' Syndrome, and first-trimester use of misoprostol (36).

The second approach, case-control surveillance, is not based on a predefined hypothesis or set of hypotheses, but is instead focused on a systematic accumulation of exposure and potential confounder information for malformed cases and controls collected over time in order to create a large repository of data suitable for testing multiple future hypotheses. This approach has been incorporated into some birth defects monitoring programs in the United States, and is the general design of the U.S. National Birth Defects Prevention Study (37). These methods are also used on an ongoing basis in programs such as the Slone Epidemiology Center's hospital-based surveillance study based at Boston University (38,39) and the Latin American Collaborative Study of Congenital Malformations (ECLAMC), which involves over 70 hospitals in several South American countries (40). These programs usually involve ascertainment of malformed cases as well as a systematic method for selection of a sample of nonmalformed infants who can be used as controls. Exposure and other risk factor information is generally gathered by postnatal maternal interview, and in some cases is supplemented by review of medical records. In addition, some designs have incorporated DNA sampling and banking from case and control children and their parents so that future hypotheses regarding genetic susceptibility or gene–environment interactions can be tested.

The primary advantage of any case-control approach in studies of rare events such as congenital anomalies is the enhanced power to detect or rule out a meaningful association for a given sample size. In contrast to pregnancy registries or other prospective designs, this method is often the only appropriate approach for detecting moderate or low-level teratogenic exposures associated with specific major malformations. Furthermore, to the extent that case-control surveillance studies collect comprehensive information on potential confounders, including vitamin, tobacco, and alcohol use, this approach can provide reassurances that more moderate effect sizes are not attributable to these other factors. Other advantages of case-control surveillance include, to a varying degree, relatively complete ascertainment of the congenital anomalies of interest within a defined population, concurrent selection of controls from the same population, and the ability to systematically validate the classification of diagnoses.

In addition, this approach provides flexibility in the ultimate use of the data, that is, based on specific research questions, subsets of cases and controls can be selected from the entire data set to test or confirm specific hypotheses. For example, this method was useful in confirming the protective effect of antenatal folic acid supplementation in reducing the incidence of neural tube defects (41), and in confirming a previously controversial finding of an association between maternal corticosteroid use and cleft lip with or without cleft palate (42). Furthermore, case-control surveillance data are amenable to hypothesis generation. For example, these data were used to first raise the question of an association between pseudoephedrine and gastroschisis (43).

The limitations of case-control studies generally relate to the use of retrospective data collection and the selection of controls. For example, maternal interviews may be conducted in some cases many months after completion of the pregnancy, which raises the possibility of limited recall of early pregnancy medication use (44). In addition, the potential for serious differential recall bias among mothers of malformed infants relative to mothers of nonmalformed controls has been cited by some (45), while the potential bias associated with the use of malformed controls has been suggested by others (46). However, case-control surveillance studies have the advantage of flexibility in selection of one or multiple control groups, malformed or not, from the larger data set as judged necessary for any specific analysis.

Because case-control surveillance programs are ongoing, they have the potential to recognize an association with a newly marketed medication; however, they may have limited sensitivity in this regard. These studies may miss an association if the medication of interest is related to a relatively unusual or uncommon congenital anomaly and/or that specific defect is not included in the range of selected anomalies for which maternal interviews are conducted. In addition, if new medications are infrequently used among pregnant women, then the case-control study may have insufficient power to detect weak or moderate associations. However, for medications that are more commonly used—e.g., by 1% or more of pregnant women—given the rarity of congenital anomalies in general, these approaches provide a relatively powerful method of hypothesis testing and hypothesis generation and can be effectively used alone and in conjunction with other methods.

Large Cohort Studies

These studies can involve open cohorts that are population based and ongoing, or can be hospital or health insurer based and/or of limited duration. For example, the Swedish Registry of Congenital Malformations in combination with the Swedish Medical Birth Registry encompasses nearly all births in Sweden and uses exposure interviews conducted by midwives during the first antenatal visit as well as data recorded prospectively in medical records (47). Large longitudinal cohort studies, each with some ability to address medication exposure in relation to risk for congenital anomalies, have been initiated in countries such as Denmark (48), and are currently being launched in the United States as part of the more than 100,000 pregnancies being recruited for the National Children's Study (http://www.nationalchildrensstudy.gov).

These studies have the advantage of large and potentially representative samples, prospective ascertainment of exposure information as well as data regarding a variety of potential confounders, and ability to collect outcome information over a long term of follow-up. In addition, women with and without the exposure of interest are concurrently enrolled as members of the cohort, facilitating the identification of one or more appropriate reference groups. Like ongoing case-control designs, studies of this type can address multiple hypotheses that need not be formulated *a priori* (49).

However, even in large cohort studies, issues of sample size can be a limitation. For example, the Collaborative Perinatal Project conducted in the 1960s, a study involving over 50,000 mother–child pairs, had inadequate power to detect weak-to-moderate associations of medication exposures with any but the most common major congenital malformations and the most commonly used drugs due to the relatively small numbers of women exposed to most specific medications of interest (50).

The Swedish Registry, with approximately 120,000 annual births that have been accumulated over more than a 25-year span of time, has enhanced power to evaluate some medication exposures. For example, using the Swedish data, investigators were able to identify over 2000 first trimester inhaled corticosteroid (budesonide-) exposed pregnancies and rule out with fairly narrow confidence intervals an increased risk in overall rate of major congenital anomalies (51). However, the numbers of exposed and affected infants were too small even in this relatively large cohort to adequately address the hypothesis of an increased risk for oral clefts, the specific major congenital malformation category that has earlier been associated with maternal systemic corticosteroid use.

Small Cohort Studies

These studies, similar in many respects to traditional pregnancy exposure registries, are focused on follow-up of a cohort of pregnant women with exposure to a target medication, and are frequently conducted by Teratology Information Services (TIS) both in North America and in other parts of the world. These studies draw on a base of callers who contact a TIS during pregnancy seeking counseling regarding the safety of a medication that has been used by the caller. For selected target exposures, women are enrolled in a cohort study, and follow-up of pregnancy outcome is obtained via repeated maternal telephone interviews and medical records review.

These studies have strengths similar to the registries described earlier with respect to the potential for rapid identification of exposed women, particularly for a new drug, as well as prospective collection of exposure and other risk factor information. However, unlike traditional pregnancy registries, TIS studies usually employ a concurrently and prospectively enrolled unexposed control group, often both a disease-matched and a nondiseased group, with extensive follow-up that is conducted in parallel with follow-up in the exposed group. This approach can help provide the most appropriate reference groups for women with a given medication exposure who have been recruited through the TIS system. Furthermore, the information collected on comparison pregnancies allows for statistical analyses that control for potential confounding. Finally, since these studies rely on the pregnant woman as the source of information, the rates of lost-to-follow-up are typically very low.

Similar to traditional pregnancy registries, the primary limitation of TIS studies relates to sample size. Published studies from TIS typically involve sample sizes with less than 200 exposed subjects (52,53). Also, similar to traditional pregnancy registry designs, TIS studies rely on spontaneous callers for recruitment of subjects which may result in selection bias.

In an effort to increase sample size and to shorten the time needed to identify a targeted number of exposed pregnancies, collaborative projects among networks of TIS sites in North America are conducted through the Organization of Teratology Information Specialists (OTIS) (31) and in Europe and beyond through the European Network of Teratology Information Services (ENTIS) (54,55). These formal collaborations can add to the variability and possibly the representativeness of subjects in the sample and increase the obtainable sample size by drawing on a larger population of potentially exposed women. However, even these studies, similar to other cohort studies with moderate sample sizes, usually are only sufficiently powered to detect or rule out high-risk associations between exposures and specific major congenital anomalies.

The primary strength of these small cohort studies is the ability to evaluate a spectrum of pregnancy outcomes following a given exposure, including major congenital anomalies, spontaneous abortion and stillbirth, preterm delivery, pre- and postnatal growth deficiency, and in some cases longer-term child development. In this context, although underpowered to evaluate rare outcomes, these studies can fill a void that better powered case-control studies cannot—that is, they can be useful in identifying increased risks for adverse outcomes that were not suspected or hypothesized *a priori*. This hypothesis-generating capacity can raise questions that can then be tested using other study methods.

A unique feature of some OTIS and individual TIS small cohort designs involves the addition of an expert team of specialty clinicians who perform blinded standardized physical exams on all live born infants involved in these studies. This physical examination protocol includes a systematic evaluation of both exposed and comparison infants for major and more subtle minor physical anomalies using a standard examination checklist. One advantage of this type of specialized examination is that more consistent and valid categorization of major congenital anomalies can be assured. In addition, this level of careful scrutiny makes it possible to identify a unique pattern of effects on fetal development, along the lines of the pattern of minor craniofacial and digital anomalies known as the anticonvulsant embryopathy, or the characteristic pattern of facial features that comprise the fetal alcohol syndrome. Alternative study designs that do not include this added layer of evaluation would be unlikely to detect syndromes with subtle presentations such as those described earlier (30,56,57).

Furthermore, with a specialized physical examination incorporated into the study design, the typical small sample sizes in TIS cohort studies can be sufficiently powered to rule out or detect specific patterns of minor structural features that occur in 10% or more of the exposed pregnancies. This is an effect size that is consistent, for example, with the incidence of the anticonvulsant embryopathy among infants born to mothers who take carbamazepine, phenytoin, or valproic acid (56).

Linked Database Studies

With the technological advances in electronic claims data and medical and pharmacy records storage, large databases that include pregnancy information can offer the advantages of large cohort studies at potentially far less cost. For example, investigators in Denmark have linked prescription database records to hospital discharge and medical birth register records for children with and without congenital anomalies to investigate the safety of a widely used antibiotic without the cost and labor of having to recruit, consent, and collect information from any individual human subject (58).

In countries where there is universal and standardized health-care delivery and record-keeping, or in countries where health-care maintenance or other extensive networks of providers serve large numbers of individuals in the population, linked prescription and birth records provide an attractive alternative method for testing hypotheses regarding drug safety in pregnancy. For example, hospital discharge data across the Canadian population has been used to evaluate adverse outcomes of pregnancies complicated by asthma (59). This approach has also been used successfully to evaluate pregnancy exposure to clarithromycin using longitudinal claims data for members from 12 geographically diverse United Health Group-affiliated insurance plans (60). Efforts are going on to develop algorithms that are acceptably accurate in identifying pregnancies when they occur, determining if and when exposure to a specific drug has occurred in that pregnancy, as well as pregnancy outcome (61).

The primary advantages of large linked databases are the availability of large numbers of possibly pregnant subjects, the ability to establish temporal relationships between exposure and outcome by constructing an historical cohort of women with dated records, and relative ease of access to previously collected administrative or claims data. This approach also avoids some of the biases involved in studies that rely on subjects to volunteer, and rely exclusively on maternal report to classify exposure, especially if that information is collected retrospectively.

On the other hand, there are limitations inherent in a study design that does not involve subject contact. For example, database studies usually cannot ensure that the medication prescribed was actually taken by the mother, taken in the dose prescribed, or taken during a specific gestational period. There are also potential limitations related to misclassification of outcome depending on the quality of the archived records used to determine the diagnosis of a congenital anomaly. However, some database studies address these limitations by validating a subset of records through other methods such as chart review or maternal interviews.

In addition, even databases covering millions of patient years may have limited power to test drug-specific hypotheses due to relatively small numbers of pregnant women exposed to any particular drug. Furthermore, databases often do not include information on potentially important confounders such as maternal tobacco or alcohol use, or exposure to over-the-counter medications. However, databases can be a relatively efficient and cost-effective method for generating and testing hypotheses related to prescription medications, and development of these as an alternative source of human data is likely to grow in importance.

WHERE DO WE GO FROM HERE

Due to the methodological challenges of studying rare exposures and rare outcomes in human populations, and the difficultly in accurately recognizing and classifying exposures and outcomes, existing sources of human data on medication-induced birth defects are limited in capacity to recognize a potential teratogenic effect with a new pharmaceutical agent, or conversely to provide reassurance that a new drug does not pose a substantial risk. However, the overriding challenge facing researchers and the public health community is the large number of drugs for which inadequate human data or no human data is available at all. Clearly, the methodological issues involved in this area call for multiple and complementary study designs to provide the most comprehensive assessment of risk. The cost implications of such a vast need may be perceived as prohibitive. Yet, the costs of lack of information may be greater. A coordinated and systematic approach to evaluating medications in human populations, both on a national and on an international basis, could contribute to more effective prevention of birth defects, and provide information that is critically and urgently needed by clinicians and pregnant women (62,63). The coordinated and integrated use of existing ongoing resources including adverse event reporting, large databases, population cohort studies, and case-control surveillance along with the additional complementary information provided by pregnancy registries and small cohort studies would require substantial efforts toward harmonization of these methods with each other and with the animal developmental toxicity research community. With the large number of prescription and over-the-counter medications used by pregnant women, a teratogen surveillance system that can adequately address these safety issues is essential.

REFERENCES

1. Rosenberg HM, Bentura SG, Maurer JD, et al. Births and deaths: United States, 1995. US Department of Health and Human Services, Public Health Service, CDC, National Center for Health Statistics, Hyattsville, MD (Monthly vital statistics report; vol. 45, no 3, suppl., 1996).
2. Yang Q, Khoury MJ, Mannino D. Trends and patterns of mortality associated with birth defects and genetic diseases in the United States, 1979–1992: An analysis of multiple-cause mortality data. Genetic Epidemiol 1997; 14:493–505.
3. Andrade SE, Gurwitz JH, Davis RL, et al. Prescription drug use in pregnancy. Am J Obstet Gynecol 2004; 191:398–407.
4. Werler MM, Mitchell AA, Hernandez-Diaz S, et al; National Birth Defects Prevention Study. Use of over-the-counter medications during pregnancy. Am J Obstet Gynecol 2005; 193:771–777.
5. Forrest JD. Epidemiology of unintended pregnancy and contraceptive use. Am J Obstet Gynecol 1994; 170:1485–1489.
6. Kessler RC, McGonagie KA, Swartz M, et al. Sex and depression in the National Cormorbidity Survey. 1: Lifetime prevalence, chronicity and recurrence. J Affect Disord 1993; 29:85–96.
7. Kwon HL, Belanger K, Bracken MB. Asthma prevalence among pregnant and child-bearing-aged women in the United States: Estimates from national health surveys. Ann Epidemiol 2003; 13:317–324.
8. Holmes LB, Wyszynski DF, Lieberman E. The AED (antiepileptic drug) pregnancy registry: A 6-year experience. Arch Neurol 2004; 61:673–678.
9. Yerby MS. Quality of life, epilepsy advances, and the evolving role of anticonvulsants in women with epilepsy. Neurology 2000; 55(Suppl 1):S21–S31.
10. Belilos E, Carsons S. Rheumatologic disorders in women. Med Clin N Am 1998; 82:77–101.

11. Goldberg HL, Nissim R. Psychotropic drugs in pregnancy and lactation. Intl J Psych Med 1994; 24:129–149.

12. Bracken MB, Triche EW, Belanger K, et al. Asthma symptoms, severity and drug therapy: A prospective study of effects on 2205 pregnancies. Obstet Gynecol 2003; 102:739–752.

13. Cohen LS, Altshuler LL, Harlow BL, et al. Relapse of major depression during pregnancy in women who maintain or discontinue antidepressant treatment. JAMA 2006; 295:499–507.

14. O'Quinn S, Ephross SA, Williams V, et al. Pregnancy and perinatal outcomes in migraineurs using sumatriptan: A prospective study. Arch Gynecol Obstet 1999; 63:7–12.

15. Lammer EJ, Chen DT, Hoar RM, et al. Retinoic acid embryopathy. New Engl J Med 1985; 313:837–841.

16. Dunn PM. Perinatal lessons from the past: Sir Norman Gregg, ChM, MC, of Sydney (1892–1966) and rubella embryopathy. Arch Dis Child Fetal Neonatal Ed 2007; 92:F513–S514.

17. Kessler DA. Introducing MedWatch: A new approach to reporting medication and device adverse effects and product problems. J Am Med Assoc 1993; 269:2765–2768.

18. Goldman SA, Kennedy DL. MedWatch FDA's medical products reporting program. Postgrad Med 1998; 103:13–16.

19. Pryde PG, Sedman AB, Nugent CE, et al. Angiotensin-converting enzyme inhibitor fetopathy. J Am Soc Nephrol 1993; 3:1575–1582.

20. Tabacova S, Vega A, McCloskey C, et al. Enalapril exposure during pregnancy: Adverse developmental outcomes reported to FDA. Teratology 2000; 61:520.

21. Spencer MJ. Fluoxetine hydrochloride (Prozac) toxicity in a neonate. Pediatrics 1993; 92:721–722.

22. Chambers CD, Johnson KA, Dick LM, et al. Birth outcomes in pregnant women taking fluoxetine. N Engl J Med 1996; 335:1010–1015.

23. Moses-Kolko EL, Bogen D, Perel J, et al. Neonatal signs after late in utero exposure to serotonin reuptake inhibitors: Literature review and implications for clinical applications. JAMA 2005; 293:2372–2383.

24. Sanz EJ, De-las-Cuevas C, Kiuru A, et al. Selective serotonin reuptake inhibitors in pregnant women and neonatal withdrawal syndrome: A database analysis. Lancet 2005; 365:482–487.

25. Andrews EB, Yankaskas BC, Cordero JF, et al. Acyclovir in pregnancy registry. Obstet Gynecol 1992; 79:7–13.

26. Preboth M. The antiretroviral pregnancy registry interim report. Am Fam Physician 2000; 61:2265.

27. Shields KE, Wilholm B-E, Hostelley LS, et al. Monitoring outcomes of pregnancy following drug exposure, a company-based pregnancy registry program. Drug Safety 2004; 27:353–367.

28. White AD, Andrews EB. The pregnancy registry program at Glaxo wellcome company. J Allergy Clin Immunol 1999; 103:5362–5363.

29. Koren G, Pastuszak A, Ito S. Drug therapy: Drugs in pregnancy. N Engl J Med 1998; 338:1128–1137.

30. Chambers CD, Tutuncu ZN, Johnson D, et al. Human pregnancy safety for agents used to treat rheumatoid arthritis; adequacy of available information and strategies for developing postmarketing data. Arthritis Res Ther 2006; 8:215–225.

31. Scialli AR. The organization of teratology information services (OTIS) registry study. J Allergy Clin Immunol 1999; 103:5373–5376.
32. Watts DH, Covington DL, Beckerman K, et al. Assessing the risk of birth defects associated with antiretroviral exposure during pregnancy. Am J Obstet Gynecol 2004; 191:985–992.
33. Honein MA, Paulozzi LJ, Cragan JE, et al. Evaluation of selected characteristics of pregnancy drug registries. Teratology 1999; 60:356–364.
34. U.S. Food and Drug Administration Office of Women's Health. Establishing Pregnancy Exposure Registries, 2992. Accessed 2.21.2006: http://www.fda.gov/womens/guidance.html
35. U.S. Food and Drug Administration Office of Women's Health. Evaluation of Human Pregnancy Outcome Data, 1999. Accessed 2.21.2006: http://www.fda.gov/womens/guidance.html
36. Pastuszak AL, Schiiler L, Speck-Martins CE, et al. Use of misoprostol during pregnancy and Möbius' syndrome in infants. New Engl J Med 1998; 338:1881–1885.
37. Carmichael SL, Shaw GM, Yang W, et al; National Birth Defects Prevention Study. Correlates of intake of folic acid-containing supplements among pregnant women. Am J Obstet Gynecol 2006; 194:203–210.
38. Mitchell AA, Rosenberg L, Shapiro S, et al. Birth defects related to bendectin use in pregnancy: 1. oral clefts and cardiac defects. J Am Med Assoc 1981; 245:2311–2314.
39. Hernandez-Diaz S, Werler MM, Walker AM, et al. Folic acid antagonists during pregnancy and the risk of birth defects. New Engl J Med 2000; 343:1608–1614.
40. Castilla E, Peters PWJ. Impact of monitoring systems: National and international efforts. In Genetic Services Provision: An International Perspective, Birth Defects: Original Article Series, Kuliev A, ed. March of Dimes Birth Defects Foundation, White Plains, NY, 1992; 28.
41. Werler MM, Shapiro S, Mitchell AA. Periconceptional folic acid exposure and risk of occurrent neural tube defects. J Am Med Assoc 1993; 269:1257–1261.
42. Carmichael SL, Shaw GM, Ma C, et al; National Birth Defects Study. Maternal corticosteroid use and oral clefts. Am J Obstet Gynecol 2007; 197:585.e1–585.e7.
43. Werler MM, Mitchell AA, Shapiro S. First trimester maternal medication use in relation to gastroschisis. Teratology 1992; 45:361–367.
44. Tomeo CA, Rich-Edwards JW, Michels KB, et al. Reproducibility and validity of maternal recall of pregnancy-related events. Epidemiology 1999; 10:774–777.
45. Khoury MJ, James LM, Erickson JD. On the use of affected controls to address recall bias in case-control studies of birth defects. Teratology 1994; 49:273–281.
46. Prieto L, Martinez-Frias ML. Response to "What kind of controls to use in case control studies of malformed infants: Recall bias versus 'teratogen nonspecificity' bias." Teratology 2000; 62:372–373.
47. Ericson A, Kallen B, Wiholm B. Delivery outcome after the use of antidepressants in early pregnancy. Eur J Clin Pharmacol 1999; 55:503–508.
48. Olsen J, Melbye M, Olsen SF, et al. The Danish national birth cohort: Its background, structure and aim. Scand J Public Health 2001; 29:300–307.
49. Irl C, Hasford J. Assessing the safety of drugs in pregnancy, the role of prospective cohort studies. Drug Safety 2000; 22:169–177.
50. Chung CS, Myrianthopoulos NC. Factors affecting risks of congenital malformations. 1. Analysis of epidemiologic factors in congenital malformations. Report from the Collaborative Perinatal Project. Birth Defects 1975; 11:1–22.

51. Kallen B, Rydhstroem H, Aberg A. Congenital malformations after the use of inhaled budesonide in early pregnancy. Obstet Gynecol 1999; 93:392–395.
52. Pastuszak A, Schick-Boschetto B, Zuber C, et al. Pregnancy outcome following first-trimester exposure to fluoxetine (Prozac). JAMA 1993; 269:2246–2248.
53. McElhatton PR, Bateman DN, Evans C, et al. Congenital anomalies after prenatal ecstasy exposure. Lancet 1999; 354:1441–1442.
54. Schaefer C, Amoura-Elefant E, Vial T, et al. Pregnancy outcome after prenatal quinolone exposure: Evaluation of a case registry of the European Network of Teratology Information Services (ENTIS). Eur J Obstet, Gynecol Reproduc Biol 1996; 69:83–89.
55. Vial T, Robert E, Carlier P, et al. First-trimester in utero exposure to anorectics: A French collaborative study with special reference to dexfenfluramine. Intl J Risk Saf Med 1992; 3:207–214.
56. Jones KL, Lacro RV, Johnson KA, et al. Pattern of malformations in the children of women treated with carbamazepine during pregnancy. New Engl J Med 1989; 320:1661–1666.
57. Chambers CD, Braddock SR, Briggs GG, et al. Postmarketing surveillance for human teratogenicity: A model approach. Teratology 2001; 64:252–261.
58. Larsen H, Nielsen GL, Sorensen HT, et al. A follow-up study of birth outcome in users of pivampicillin during pregnancy. Acta Obstet Gynecol Scand 2000; 79:379–383.
59. Wen SW, Demissie K, Liu S. Adverse outcomes in pregnancies of asthmatic women: Results from a Canadian population. Ann Epidemiol 2001; 11:7–12.
60. Drinkard CR, Shatin D, Clouse J. Postmarketing surveillance of medications and pregnancy outcomes: Clarithromycin and birth malformations. Drug Safety 2000; 9:549–556.
61. Hardy JR, Holford TR, Hall GC, et al. Strategies for identifying pregnancies in the automated medical records of the General Practice Research Database. Drug Safety 2004; 13:749–759.
62. Olsen J, Czeizel A, Sorensen HT, et al. How do we best detect toxic effects of drugs taken during pregnancy? A Euromap paper. Drug Safety 2002; 25:21–32.
63. Mitchell AA. Systematic identification of drugs that cause birth defects—a new opportunity. N Engl J Med 2003; 349:2556–2559.

Index

T - #0361 - 101024 - C8 - 229/152/22 - PB - 9781138372443 - Gloss Lamination